Managing E-Commerce and Mobile Computing Technologies

edited by

Julie Mariga
Purdue University, USA

IRM Press
Publisher of innovative scholarly and professional information technology titles in the cyberage

Hershey • London • Melbourne • Singapore • Beijing

Acquisitions Editor:	Mehdi Khosrow-Pour
Senior Managing Editor:	Jan Travers
Managing Editor:	Amanda Appicello
Copy Editor:	Julie Randall
Typesetter	Amanda Lutz
Cover Design:	Michelle Waters
Printed at:	Integrated Book Technology

Published in the United States of America by
IRM Press (an imprint of Idea Group Inc.)
701 E. Chocolate Avenue, Suite 200
Hershey PA 17033-1240
Tel: 717-533-8845
Fax: 717-533-8661
E-mail: cust@idea-group.com
Web site: http://www.irm-press.com

and in the United Kingdom by
IRM Press (an imprint of Idea Group Inc.)
3 Henrietta Street
Covent Garden
London WC2E 8LU
Tel: 44 20 7240 0856
Fax: 44 20 7379 3313
Web site: http://www.eurospan.co.uk

Library of Congress Cataloging-in-Publication Data

Mariga, Julie R.
Managing e-commerce and mobile computing technologies / Julie Mariga.
p. cm.
Available also in electronic form.
Includes bibliographical references and index.
ISBN 1-931777-46-2 (soft cover) -- ISBN 1-931777-62-4 (ebook)
1. Electronic commerce--Technological innovations. 2. Mobile commerce. 3. Mobile computing. I. Title.
HF5548.32.M3734 2003
658.8'4--dc21

2002156232

British Cataloguing in Publication Data
A Cataloguing in Publication record for this book is available from the British Library.

Other New Releases from IRM Press

Managing E-Commerce and Mobile Computing Technologies

Table of Contents

Preface

One of the hottest topics in the information technology world today is mobile computing and mobile commerce (m-commerce). The rate at which technology impacts our professional and personal lives, businesses, and society is remarkable. The information revolution has transformed global economies and how companies do business.

The client/server paradigm moved the computing industry into new areas, both from a technical and business standpoint. As the Internet paradigm continues to evolve, one important area that is growing is the wireless and mobile areas. The pace at which the computing and communication industries are converging is very rapid. Wireless phones and personal digital assistants (PDAs) are quickly evolving into powerful multipurpose devices. With the continued evolution of m-commerce and electronic commerce (e-commerce) companies are doing more and more business electronically and in mobile environments.

This book discusses many of the issues involved in developing mobile and electronic architectures, deciding on what mobile devices and mobile operating systems should be implemented on those architectures, mobile security is addressed, various frameworks and business models are discussed, as well as factors that influence decisions that are made regarding the technology. The book introduces a number of important topics that need to be understood before making any large investments in this technology area.

The book is organized into 19 chapters. A description of each of the chapters follows:

Chapter 1 entitled, "A Concept for the Evaluation of E-Commerce-Ability" by Ulrike Baumöl, Thomas Stiffel and Robert Winter develops a concept to evaluate the e-commerce-ability of a corporation and applies the framework to basic roles of the e-commerce environment. The concept comprises two components: A four dimensional framework is proposed which can be used to represent the degree of external coordination, the degree of alignment of business towards organizational and cultural rules of the networked economy, the degree of orientation towards customer needs and the degree of systematic and integral use of information and communication technology (ICT). Based on this framework, an evaluation approach is presented which supports a maturity analysis.

Chapter 2 entitled, "M-Commerce: A Location-Based Value Proposition" by Nenad Jukic, Abhishek Sharma, Boris Jukic and Manoj Parameswaran defines Mobile Commerce (m-commerce) as a process of conducting commercial transactions via a "mobile" telecommunications networks using communication, information, and payment devices such as a mobile phone or a palmtop unit. They analyze the potential ramifications to the field of marketing and changes in the market due to the advent of m-commerce. In particular, they analyze the opportunities that various characteristics of the m-commerce model can bring to the field of marketing. They investigate the likelihood of emergence of mall-like zones that are based both on the geographical proximity of services and goods providers and the use of mobile communication devices. Such zones have a potential of becoming the basic units for any analysis of m-commerce scenarios. As m-commerce attains maturity, the zones could become the fundamental parameter in marketing evaluation.

Chapter 3 entitled, "E-Commerce Adoption in Small Firms: A Study of Online Share Trading" by Pak Y. P. Chan and Annette M. Mills examines adoption of an electronic commerce technology (namely, order-execution online trading technology) by six small brokerage firms at various stages in the evaluation and adoption process. The study is informed by innovation theory and prior research, and seeks to identify the key factors influencing the adoption process. Consistent with innovation theory, the case findings suggest that three classes of factors influence adoption: innovation factors, innovation factors, organisational factors, and environmental factors. The key factors within each of these classes were identified as compatibility and perceived benefits (innovation factors), IT sophistication, internal and external IT support, and management support (organisational factors); and pressure from e-commerce-able competitors and clients (environmental factors). Of these variables, compatibility and perceived benefits were found to be the most significant, impacting e-commerce adoption.

Chapter 4 entitled, "Conceptualizing the SMEs' Assimilation of Internet-Based Technologies" by Pratyush Bharati and Abhijit Chaudhury conceptualizes a model for the assimilation of internet-based technologies in small and medium enterprises (SMEs). The research examines the factors influencing the assimilation of Internet based technologies and the penetration of these technologies in SMEs. Internet based technologies are complex organizational technologies. The model uses the learning related scale, related knowledge, and diversity together with several control variables like host size, IT size, specialization, education. Several external factors, such as influence of customers, suppliers, vendors and competitors that have been suggested in studies of SMEs have also been included to explain the assimilation and diffusion of innovation. The results of an exploratory survey are presented and future research is discussed.

Chapter 5 entitled, "Strategic Issues in Implementing Electronic-ID Services: Prescriptions for Managers" by Bishwajit Choudhary examines e-security solutions (e.g., digital certificates, e-signatures, e-IDs) that have gained tremendous attention as they promised to plug the security loopholes and create trusted electronic markets. Implementation of such critical, complex and costly security solutions demands their thorough assessment at technical as well as business levels. Based on the author's experience at one of Scandinavia's leading vendors of banking solutions and infrastructure, the chapter develops basic concepts, discusses strategic (product, market and technical) concerns and finally summarizes the contemporary challenges facing the implementation of e-ID schemes.

Chapter 6 entitled, "Potential Roles for Business-to-Business Marketplace Providers in Service-Oriented Architectures" by Markus Lenz and Markus Greunz introduces the concept of a service oriented architecture and assesses its applicability to business-to-business (B2B) marketplaces. In order to provide substantial value to their members marketplaces have to offer a comprehensive service offering that aims to support all phases of the transaction process. Building up such a service offering is not a one-time effort, but electronic B2B marketplaces will have to continuously evolve their service offerings and to adapt it to ever-changing needs of the companies they serve. Therefore, instead of trying to create such an extensive service offering on their own, we argue that B2B marketplaces have to make use of partnerships with specialized service providers. Only by partnering they can make use of speed and scale of such a collaboration of specialists (Ernst et al., 2001). This view of B2B marketplaces as integrators of services from multiple parties puts some new demands on architectural issues of the marketplaces. A service-oriented architecture may help to cope with these new demands. To test this view a survey among European B2B marketplaces has been carried out in order to match their service development and their needs with the characteristics of a service-oriented architecture and the potential roles that marketplace providers can play in such a concept.

Chapter 7 entitled, "An Evaluation of Dynamic Electronic Catalog Models in Relational Database Systems" by Kiryoong Kim, Dongkyu Kim, Jeuk Kim, Sang-uk Park, Ighoon Lee, Sang-goo Lee and Jong-hoon Chun discusses electronic catalogs that are electronic representations of required to manage diverse and flexible schemas of products in electronic commerce. Although relational database systems seem an obvious way for their storage, traditional designs of relational schemas do dot support electronic catalogs. Therefore, new models for managing diverse and flexible schemas in relational databases are required for such systems. In this chapter, several models for electronic catalogs using relational tables, and an experimental evaluation of their efficiency. Some of the proposed models were more efficient than currently used ones, and they can be put to practical use.

Chapter 8 entitled, "Extending Client-Server Infrastructure Using Middleware Components" by Qiyang Chen and John Wang research the embracing of inapt infrastructure technology which is a major threat in developing extensive and efficient Web-based systems. The architectural strength of all business models demands an effective integration of various technological components. Middleware, the center of all applications, becomes the driver—everything works if middleware does. In the recent times, the client/server environment has experienced sweeping transformation and led to the notion of the "Object Web." Web browser is viewed as a universal client that is capable of shifting flawlessly and effortlessly between various applications over the Net. This paper attempts to investigate middleware and the facilitating technologies and point toward the latest developments, taking into account the functional potential of the on-market middleware solutions as well as their technical strengths and weaknesses. The chapter describes various types of middleware including database middleware, Remote Procedure Call (RPC), application server middleware, message-oriented middleware (MOM), Object Request Broker (ORB), transaction-processing monitors, and Web middleware, etc., with on-market technologies.

Chapter 9 entitled, "E-Business Experiences with Online Auctions" by Bernhard Rumpe discusses how online auctions are among the most influential e-business applications. Their impact on trading for businesses as well as consumers is both remarkable and inevitable. There have been considerable efforts in setting up market places, but with respects to market volume online trading still lays in its early stages. This chapter discusses the benefits of the concept of Internet marketplaces, with the highest impact on pricing strategies, namely, the conduction of business online auctions. It discusses their benefits, the problems occurring, and possible solutions. In addition, sketch actions for suppliers to achieve a better strategic position in the upcoming Internet market places.

Chapter 10 entitled, "Implementing Privacy Dimensions within an Electronic Storefront" by Chang Liu, Jack Marchewka and Brian Mackie research how electronic businesses will attempt to distinguish themselves from their competition and gain competitive advantage by customizing their Web sites in order to build a strong relationship with their customers. This will require the collection and use of personal information and data concerning the customer's online activities. Although new technologies provide an opportunity for enhanced collection, storage, use, and analysis of this data, concerns about privacy may create a barrier for many electronic businesses. For example, studies suggest that many people have yet to shop online or provide personal information due to a lack of trust. Moreover, many others tend to fabricate personal information. To this end, many electronic businesses have attempted to ease customers' concerns about privacy by posting privacy policies or statements or by complying with a particular seal program. Recently, the Federal Trade Commission has proposed four privacy dimensions that promote fair information practices. These dimensions include: (1) notice/awareness, (2) access/participation, (3) choice/consent, and (4) security/integrity. An electronic storefront was developed to include these privacy dimensions as part of study to learn how privacy influences trust and, in turn, how trust influences behavioral intentions to purchase online. The empirical evidence from this study strongly suggests that electronic businesses can benefit by including these privacy dimensions in their Web sites. This chapter will focus on how these dimensions can be implemented within an electronic storefront.

Chapter 11 entitled, "An E-Channel Development Framework for Hybrid E-Retailers" by In Lee researches the profound impact of e-commerce on organizations, e-channel development emerged as one of the most important challenges that managers face. Unfortunately, studies indicate that managers in most large companies are still unclear about an e-commerce strategy, and tend to lack adequate e-commerce development expertise. Poorly planned and developed e-commerce channels add little value to organizations. Furthermore, these poorly developed e-channels may even have negative impact on their organizations by confusing and disappointing customers who value a seamless cross-channel experience. To develop an e-channel that delivers higher utility to customers and generates sustainable long-term profits, managers need to analyze how an e-commerce channel affects the performance of existing channels and develop a company-wide e-channel development program. Based on a number of e-commerce case studies, we developed an e-channel development framework which consists of five step-by-step phases: (1) strategic analysis, (2) e-channel planning, (3) e-channel system design, (4) e-channel system development, and (5) performance evaluation and refinement. This framework helps managers evaluate the impact of e-commerce channels on organizational performance and determine the most appropriate channel design and integration mechanisms for the achievement of business strategies. This chapter also discusses impact of e-channel structures on organizational performance.

Chapter 12 entitled, "Organization, Strategy and Business Value of Electronic Commerce: The Importance of Complementarities" by Ada Scupola describes how many corporations are reluctant to adopt electronic commerce due to uncertainty in its profitability and business value. This chapter introduces a business value complementarity model of electronic commerce. The model relates high level performance measures such as business value first to intermediate performance measures such as value chain and company strategy and then to the e-business performance drivers as business processes and complementary technologies. The model argues that complementarities between the different activities of the value chain, corresponding business processes and supporting technologies should be explored to reach a better fit between strategy, business model and technology investments when entering the electronic commerce field. The exploration of such complementarities should lead to investments in electronic commerce systems that best support the company strategy, thus minimizing failures. From a practical point of view, managers could use this framework as a methodology to increase the business value of electronic commerce to a corporation.

Chapter 13 entitled, "Continuous Demand Chain Management: A Downstream Business Model for E-Commerce" by Merrill Warkentin and Akhilesh Bajaj discusses the demand side of supply chain management which has drawn considerable research attention, with focus on disintermediation and syndication models. In this chapter, we evaluate new business models for establishing a continuous demand chain structure to streamline the logistics between the vendor and its direct consumers. The Continuous Demand Chain Management (CDCM) model of E-Commerce is one in which the physical products for sale are delivered directly to the customer without the use of a third party logistics provider, such as a common carrier, and in which the physical product may be continuously "pulled" from the seller. We present three submodels of CDCM: The CDCM Model A applies to business-to-consumer

(B2C) online sellers of physical goods who own or control their own delivery vehicles and may provide further services to extend the value proposition for the buyer. The online grocer is a typical example of businesses in this category. The CDCM Model B applies to business-to-business (B2B) sellers of physical goods, who also own a significant portion of their delivery fleet and deliver goods on demand to local distributors or business customers. Office supply E-Merchants provide an example of this model. The CDCM Model C applies to businesses that typically provide virtually instantaneous delivery of third party goods to consumers or businesses. Businesses in this category own or control their own delivery fleet and add value by delivering items within very short periods of time, usually one-hour delivery.

In order to analyze these models we conducted structured interviews with key senior managers of one representative business each in the CDCM Model A and Model B categories. We extensively surveyed recent literature on companies in the CDCM Model C category. We use the results of our study to analyze different aspects such as revenue streams, cost structure, and operational peculiarities of businesses following the CDCM model, and finally discuss the long-term viability of the sub models.

Chapter 14 entitled, "Web-Based Supply Chain Integration Model" by Latif Al-Hakim discusses various business process supply-chain models and emphasises the need for organisations to apply CRM concepts and to integrate the Internet within the functions of the supply chain in order to be able to gain good customer expectations in the era of e-commerce. This chapter outlines a framework for developing a web-based supply chain integrating model based on SCOR and key features of CPFR and attempts to link this model with Business Process Reengineering and with traditional productivity improvement programs. The development of a website at two levels is suggested. The first level is within the public domain and the other is limited to supply chain partners. The chapter incorporates fuzzy set theory into the dynamic of production scheduling to allow the integrating model to deal with vague constraints and to enable conflicting multi-criteria objectives to be managed effectively in the production environment.

Chapter 15 entitled, "A Cooperative Communicative Intelligent Agent Model for E-Commerce" by Ric Jentzsch and Renzo Gobbin researches the complexities of business continues to expand. First technology, then the World Wide Web, ubiquitous commerce, mobile commerce, and who knows. Business information systems need to be able to adjust to these increased complexity while not creating more problems. Here we put forth a conceptual model for cooperative communicative intelligent agents that can extent itself to the logical constructs needed by modern business operations today and tomorrow.

Chapter 16 entitled, "Supporting Mobility and Negotiation in Agent-Based E-Commerce" by Ryszard Kowalczyk and Leila Alem presents recent advances in agent-based e-commerce addressing the issues of mobility and negotiation. It reports on selected research efforts focusing on developing intelligent agents for automating the e-commerce negotiation and coalition formation processes, and mobile agents for supporting deployment of intelligent e-commerce agents and enabling mobile e-commerce applications. Issues such as trade-off between decision-making in negotiation and mobility capabilities of the agents are also discussed in this chapter.

Chapter 17 entitled, "Deploying Java Mobile Agents in a Project Management Environment" by F. Xue and K. Y. R. Li introduces mobile agent technology and explains how it can help businesses to implement client-server enterprise computing solutions. A Java mobile agent-based project management system prototype is presented to demonstrate the main features of mobile agents (mobility, functionality, intelligence and autonomous), and how they help to enhance communication processes and facilitate security within the project environment. It suggests a practical way to isolate all host resources from all visiting agents using host agents and exported host functions. It also proposes a communication infrastructure to support intelligent dialogue between agents.

Chapter 18 entitled, "Factors Influencing Users' Adoption of Mobile Computing" by Wenli Zhu, Fiona Fui-Hoon Nah and Fan Zhao introduces a model that identifies factors influencing users' adoption of mobile computing. It extends the Technology Acceptance Model (TAM) by identifying system and user characteristics that affect the perceived usefulness and perceived ease of use of mobile computing, which are two key antecedents in TAM. Furthermore, it incorporates two additional constructs, trust and enjoyment, as determinants in the model, and proposes specific factors that influence these two constructs. The long-term goals of this work are to gain an increased understanding of adoption issues in mobile computing, and to explain how specific HCI design issues may affect adoption by users.

Chapter 19 entitled, "Mobile Computing Business Factors and Operating Systems" by Julie R. Mariga introduces the enormous impact mobile computing is having on both companies and individuals. Companies face many issues related to mobile computing. For example, which devices will be supported by the organization, which devices will fulfill the business objectives, which form factor will win, which features and networks will future devices offer, which operating systems will they run, what will all this cost, what are the security issues involved, what are the business drivers? This chapter will discuss the major business drivers in the mobile computing field and provide an analysis of the top two operating systems that are currently running the majority of mobile devices. These platforms are the 1) Palm operating system (OS) and 2) Microsoft Windows CE operating system. The chapter will analyze the strengths and weaknesses of each operating system, discuss market share and, future growth.

Acknowledgments

The editor would like to acknowledge the help of all involved in the collation and review process of the book, without whose support the project could not have been satisfactorily completed. A further special note of thanks goes also to all the staff at Idea Group, Inc., whose contributions throughout the whole process from inception of the initial idea to final publication have been invaluable. In particular to Amanda Appicello, Managing Editor, who continuously helped me via e-mail to keep the project on schedule and to Mehdi Khosrow-Pour, whose enthusiasm motivated me to initially accept his invitation for taking on this project.

To my family - Sharon, Michelle, Anthony, and Bob — for their support and encouragement.

And last, but not least, my mentor, James E. Goldman, whose leadership and guidance has been greatly appreciated.

Julie R. Mariga, Editor
Purdue University, USA
November 2002

Chapter I

A Concept for the Evaluation of E-Commerce-Ability

Ulrike Baumöl, Thomas Stiffel and Robert Winter
University of St. Gallen, Switzerland

ABSTRACT

This chapter develops a concept to evaluate the e-commerce-ability of a corporation and applies the framework to basic roles of the e-commerce environment. The concept comprises two components. A four-dimensional framework is proposed which can be used to represent the degree of external coordination; the degree of alignment of business toward organizational and cultural rules of the networked economy; the degree of orientation toward customer needs; and the degree of systematic and integral use of information and communication technology (ICT). Based on this framework, an evaluation approach is presented that supports a maturity analysis.

INTRODUCTION

While many companies are still implementing or improving their Internet-facilitated e-commerce activities, a new wave of technology-driven innovation has arrived: Mobile (M-commerce). Enabled by the progress in wireless technology and the increasing number of mobile devices, expectations are high again. We deduced from our current research that electronic forms of buying/selling (i.e., e-commerce including m-commerce) can only be successful if the corporation is structured according to specific requirements. That means that e-commerce and m-commerce projects are at failure risk if realized solely based on the Internet as a new distribution channel without changing the internal view on customer processes and without restructuring certain elements such as internal processes and structures and inter-business networking.

However, corporations that want to implement successful e-commerce activities first of all need to have a framework for reflecting and analyzing their current status before measures can be defined to achieve e-commerce-ability.

The first step of such a systematic approach has to include not only a framework of dimensions which allows the reflection and analysis of patterns of e-commerce business models or roles, respectively, but also a set of parameters which represents measurable success. We therefore developed a concept consisting of two "pillars":

1. A four-dimensional framework is proposed that can be used to represent the degree of external coordination; the degree of alignment of business toward organizational and cultural rules of the networked economy; the degree of orientation toward customer needs; and the degree of systematic and integral use of information and communication technology (ICT). Each of the framework's dimensions is described by a set of characteristics which are used as a metric to render the creation and, moreover, the comparison of the patterns possible. Our hypothesis is that there are success patterns which depend on the specific roles existing in an e-commerce environment.
2. Based on this framework, an evaluation approach is presented that supports the analysis of the corporation based on value-driven quantitative and qualitative parameters reflecting economic success. The concept can be put to use by visualizing the pattern of the respective corporation to be analyzed; comparing it with the success pattern of the role; analyzing the status regarding the important value drivers; identifying the gap; and, finally, defining measures to close the gap.

But before we can start to develop the concept of e-commerce-ability, we have to take a closer look at the terms and models that represent a basis for this chapter. On the one hand, we reflect the understanding of the terms e-commerce and m-commerce. On the other, hand we have to look at existing e-commerce maturity concepts and decide whether they can be used for developing our e-commerce-ability concept.

DEFINITIONS OF E-COMMERCE

The understanding of e-commerce is widespread. A common definition is difficult to give because of many inconsistent approaches (Wigand, 1997). Therefore, a discussion of an appropriate definition is necessary.

Many definitions do not strictly separate e-commerce and e-business. However, the definition space of e-business is more complex and inconsistent. In this study, e-business is interpreted as a superset of e-commerce. E-business are those business activities that are a part of a value network; address the customer process; and use information and communication technologies (ICT) in an integrative way based on the organizational and cultural rules of the networked economy.

Most definitions assume that e-commerce is enabled by the development and implementation of electronic media such as the Internet, whereby it is not uniform in how far "old" electronic media, like telephone, telex and television, are included.

The definition of e-commerce as "doing business electronically" (European Commission, 1997) is too broad and interpretable, whereas Gartner Group's definition (1999) "e-commerce is a dynamic set of technologies, applications and business processes that link corporations, consumers and communities" is too narrowly focused on the transactional aspect. This focus is more explicitly followed by Timmers (1998) who defines e-commerce as "any form of business transaction in which the parties interact electronically rather than by physical exchange or direct "physical contact". Other approaches in this direction mostly differ in the degree of detail of the trade / transaction process or in the selection of specific processes such as procurement or distribution (Aldrich, 1999; Morasch et al., 2000). A further approach stresses the enhancements evoked by the enabling technologies in the form of more effective and efficient processes (cf. Baldwin et al., 2000).

Resuming this discussion, Kalakota et al. (1997) can be cited: "depending on whom you ask, electronic commerce has different definitions," particularly with regard to communications, business process, service and online necessity.

The definition used for the purposes of this article focuses on the transactional approach and uses the definition of Kalakota et al. (1997) as a basis: "buying and selling over digital media," whereas buying can be left out if the buying process is electronic, the selling process is electronic as well. To be more precise: goods can also be services and the selling process can be sale, commerce or distribution, as digital goods can be sold for free. So e-commerce is the trade (sales, commerce, distribution) of goods and services, i.e., products, by electronic means.

As a consequence, e-commerce activities are mostly objectively observable activities of corporations. Implicitly, it is also assumed that those activities have a deep impact on the structure of the corporation.

As already mentioned, m-commerce can be interpreted as a subset of e-commerce by referring to those e-commerce activities which involve wireless technologies (i.e., mobile devices like handphones, personal digital assistants or handheld computers) (cf. Shih, 2002; Siau, 2001; Kalakota, 2002).

The conceptual differences between e-commerce and m-commerce are primarily based on the mobility or the location, respectively. Beyond transactions and information access, the location of m-commerce users can be determined, which opens new forms of services and transactions. On the other side, corporations can reach specific users anytime and anywhere, not only with regard to a specific person, but also with regard to a specific geographical region which again enables new forms to disseminate information to consumers (Siau, 2001).

Despite these differences, we believe the e-commerce-ability framework to be appropriate for m-commerce as well.

DIFFERENT APPROACHES AND FRAMEWORKS TO MEASURE E-COMMERCE-ABILITY

E-commerce-ability, or readiness, the focus of this chapter, describes the capability of a corporation to perform e-commerce successfully. Generally, to analyze the capability, two questions should be answered: (1) What are the impacts or opportunities of e-commerce and (2) How can e-commerce activities be analyzed and evaluated with regard to the specific situation and the strategy of the corporation? Numerous approaches give answers to those questions in different ways.

Many models that reflect or, explain factors or forecast their impact on corporations, often have a special and narrow focus. Schwartz (1998), for example, claims essential principles to grow the business, especially on the World Wide Web, for example, include the offer of experiences, compensation for personal data, high customer comfort, continual adaptation to the market or establishment of brands. Zerdick et al. (2000) and Kelly (1998), in contrast, are more general, and describe 10 theses for the new economy, i.e., digitalization of the value chain, attainment of critical mass or competition and cooperation by value networks.

Approaches of that kind are numerous, but although most of them seem to be valid, the transformation into business is difficult. Reasons include a missing methodology, as well as, the absence of a holistic view or framework in which the interconnection and most important dimensions are considered.

An approach given by the European Commission (1998) partly considers those aspects and distinguishes several levels of e-commerce-activities in a continuum from easy to complex to implement into business respectively a continuum from standard to custom applications. Examples are electronic presences, national payment and international electronic distribution. The latter usually implies a high risk and high financial investments, whereas the others are mostly covered by standard and easy-to-realize applications. This model may give a clue for corporations on what they can do and which standard applications might exist, but a statement concerning their ability to do so or to know what to do cannot be made.

A model which takes this into account is proposed by Canter (2000). Starting point is the need for agility in the information age because corporations have to react faster in their quickly changing environment. Bases of the framework are the capability maturity model (CMM) of the Carnegie Mellon University's Software Engineering Institute (cf. Paulk et al., 1993) and the decision cycle OODA (Observation-Orientation-Decision-Action). The five levels of the CMM are transformed into five levels with different characteristics of changeability and agility also known as the Change Proficiency Maturity Model (CPMM). This model can be used for business areas such as vision and strategy, innovation management or relationship management. The aim is to assist the OODA-process supported by given tools to develop and reach a better maturity.

Even if the need for agility is accepted, the strong internal perspective, which excludes external areas like external relationships or customer needs, is the disadvantage of this model.

For the same reason, approaches of Whitely (1998) and Grant (1999) cannot be used in this study. Likewise, the maturity of businesses is treated by Anghern (1997), but with an external focus on business activities on the Internet such as: information, communication, distribution and transaction. With those, he distinguishes levels concerning the customization and sophistication. Additionally, the author identifies a set of core competencies that are necessary to succeed within those levels. To summarize, this framework can be used for the "analysis of business-related Internet strategies, as well as a systematic approach to guide the strategic building process..." (cf. Anghern, 1997).

However, the neglect of internal matters also excludes this and the similar approach of Burgess et al. (1999). Indeed, those implementations demonstrate that multidimensional models are more adequate for the complex e-commerce environment, but a holistic approach is missing.

Although PriceWaterhouseCoopers (PWC, 2000) and the similar Bain and Company (2000), present such an integrated approach, they both refer to e-business. The PriceWaterhouseCoopers approach consists of nine dimensions:

1. E-business strategy;
2. Organization and competencies;
3. Business processes on the e-business value chain (advertising, ordering, delivery, billing, debt collection, customer care);
4. Web site performance (financially or otherwise);
5. Taxation issues;
6. Legal and regulatory aspects;
7. Systems and technology used;
8. ICT and logistics processes; and
9. E-business security.

These dimensions are used, on the one hand, to evaluate the readiness of the corporation to develop and, on the other hand, to benchmark e-business performance.

Comparing those approaches with the proclaimed aim of the study we believe only the latter approaches of PriceWaterhouseCoopers (PWC 2000) and Bain and Company (2000), respectively, to be appropriate. Nevertheless, they both have a scope that is too wide and complex for a quick and effective analysis of e-commerce activities.

Consequently, a framework has to be developed that offers an integrative approach to the corporation and its capabilities, as well as the coverage for the most important success-defining factors.

DEDUCTION OF A CONCEPT FOR THE EVALUATION OF E-COMMERCE-ABILITY

The analysis concept for e-commerce-ability consists of two essential elements. First, the framework for reflecting and comparing the patterns of e-commerce role profiles and second, a valuation approach for analyzing the economic success of the e-commerce activities with regard to the value drivers of the corporation. Both elements combined enable the evaluation of the e-commerce-ability of a corporation. Moreover, after analyzing the current status of the e-commerce activities, management can decide on further steps to be taken to render them e-commerce-able in a successful way.

First Pillar of the Concept: A Four-Dimensional Framework for Creating Patterns

This basic framework consists of four dimensions that represent characteristics we think to be important for a corporation of the net economy (also cf. Figure 2).

Degree of orientation toward customer needs

All ideas presented in this paper are based on the strong belief that, in the information age, every corporation has to align its activities with its customers' needs. Information and communication technology (ICT) has been an enabler for the extensive gathering and processing of customer data. On the other hand, ICT made possible the adaptation of a corporation's range of services to customer needs. Therefore, intensive consideration of customer needs seems to be an important means of differentiation. Often, customers do not want to purchase a product or service, but instead intend to get a solution to a certain problem. This solution can be single goods and services or, more likely, bundles of goods and services. Successful e-commerce corporations offering those problem solutions therefore might be able to differentiate from their competitors (cf. Österle, 2000, pp. 45-47).

Another means of differentiation might be the ability of a corporation to provide highly customizable products and services to their customers, i.e., to a certain extent enabling them to modify product characteristics.

Therefore, we define the following categories for the measure "Degree of orientation toward customer needs":

- Standard products: a corporation offers a range of products whose specifications are not modifiable.

- Mass customization: after intensively analyzing the customer needs, a set of customized products will be developed (based on standard products). Corporations offering those variations pursue the aim of making the product more attractive to groups of customers while at the same time limiting the costs of individualization.
- Full coverage of needs: this category is an extension of the previous one. A corporation does not offer only goods and services that are highly customizable (as far as the individual customer needs are concerned), but also individual problem solutions consisting of a bundle of goods and services.

Degree of systematic and integral use of ICT

Throughout the past 10 years, the role of information and communication technology has changed from a mere support function to a strategic tool for the generation of competitive advantages. We believe that e-commerce corporations need to pay special attention to the way of using ICT because they cannot – like in "normal" shopping marts – differentiate from their competitors by individualized face-to-face customer care. Instead, they have to use electronic means for establishing the best possible customer relationship.

Venkatraman (1991) describes five levels of ICT-induced business reconfiguration:

1. Localized exploitation; at this level, corporations use ICT to support selected business functions. Thus, they can realize efficiency gains in isolated functions, without any influence on daily operative business. However, localized exploitation does not lead to gains in effectiveness. If ICT is used only localized, i.e., without any strategic vision, a corporation's competitors can easily copy it. Therefore no long-term competitive advantages can be generated from its usage.
2. Internal integration; this level is based on the localized exploitation of ICT. Internal integration enhances the benefits of localized exploitation by building an internal electronic infrastructure that enables the integration of tasks and processes, as well as functions, and therefore links all local ICT "islands." Internal integration aims at enhancing both efficiency (by reducing time and distance) and effectiveness (by improving information sharing between business processes). It has the potential to be an effective means of differentiation from a corporation's competitors, because it enables the creation of unique business processes.

According to Venkatraman (1991) those two levels form the basis for the purposeful deployment of ICT to (3) redesign business processes or (4) business networks and (5) redefinition of the business scope. The potential of ICT can be exploited fully only by performing these activities.

Figure 1: Five levels of ICT-induced reconfiguration (cf. Venkatraman 1991, p.127)

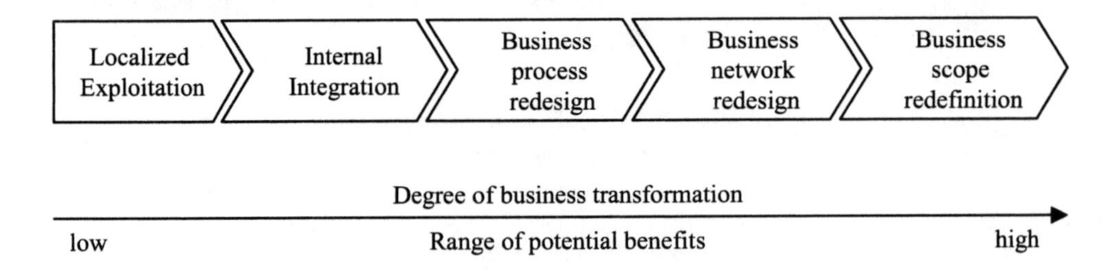

For the dimension "Degree of systematic and integral use of ICT" this results in the following three measures:

- Support of internal or external business processes; this degree corresponds to Venkatraman's level of internal integration. Moreover, integrated internal or external workflows become possible by integrating ICT solutions that support local business functions.
- Support of internal and external business processes; this degree enhances the first one by enlarging the extent of supporting business processes not only internally or externally, but both at the same time.
- Support of internal and external as well as inter-organizational business processes; ICT in this case is used to support inter-business processes and thus enables the design of more efficient and effective processes throughout a business network.

Degree of the alignment of the business toward organizational and cultural rules of the net economy

Besides answering the question of the optimal degree of external coordination, a successful e-commerce corporation also has to think about how to align its internal organization toward organizational and cultural rules of the net economy. It has to review whether its organizational structure corresponds to the requirements resulting from the fast-changing environment.

Basically, there are two extreme values within this dimension: structures characterized by a high degree of organizational rules and hierarchy (e.g., matrix organization, divisional organization), and very flexible organizational forms (e.g., project organization) where durable structures only exist during a certain project or task. The question of whether one or the other form of internal organization is more appropriate for a successful e-commerce corporation cannot be answered in general terms.

The fast-changing environment (technological innovations, new competitors, etc.) and an increased necessity to satisfy the ever-rising customer needs requires the ability to flexibly make changes in the internal organization. However, there might be a critical size where corporations cannot exist anymore without a set of organizational rules (i.e., hierarchies).

Degree of external coordination

This dimension has been added to our framework since we recognized that, in the information age, some general conditions have been changed that allow new efficient forms of interorganizational coordination. Our ideas are based on the relation established by Coase (1988) between the cost of using the price mechanism of a market and the existence of firms. The main reason for the existence of a firm is that "... the operation of a market costs something and that, by allowing some authority (an 'entrepreneur') to direct the resources, certain marketing costs are saved" (Coase, 1988, p. 40). Examples for those marketing (or transaction) costs are the costs of information procurement and processing or the costs of contracting.

The emergence of Internet technologies has resulted in a dramatic decrease of transaction costs. Examples are the costs of looking up business partners in electronic directories or the costs of exchanging data with them.

At the same time, Internet technologies also contributed to decreasing costs of intraorganizational activities, e.g., internal workflow management systems and document management systems dramatically reduced the processing time of certain business transactions and, therefore, increased the throughput.

Depending on the individual corporation, the decrease in transaction costs may exceed the decrease in intraorganizational costs or vice versa. Therefore, forms of external coordination via markets may become more efficient than the currently prevalent coordination within a firm.

Successful e-commerce corporations need to observe these trends, analyze their coordination of production and possibly adapt it to the new situation.

One possible form of adaptation might be to form alliances with certain corporations and to split the value chain into single steps, each of them to be provided by one corporation. An even more ambitious approach could be to participate in a value network, where each partner performs a dedicated role and is responsible for maintaining and driving both communication and cooperation concerning his "node" in the network.

ROLES IN THE E-COMMERCE ENVIRONMENT AND THEIR PATTERNS

A large number of classification approaches for e-commerce have already been proposed (cf. Dubosson-Torbay et al., 2002). The most simple is based on the reflections of Coase (1988) or the intermediation and disintermediation thoughts of Benjamin et al. (1995). Basically, there are two roles in the e-commerce environment

Figure 2: Framework for creating patterns of roles in the e-commerce environment

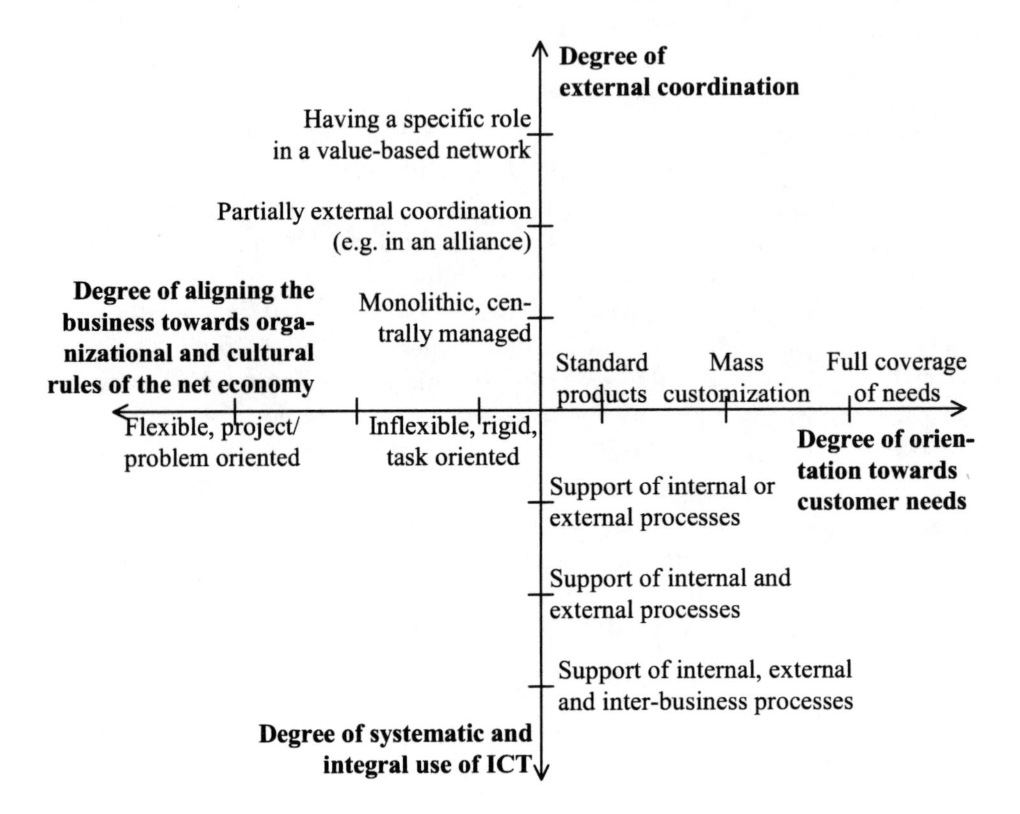

that we included in our framework. This is, first of all, the role "service provider" (SP), who is specialized in the production of very specific goods or services that correlate to its core competencies. Secondly, there is the "service integrator" (SI), whose main goal is to fulfil a certain need of a customer that mostly consists of many different goods and services which must be combined to one solution. The SI integrates all the necessary products and sells this individual solution to the customer. Thus, the definition of these roles is based on different requirements.

The SP role requires very good logistics; i.e., deliver the product at the right time, in the right amount, in the right quality, to the right addressee. The addressees or customers, respectively, can be both different SIs needing the product for creating the individual customer solution and the end customer him or herself, who only expects a certain product, but not a specific solution or service. Because the performance of this role is very specific, it mainly targets the effectiveness and efficiency of the process, thus also focusing on economies of scale. For example, it shows a pattern with the following characteristics regarding the different dimensions.

Table 1: Characteristics of the SP regarding the dimensions of the e-commerce-ability framework

Dimension	Characteristic
Degree of orientation toward customer needs	The SP only produces standardized or mass customizable products due to reasons of efficiency, e.g., to be able to realize a very short "time to customer."
Degree of systematic and integral use of ICT	In this case, it is not necessary to realize the highest possible degree, because the external processes are fixed and do not request a fully flexible integration or coordination of constantly changing services, for example.
Degree of the alignment of the business toward organizational and cultural rules of the net economy	Due to reasons of efficiency and a short time to market, the organization should be task-oriented and have a hierarchical structure. This leads to standardized processes which seem to be helpful as far as the implementation of a short lead time of the core processes is concerned.
Degree of external coordination	Because the SP has a very specific and clearly defined role in a value-based network, the degree of external coordination is high, although with a limited set of partners.

The SI role, however, requires a maximum amount of flexibility in regards to the creation of the performance needed. Therefore, effectiveness and efficiency — although of course still being important, due to economic reasons — are not considered main targets. Much more important is the good and lasting customer relationship that is established by the ability to coordinate many different partners to establish a large "pool" of goods and services, from which the most eligible ones can be chosen dynamically to create the best possible solution for the end customer. Taking this into account, we can deduce the following characteristics in regards to the dimensions of the framework for the SI, also depicted in Figure 3.

Table 2: Characteristics of the SI regarding the dimensions of the e-commerce-ability framework

Dimension	Characteristic
Degree of orientation toward customer needs	The SI provides full coverage of any need the customer states.
Degree of systematic and integral use of ICT	The maximum degree of this dimension should be reached in order to establish the most efficient integration process possible.
Degree of aligning the business toward organizational and cultural rules of the net economy	Since each need is individual, the organization must be very flexible and problem-oriented. Creativity for problem solving is mandatory and is difficult to create in a rigid environment.
Degree of external coordination	The SI also has a very specific and clearly defined role in a value-based network, thus the degree of external coordination is very high.

Integrated into the framework, the following patterns evolve as shown in Figure 3.

Having identified the patterns the different roles in an e-commerce environment should have, we now have to define the second part of being successful; i.e., we have to develop a concept for evaluating the economic success of e-commerce activities.

VALUE AS SECOND PILLAR OF EVALUATING THE SUCCESS OF E-COMMERCE ACTIVITIES

The success of a business model can be measured in many different ways. The most common way is to quantify it using financial figures such as the Return on Investment (ROI), Earnings before Interest and Taxes (EBIT) and Cash Flow (CF). The shareholder value discussion focuses on long-term sustainability and thus uses ratios such as the Discounted Cash Flow (DCF) or Economic Value Added (EVA) (Stern et al., 2000). Business models of the net economy, however, have proven that

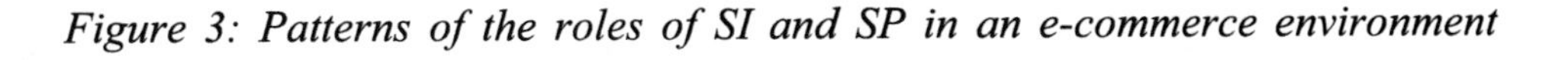

Figure 3: Patterns of the roles of SI and SP in an e-commerce environment

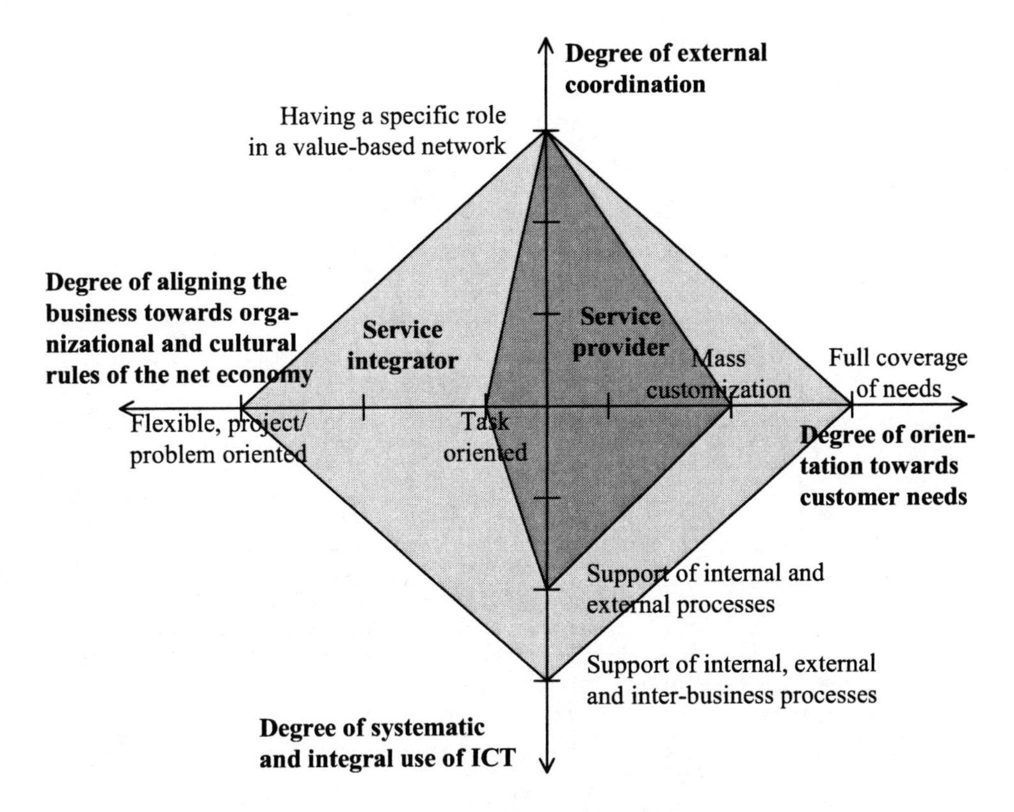

the mere consideration of financial figures is not sufficient. Especially in the first stages of a start-up (i.e., the lowest point of the so-called "hockey stick," where the main financial figures take a dive), financial ratios are not sufficient for reflecting the real value of a corporation. Therefore, we think that although financial figures still play an important role, they have to be complemented by qualitative parameters. Moreover, we base on the shareholder value approach for evaluating success. The term "value" implies that we drop the short-term perspective and analyze the medium to long-term substance and sustainability of a corporation. Value is often defined using operating performance and long-term growth. Because this does not represent all perspectives of shareholder value, but only the financial ones, the important elements of creating shareholder value are depicted in Figure 4 (Copeland et al., 2000, p. 91).

The main reason for our decision to choose this approach is our definition of "successful" as the creation of value over a medium- to long-term period of time.

Applying the shareholder value approach for these kinds of business models we base on Copeland, Koller and Murrin (cf. here and in the following Copeland et al.,

Figure 4: Important elements of value creation

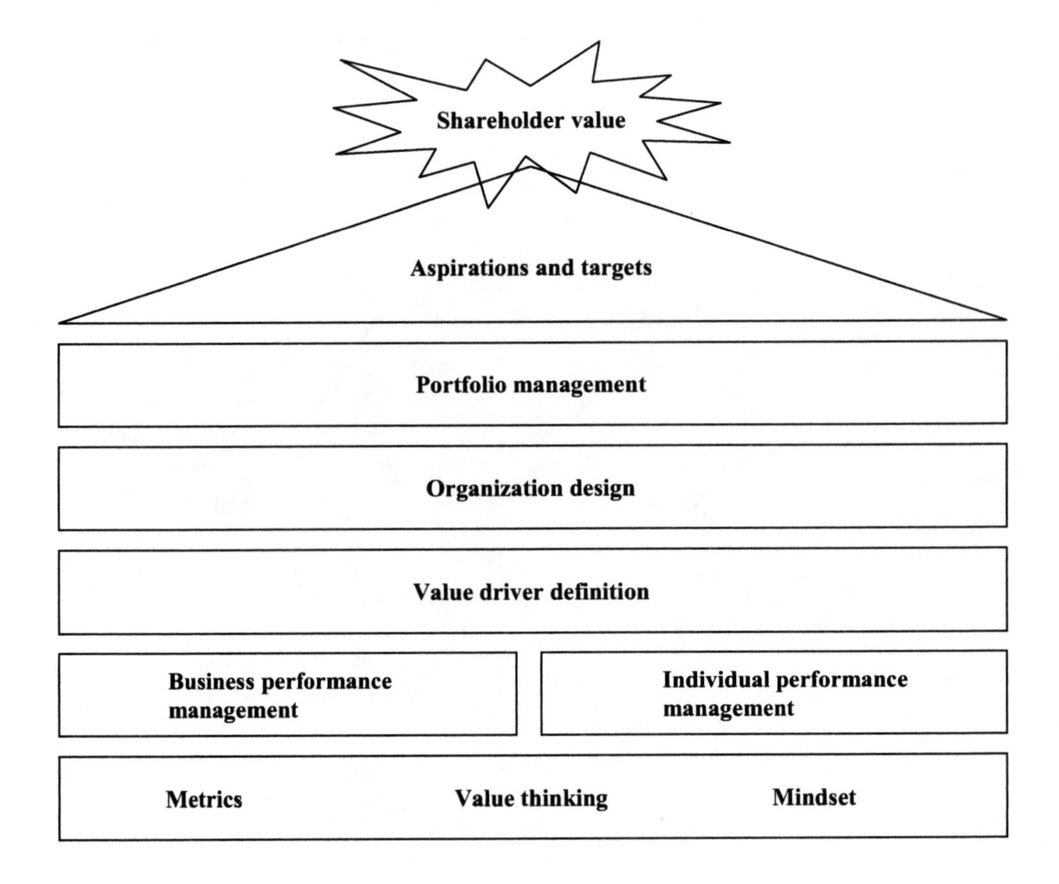

2000) and thus choose the quantitative as well as the qualitative parameters from the value drivers discussed there. Value drivers are those factors that have the highest impact on value both in regards to day-to-day business, as well as investment decisions having a medium- to long-term horizon. Thus, value drivers are performance indicators that differ from corporation to corporation, but have to be fully understood on each management level to be able to control value.

Copeland et al. suggest the drawing of a value tree for the representation of the value drivers chosen. We now adapt this for our purposes of valuating e-commerce activities of a corporation.

The e-commerce value tree is built by the following value drivers:

- *Contacts over electronic channels* with the "branches" being *types of information requested*, *customer segments attracted*, and *time to reaction*
- *Goods and services traded over electronic channels* with the "branches" being *types of products traded*, *lead time of process* and *customer satisfaction*

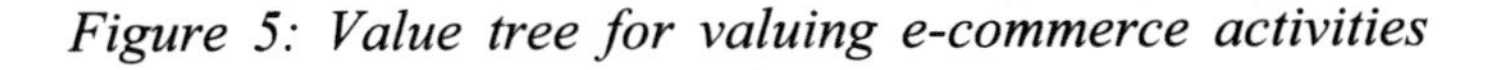

Figure 5: Value tree for valuing e-commerce activities

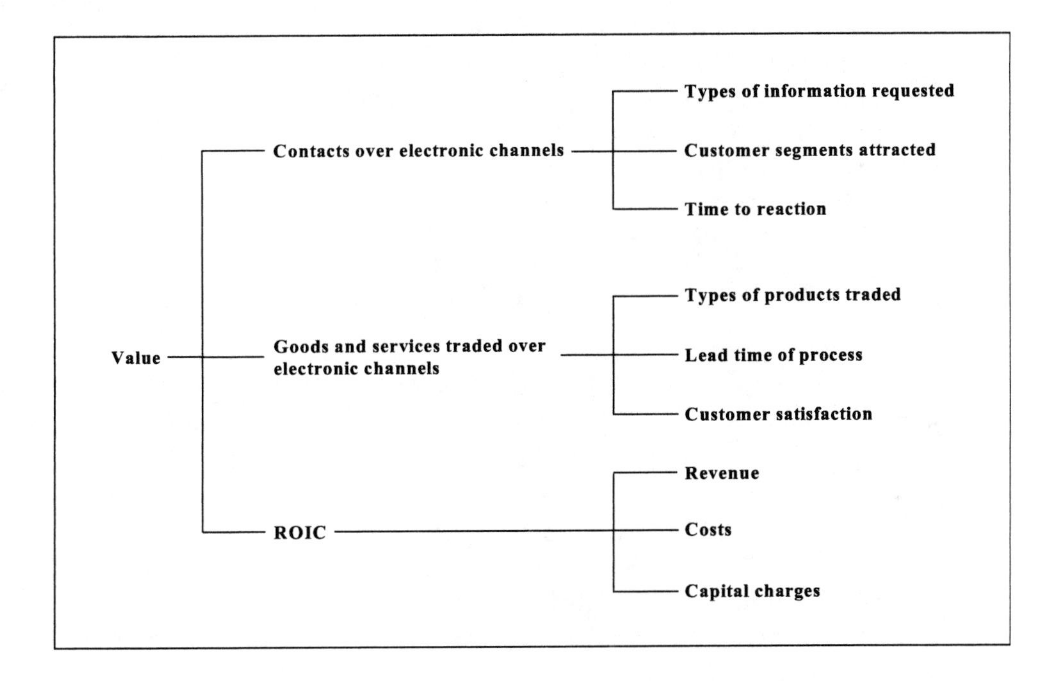

- *Traditional profit and loss tree* with *ROIC (Return on Invested Capital)* with the "branches" being *revenue*, *costs* and *capital charges*

The value tree can be enlarged or modified according to the analysis needs of management. Moreover, key performance indicators (KPI), which serve as metrics for operationalizing the value drivers, have to be defined (e.g., for the value driver, "customer satisfaction," a KPI "# of returned products" and/or "% of returned feedback questionnaires below the evaluation <satisfied>" could be defined). There are a few more points which ought to be observed for successfully using the value tree: First, it must be designed according to the specific analysis needs, the operative and strategic requirements and the environment of the corporation. Second, it must be appropriately cascaded down to each management level in order to ensure that its implications and the individual contributions to success become clear. Third, it must be analyzed during several consecutive periods to be able to observe changes and decide on actions in case of unwanted deviations. Moreover, it must be regularly adapted in order to be able to take the current requirements of e-commerce and competition into account.

CONCLUSIONS

The above developed concept is a first step toward analyzing e-commerce activities in a broader context of qualitative and quantitative parameters. It is designed to support management decisions regarding the medium and long-term strategies focusing on e-commerce activities and, moreover, to enable the evaluation of the current status, or position, as far as e-commerce-ability, i.e., the ability to perform successful e-commerce, is concerned.

The next step must be the validation of the concept by applying it to e-commerce corporations and enhancing it by developing a rule base for e-commerce success patterns regarding different business models within the role profiles.

REFERENCES

Aldrich, D. F. (1999). *Mastering the Digital Marketplace: Practical Strategies for Competitiveness in the New Economy.* New York: Wiley.

Angehrn, A. (1997). Designing mature Internet business strategies: The ICDT model. *European Management Journal,* 15, 361-369.

Bain and Company. (2000). Bain and Company eQ – Test. Retrieved October 15, 2000, from the World Wide Web: http://www.bain.de.

Baldwin, L. P. & Currie, W. L. (2000). Key issues in electronic commerce in today's global information infrastructure. *Cognition, Technology & Work, 2* (1), 27-34.

Baumöl, U., Fugmann, T., Stiffel, T. & Winter, R. (2001). A concept for the evaluation of e-commerce-ability. In A. Gangopadhyay (Ed.). *Managing Business with Electronic Commerce: Issue and Trends* (pp.199-213). Hershey, PA: Idea Group Publishing.

Benjamin, R.I., & Wigand, R.T. (1995). Electronic markets and virtual value chains on the Information Superhighway. *Sloan Management Review*, Winter, 62-72.

Burgess, L. & Cooper, J. (1999). A model for classification of business adoption of Internet commerce solutions. In S. Klein, J. Gricar & A. Pucihar (Eds.). *Global Networked Organizations: 12th International Bled Electronic Commerce Conference* (pp.46-58). Kranj, Slovenia: Moderna Organizacija.

Canter, J. (2000). An agility-based - ODDA model for the -commerce/e-business corporation. /- Retrieved October 15, 2000 from the World Wide Web: http://www.belisarius.com/canter.htm.

Coase, R. H. (1998). *The Firm, the Market, and the Law.* Chicago: The University of Chicago Press.

Copeland, T., Koller, T., & Murrin, J. (2000). *Valuation - Measuring and Managing the Value of Corporations*. New York: Wiley.

Dubosson-Toray, M., Osterwalder, A. & Pigneur, Y. (2002). E-business model design, classification and measurements. *International Business Review, 44* (1), 5-23.

European Commission (1997). A European Initiative in Electronic Commerce Retrieved May 29, 1997 from the World Wide Web: http://www.cordis.lu/esprit/src/ecomcom.htm.

European Commission (1998). Electronic commerce - An introduction. Retrieved October 15, 2000 from the World Wide Web: http://www.ispo.cec.be/ecommerce/answers/introduction.html.

GartnerGroup. (1999). Electronic Commerce Platforms and Applications (Report No. R-07-1624). GartnerGroup RAS Services.

Grant, S. (2000). E-commerce for small businesses. Retrieved October 15, 2000 from the World Wide Web : http://www.inst.co.uk/papers/small.html.

Kalakota, R. & Robinson, M. (2002). *M-Business: The Race to Mobility*. New York: McGraw-Hill.

Kalakota, R. & Whinston, A. B. (1997). *Electronic Commerce: A Manager's Guide*. Reading, MS: Addison-Wesley.

Kelly, K. (1998). *New Rules for the New Economy*. New York: Viking.

Morasch, K. & Welzel, P. (2000). Emergence of electronic markets: Implication of declining transport costs on firm profits and consumer surplus . In F. Bodendorf & M. Grauer (Eds.). *Verbundtagung Wirtschaftsinformatik*. 2000. Aachen, Germany: Shaker.

Österle, H. (2000). Enterprise in the information age. In H. Österle, E. Fleisch, & R. Alt (Eds.). *Business Networking* (pp.17-54). Berlin: Springer.

Paulk, M. C., Chrissis, M. B. & Weber, C. V. (1993). Capability Maturity Model, Version 1.1. *IEEE (Software)*. 10, 18-27.

PriceWaterhouseCoopers (2000). PriceWaterhouseCoopers E-Business: About PWC. Retrieved October 15, 2000 from the World Wide Web: http://208.203.128.56/external/ebib.nsf/docid/8149EB15B63E21508025691F004623AB?opendocument.

Schwartz, E. (1998). *Webonomics: Nine Essential Principles for Growing Your Business on the World Wide Web*. New York: Broadway Books.

Shih, G. & Shim, S. S. Y. (2002). A service management framework for m-commerce applications. *Mobile Network and Applications, 7*(3), 199-212.

Siau, K., Lim, E.-P. & Shen, Z. (2001). Mobile commerce: Promises, challenges, and research agenda. *Journal of Database Management, 12*(3), 4-13.

Stern Stewart & Co. (2000). Evanomics. Retrieved October 15, 2000 from the World Wide Web: http://www.evanomics.com.

Timmers, P. (1998). Business models for electronic markets. *Electronic Market, 8* (2), 3-8.

Venkatraman, N. (1991). ICT-induced business reconfiguration. In M. S. Scott Morton (Ed.) *The Corporation of the 1990s* (pp. 122-158). Oxford: Oxford University Press.

Whiteley, D. (1998). EDI maturity: A business opportunity . In C. T. Romm & F. Sudweeks (Eds.). *Doing Business Electronically: A Global Perspective of Electronic Commerce* (pp. 139-150). Berlin: Springer.

Wigand, R. T. (1997). Electronic commerce: Definition, theorie, and context. *The Information Society,* 13, 1-16.

Zerdick, A., Picot, A., Schrape, K., Artope, A., Goldhammer, K., Lange, U. T., Vierkant, E., Lopez-Escobar, E. & Silverstone, R. (Eds.). (2000). *E-conomics. The Economy of E-commerce and the Internet.* Berlin: Springer.

ENDNOTES

[1] An earlier version of this work has been published under the title "A Concept for the Evaluation of E-Commerce-Ability" in (Baumöl et al., 2001).

Chapter II

M-Commerce: A Location-Based Value Proposition

Nenad Jukic and Abhishek Sharma
Loyola University, Chicago, USA

Boris Jukic
George Mason University, USA

Manoj Parameswaran
University of Maryland, USA

ABSTRACT

Mobile Commerce (m-commerce) has been defined as a process of conducting commercial transactions via "mobile" telecommunications networks using a communication, information, and payment devices such as a mobile phone or a palmtop unit. Here, we analyze the potential ramifications in the field of marketing and changes in the market due to the advent of m-commerce. In particular, we analyze the opportunities that various characteristics of the m-commerce model can bring to the field of marketing. We investigate the likelihood of the emergence of mall-like zones that are based both on the geographical proximity of services and goods providers and the use of mobile communication devices. Such zones have a potential for becoming the basic units for any analysis of m-commerce scenarios. As m-commerce attains maturity, the zones could become the fundamental parameter in marketing evaluation.

INTRODUCTION

The advent of wireless and mobile technology has created both new opportunities and new challenges for the business community. Our aim here is to examine the potential impact of mobile commerce (m-commerce) on the relationship between customers and the providers of goods and services.

In its present state, m-commerce can be viewed as an extension of conventional, Internet-based e-commerce, which adds a different mode of network and accommodates different end-users' characteristics. However, if the predictions stating that mobile and wireless computing will dominate the Internet industry in the future (Vetter, 2001) materialize, the e-commerce and m-commerce could become a singular blended entity. Moreover, in the long run, m-commerce has the ability to emerge as a separate and different model. The magnitude of the mobile revolution has the potential to create a substantial pressure on the current e-commerce business models. It could generate apertures for new mobile Internet companies, engendering a stream of change among established e-commerce paradigms and leading to a reconfiguration of value propositions in many industries. Presently, m-commerce is still bugged by its limitations. The issues that are still being dealt with include: uniform standards, ease of operation, security of transactions, minimum screen size, display type and bandwidth, billing services, and the relative impoverishment of web sites. Due to current technological limitations, limited service availability and varying mobile consumer behavior patterns, business strategies developed for m-commerce applications will find it necessary to emphasize differing characteristics than traditional e-commerce strategies. At the same time, a successful m-commerce provider must understand that consumers are unwilling to spend long periods "surfing" on these inherently less user-friendly wireless devices. Wireless users demand packets of hyper-personalized information, not scaled-down versions of generic information. Therefore, technology-focused wireless Internet business models could be replaced by models that best integrate the unique characteristics of wireless m-commerce (Clarke, 2001).

To summarize, m-commerce is not a new distribution tool, a mobile Internet or a substitute for personal computers. Rather, it is a new aspect of consumerism and a much more powerful way to speak with consumers. Unleashing the value of m-commerce requires understanding the role that mobility plays in people's lives. That calls for a radical shift in thinking (Nohria & Leestma, 2001).

M-commerce, as defined by Muller and Veerse (1999), stands for conducting commercial transactions via a "mobile" telecommunications network using a communication, information and payment (CIP) device such as a mobile phone or a palmtop unit. In a broader sense, m-commerce can be defined simply as exchanging products, ideas and services between mobile users and providers.

The following section gives an overview of the characteristics of m-commerce. We discuss the fundamental characteristics of m-commerce that have the potential to influence the basic marketing orientation of both sellers and buyers, and above all, alter the general dynamics of the market. Subsequent sections discuss various ways in which the emergence of m-commerce could impact basic marketing orientation factors and overall marketing dynamics.

BACKGROUND

The terms mobile and wireless often are thought of as synonyms, but this is not always entirely accurate. The mobile user does not necessarily need to use wireless interfaces and wireless interfaces do not necessarily support mobility (Varshney, 2000; Vetter, 2000). For example, a user can browse through a store with a hand-held device, scanning items for inclusion into a wedding-registry database. The information will be transferred into the database once the user has finished browsing through the store and returned the hand-held device to the store clerk. The user was mobile, even though no wireless communication technology was used. On the other hand, a user sitting in his home could be using a wireless network as a part of his Internet connection, and that still would not make him a mobile user.

A mobile network is characterized by two main capabilities. The first is the ability to maintain communication between non-static locations. The second critical feature is the ability to keep track of the location. This characteristic has been conspicuous in new developments like Location-Based Services. It is very likely that a key driver of m-commerce growth will be location-based technology and applications. When operators know where their customers are, all sorts of commerce opportunities become possible (Secker, 2001).

Location-Based Services, or LBS, can be described as applications that react according to a geographic trigger. A geographic trigger might be the input of a town name, zip code or street into a web page, the position of a mobile phone user or the precise position of a vehicle. Using the knowledge of where someone is or where they intend to go is the essence of LBS.

LBS have been around for a long time. The arrival of high bandwidth mobile networks has brought the spotlight onto their potential. Large amounts of information that can be re-purposed for the wireless Internet, coupled with the ability to filter and personalize content by reference to a user's physical location, provide compelling businesses opportunities.

The following describes different components of Location-Based Services.

Geo-coding

Geo-coding is the task of processing textual addresses to add a positional coordinate (latitude and longitude) to each address. These coordinates then are indexed to enable the addresses to be searched geographically in ways such as "*find me a nearest* ___."

Map content

Map content can either be in a Raster or Vector format. Raster images are pre-rendered pictures such as aerial photographs. Vector data is a series of matching layers, each layer containing a specific type of information (a layer for parks, motorways, streets, rivers, etc.). Both formats can be used to display maps on a screen.

Proximity searching

Proximity searching is the method of finding relevant information to meet the user's specific request. Examples include: "*find everything within a radius of X,*" "*select all Y's I will drive past in the next hour*" or "*show me where I am.*"

Routing and driving directions

This refers to the interaction between the user location (origin) and a planned destination. Routes can be calculated and displayed onto a map and driving directions can be provided according to the "*shortest distance*" or the "*fastest route.*" In case of changing traffic or road situations, traffic data can be merged with the static map content to provide real-time alternative route suggestions and give you notice of the adjusted travel time.

Rendering

Rendering refers to the production of maps for display on the screen of a device. Rendered images typically are personalized according to the specific LBS request.

These location-based services will be an essential part of future mobile packages. Companies will need technology platforms in order to facilitate these services. Global and scalable platforms that are easy to integrate in the existing infrastructure will be essential. The technologies that satisfy these criteria are progressing at a fast pace in the form of wireless local area networks (WLANs), satellite-based networks, wireless local loops (WLL), mobile Internet Protocol (IP), and wireless asynchronous transfer mode (ATM) networks (Varshney, 2000; Vetter, 2000). However, just the extent of basic cellular phone usage is a good enough indicator of the magnitude of wireless potential. Already, the number of cellular subscribers in Europe, the U.S. and Asia is far greater than the number of users of the Internet. According to wireless service providers, m-commerce is viewed as the primary means of creating more value per subscriber and it has the greatest chance of success in the situations where it provides the three C's: convenience, low-cost, and compulsive usage (Swartz, 2001).

In addition to providing users with mobility and the ability to be tracked, m-commerce applications have the ability to achieve a high (i.e., detailed and accurate) level of personalization of the interaction with the customer. For example, as an addition to a web database, an m-commerce engine can include a real-time personalization engine that contains personal profiles of individuals, preferences, location marks for home or office and secure online payment details (Waddington, 2001). Because m-commerce provides a highly interactive environment, personalization through techniques like collaborative filtering can be highly accurate. One of the more disruptive aspects of e-commerce has been the inability to verify the accuracy of much of the data passed into the system. In other words, the data gathered via interactive web sites is as accurate as the input skills (or intentions) of the user who entered it. On the other hand, due to the simultaneous automatic interaction between the consumer and the service provider in the m-commerce setting, accurate individual data can be collected in an easy manner.

In addition to the obvious ability to influence the consumer behavior, m-commerce has the opportunity to create utilities of various types. The following sections discuss such opportunities.

The Impact of Mobile Commerce – Creation of Zones

The foremost perceptible influence of m-commerce, as it assumes significant dimension, could be the systematic but rapid generation of wireless shopping areas or zones, similar to real or virtual malls. These emerging market entities could be more succinct than their predecessors. A zone would be characterized by a renewed focus of the sellers and suppliers on the prospective clients in the zone and by a highly specific conversation between sellers and the zone-specific customers. The arrival of m-commerce devices that dynamically track the best deals on commodities and their respective locations could further facilitate the formation and subsequent distinction of zones based on the extent of the mobility of their customers and the degree of communication of the seller.

In addition, within a zone, customers' movements will not create or deplete the place utility of a product, but a seller — who is aware that the customer is equipped with the information on the best options for purchase — might be forced to offer the best price to attract customers for its general commodities. This realization could lead to a rapid reduction in prices within limited zones that eventually culminates as a terminal value of prices for every commodity. These effects can be summarized as the development of zones into wide and diversified (but integrated) supermarkets that are easily accessible to an associated population of customers and operate upon a strong, constant, two-way conversation between the seller and the buyer. Another ramification could be price differences (premium and discount) *between* zones based upon the place utility of the respective zones. However, as stated earlier, prices *within* zones would likely remain uniform. Furthermore, there is a possibility of price wars between zones and/or existing supermarkets. Under these conditions, the suburban supermarkets could be forced to maintain lower prices to compensate for the lack of place utility for their services.

As a synonym of accentuated and prompt interaction between the seller and the buyer, a zone may adversely affect the effectiveness of a discount as a marketing tool. The most common inspiration behind offering a discount is the tendency of the buyer to buy more than what he needs or to buy a product that does not match exactly his needs. Under the new circumstances, a discount could quickly propagate throughout the zone and this new price could leave no incentive for the buyer to stock for the future or needlessly deviate from his usual course of purchase.

The Impact of Mobile Commerce – Place Utility

Due to the constant tracking of the goods an individual needs and equally vigilant tracking of the customers, individual purchase opportunities can be optimized. This scenario can be compared with one in which a good is always available to the customer at a fixed distance equivalent to the extent of a zone. For example, a person who issues and transmits a query in the middle of a zone, seeking a closest Pizza-Hut® restaurant, can be considered an individual who carries proximity to the store along with him. This association of quick and ubiquitous availability of desired goods in a zone can lead to other subtle but important impacts on the marketing strategy, pricing decisions and consumers' orientation.

As a reverse trend in a consumer and seller relationship, the stores could come closer to the customers or to the places where their chances of intersecting the customers are greater. This further implies that the suppliers could be forced to spread their operations across wider regions to increase interaction with customers. This expansion should be directed toward having presence in more than one zone, rather than toward a heavy concentration in one particular zone. Direct implication of this reshuffle could be specifically borne by the bulky retail entities, like supermarkets, that are characterized by suburban location, huge inventory stock, and attractive one-stop–shopping-appeal. With the advent of the facility that could provide dynamic information regarding the best bargain nearest to the customer, the appeal of one-stop-shopping could face a serious challenge as the motive behind the journey to a suburban superstore or mart fades (since it could be replaced by a short sprint to the nearest zone).

Location-based marketing, via global positioning technology, will soon be available in mobile devices. Through this technology, m-commerce providers will be better able to receive and send information relative to a specific location (Clarke, 2001). All this facilitates the creation of a place utility as discussed above.

The Impact of Mobile Commerce – Personalization

Personalization has been one of the foremost aims of any marketing tool ranging from the traditional advertisement to modern e-commerce technologies. The personalization of content and services that help consumers make their purchasing decisions is pivotal and information that facilitates this goal is key to the overall success of m-commerce (Raisinghani, 2001). Typically, an individual uses one mobile device. This makes mobile devices ideal for individual-based target marketing. Additionally, personalized content assumes greater importance in mobile devices, because of the limitation of the user interface. Relevant information must always be only a single "click" away, because web access with any existing wireless devices is not comparable to a PC screen either by size, resolution or "surfability" (Clarke, 2001). M-commerce can utilize such functionality to realize its basic goals. For example, a retail chain can target all females ages 14 to 21 who are within 400 meters of its stores anywhere in the country with its latest bargain. In this context, two relevant features of the m-commerce architecture are the ability of the system to track the individual and the quality of personalization engines.

The ability of the system to track implies, essentially, a robust platform that allows triangulation between different network base stations which, in turn, allows the engine and mobile devices users to track the location of users within the range of the zone. The tracking ability can determine the exact coordinates of the individual and henceforth suggest the nearest commodities that suit his profile, providing those coordinates as well. A store or a group of stores of any size can mimic a carefully arranged grocery store that considers the buying pattern of the individual. This extreme level of target marketing is difficult to achieve by traditional methods.

At the same time, new personalization methods (especially implicit techniques like collaborative filtering) could raise the personalization to a level that is far greater than a simple sum of a few patterns computed by traditional rule-based or segmentation-based software. Instead of just figuring out what product a customer will be interested in, these personalization techniques operating in the m-commerce scenario could even arrive at conclusions about the psychology of the individual. The strength of the personalization techniques in m-commerce emanates from the ability of the system to keep a close watch on the activities and behavior of the individual and provide a rich, implicit input to the collaborative part of the personalization engine, which makes the filtering and extrapolation part of the process more accurate.

The Impact of Mobile Commerce – Promotional Utility

As an aftermath effect, the formation of zones and high level of personalization could make m-commerce an ideal tool of promotional utility. As mentioned earlier, m-commerce as a gateway to immense personalization could translate well into the promotion of a good or service. It is highly probable that personalized promotion efforts could materialize into more successful market conversations. In a hypothetical situation, where a restaurant sends its menu for the day to all the prospective customers in a 500- meter radius, due to personalization techniques, every individual can receive his favorite dish in highlighted or different format. Because promotional messages to customers will be based on their respective coordinates, these wireless advertisements could be considered as billboards on the top of the stores, which are visible only when you step into that particular zone. With the combination of tracking ability and high level of personalization, m-commerce promotions can guide the target customer to the doorstep of the store, an attribute absent in traditional promotional efforts.

Despite this rearrangement and the creation of new stores in the market, some of the restrictions could still hold. There could be little "zone effect" on the goods like cars, expensive electronics and jewelry, where place utility does not match price utility. In addition, even easy access to goods might not dissolve brand equity of a good or service with lower level of presence in an immediate zone.

The Impact of Mobile Commerce – Interaction

Another major influence of m-commerce on market dynamics could be related to the level of interaction between the seller and the buyer. M-commerce can expedite the interaction between the store and the consumer and bridge the associated gaps of e-commerce.

In m-commerce, the target customer is highly mobile and moves in close surroundings to the seller. Under this scenario, there could be a perceptible reduction in the time between advertisement and the actual transaction. This can be compared to the market of yesteryear, where a seller shouted his offer to the customer and the customer gave an immediate reply. M-commerce has facilitated a geographical proximity between the seller and the customer, a characteristic absent in e-commerce. E-commerce has faced challenges in the individual-specific goods like clothes and fragrances, which are not bought very often without personal involvement. In many cases, personal goods simply require physical presence of the end user. As discussed earlier, the entry of m-commerce can affect the general characteristics of marketing. These effects can be different for different market segments and they can vary in intensity. A regular e-commerce customer, who is mainly concerned with the cost part of the transaction, may not be highly influenced by m-commerce in the initial days. But, as zones start forming, m-commerce can become a competitive and cheap alternative to e-commerce. During the infancy of m-commerce (pre zone-time), the target market for m-commerce is those individuals who care more for the time utility created by the service rather than the price utility. At this time, a typical m-commerce customer is a highly mobile individual who is explicitly willing to pay for the place utility generated. However, the repercussions of m-commerce can raise significant challenges for the marketing appeal of an e-commerce model based on potential for cost reductions, just in time possibilities and easy access to major alternatives.

CONCLUSION

This text analyzed and described a conceptual framework for the progression of the initial stages of m-commerce, with special emphasis on how the target market may change as the technology matures. Recent trends illustrate a rapidly changing environment, where hundreds of millions of subscribers have moved to different forms of wireless technology to acquire mobility and to achieve a quick communication network (Standard & Poor's, 2001). At the same time, liquid crystal display technologies, and electric and digital papers are indicating the arrival of wireless communication as a way of life (Vetter, 2001). Under these circumstances, the discussed aspects of m-commerce assume unprecedented importance.

REFERENCES

Clarke, I. (2001). Emerging value propositions for m-commerce. *Journal of Business Strategies, 18*(2), 133-148.

Lewis, S. (2001). M-commerce: Know your customer. *Asian Business, 37*(5), 34.

Muller-Veerse F. (1999). Mobile Commerce Report, *Durlacher Reports*. Retrieved from the World Wide Web: http://www.durlacher.com/bbus/resreports.asp

Nohria, N. & Leestma, M. (2001). A moving target: The mobile-commerce customer. *MIT Sloan Management Review, 42*(3), 104.

Raisinghani, M. S. (2001). WAP: Transitional technology form-commerce. *Information Systems Management, 18*(3), 8-16.

Secker, M. (2001). Does m-commerce know where it's going? *Telecommunications, 35*(4), 85-88.

Standard and Poor's. (2001). *Industry Surveys*. Telecommunications: Wireless. 169, 44(2): Shere C. & Abreu K.

Swartz, N. (2001). Hot & cold: M-commerce opportunities. *Wireless Review, 18*(6), 32-38.

Varshney, U. & Vetter, R. (2000). Emerging mobile and wireless networks. *Communication of the ACM, 43*(6), 73-81.

Vetter, R (2001). Wireless Web. *Communication of the ACM, 44*(3), 60-61.

Waddington, P. (2001). Wireless odyssey. *Oracle Magazine, 15*(1), 81-108.

Waters, J.K. (2000). Getting personal on the Web. *Application Development Trends,* 25-32.

Chapter III

E-Commerce Adoption in Small Firms: A Study of Online Share Trading

Pak Yuen P. Chan
ATC Hong Kong University of Science and Technology, ROC, Hong Kong

Annette M. Mills
University of Canterbury, New Zealand

ABSTRACT

This research examines the adoption of e-commerce technology (namely, order-execution online trading technology) by six small brokerage firms at various stages in the evaluation and adoption processes. The study is informed by innovation theory and prior research, and seeks to identify the key factors influencing the adoption process. Consistent with innovation theory, the case findings suggest that three classes of factors influence adoption: innovation factors, factors, organizational factors and environmental factors. The key factors within each of these classes were identified as compatibility and perceived benefits (innovation factors); IT sophistication, internal and external IT support and management support (organizational factors); and pressure from e-commerce-able competitors and clients (environmental factors). Of these variables, compatibility and perceived benefits were found to be the most significant in impacting e-commerce adoption.

INTRODUCTION

Electronic commerce (e-commerce) is significantly changing the dynamics of the business environment and the way in which people and organizations do business with one another. E-commerce addresses many of the needs within organizations. For example, e-commerce enables and facilitates electronic markets: it allows firms to cut service costs while improving speed of delivery and to simplify and streamline business processes; it supports the delivery of information products, services and payments; and it improves information exchange with customers. E-commerce also can enhance company image, enable access to new customers, and generate new business opportunities (Applegate et al., 1996; Nath et al., 1998; Turban et al., 2000). For these and many other reasons, firms are adopting e-commerce technology.

E-commerce adoption is particularly relevant in the financial services sector. Rapid advances in information and communication technology and increased public awareness have allowed e-commerce to evolve into a major distribution and service channel for financial services (e.g., online banking and online share trading), changing business processes as well as the dynamics of the financial market (e.g., Allgood, 2001). Although the financial services sector traditionally has lead the way in information technology (IT) adoption, there are few studies in the literature on e-commerce adoption in this sector (Ngai & Wat, 2002). Furthermore, because small firms often possess characteristics that are atypical of large firms (Fink, 1998), the relevance of existing research to small firms is not always apparent. It is therefore expected that this study will extend current understanding of e-commerce adoption in the financial services sector, and of small firm adoption. Understanding the factors that influence adoption is key to the success of e-commerce initiatives in any industry. For example, diffusion theory suggests that a technology would have little utility unless a critical mass of adoption is achieved (Rogers, 1989).

Drawing on existing research and innovation literature, this chapter seeks to identify key factors that motivate or inhibit e-commerce adoption among small firms. The contents are organized as follows. The following section introduces innovation theories that have been successfully used to explain innovation adoption and reviews previous findings. Innovation theory is then used to inform an investigative study of e-commerce adoption in small firms; more specifically, the adoption of order-execution online share trading technology by small brokerage firms. The research methodology is discussed and the research model assessed using case findings. The chapter concludes with a discussion of the findings, limitations, and directions for future research.

BACKGROUND

Innovation theory suggests that adoption of an innovation may be viewed in terms of stages of adoption (e.g. Rogers, 1983; Zaltman et al., 1973). Zaltman et al. (1973) proposes two stages of innovation adoption: the *initiation* stage involving knowledge and awareness, formation of attitudes toward the innovation and decision-making (to adopt or not adopt an innovation); and the *implementation* stage (i.e., the actual implementation of the technology). Rogers (1983) proposes a similar theory, namely the *innovation-decision* process which comprises the stages of *knowledge*, *persuasion*, *decision*, *implementation* and *confirmation*. It is at the *decision* stage that the organization determines whether to adopt (accept) or not adopt (reject) the innovation. This study focuses on the pre-implementation stages of adoption. The objective is to identify factors that enable or inhibit adoption of an e-commerce innovation in small firms.

Much of the prior adoption research focuses on three classes of factors as influencing adoption (Kuan & Chau, 2001; McGowan & Madey, 1998), namely, *innovation (technological) factors* (e.g., perceived benefits, complexity, compatibility), *organizational factors* (e.g., size, management support, IT support) and *environmental factors* (e.g., pressure from clients and competitors). This basic framework has been successfully used to identify factors that influence organizational adoption of various information technologies, including EDI by small firms (Chwelos et al., 2002; Iacovou et al., 1995; Kuan & Chau, 2001) and large firms (McGowan & Madey, 1998), the Internet by small firms (Mehrtens et al., 2001), open systems (Chau & Tam, 1997) and communication technologies (Prekumar & Roberts, 1999). In this research, this framework is used to guide the study of an e-commerce adoption; hence this research will seek to identify and examine these three types of factors considered significant for the adoption.

Prior research suggests a number of individual factors (within each of the adoption factor classes identified above) as driving the adoption of various information and communication technologies. These include perceptions of advantages, compatibility, complexity, external pressure, pro-activity toward technology, organizational support, financial readiness, and internal and external IT support (Duxbury & Corbett, 1996; Kendall et al., 2001; King & Teo, 1996; Lederer et al., 1997; Rai & Bajwa, 1997; Raymond & Bergeron, 1996; Salehi, 1997). Lack of employee knowledge and skills, cost of development, maintenance, management resistance, incompatibilities with existing systems, lack of infrastructure to support the IT innovation, lack of IT drivers and low economies of scale are factors that may slow adoption (AC Nielsen, 2001; Duxbury & Corbett, 1996; King & Teo, 1996).

Firm size is also a key adoption factor; prior research suggests that smaller firms may be less likely to adopt e-commerce (AC Nielson, 2001; Premkumar & Roberts, 1999). For small firms that do adopt information technologies, research suggests relative advantage, perceived benefits, financial readiness, technological readiness, IT knowledge of non-IS professionals, management support, CEO attitudes, internal and external IS support and external pressure are factors influencing adoption (Chwelos et al., 2001; Cragg & King, 1993; Fink, 1998; Iacovou et al., 1995; Kendall et al., 2001; Kuan and Chau, 2001; Mehrtens et al., 2001; Thong & Yap, 1985). On the other hand, lack of internal IT support and inadequate financial resources inhibit adoption among small firms (Cragg & King, 1993).

THE RESEARCH MODEL

This section reviews prior research on individual factors that influence information technology (IT) adoption. The research model considers the role of individual factors such as perceived benefits and compatibility, financial and technological readiness, management and staff support, and environmental pressures from competitors and clients on e-commerce adoption. The proposed model is then tested using case data drawn from small brokerages adopting order-execution online share trading technology.

Innovation (Technological) Factors

Innovation factors include *perceived benefits* and *compatibility* (McGowan & Madey, 1998). *Perceived benefits* refers to the direct (operational savings related to internal efficiency of the organization) and indirect (opportunities derived from the impact of EC on the business processes and relationships) benefits that a technology can provide the firm (e.g., Iacovou et al., 1995). Research suggests that these benefits include cost leadership, assistance in marketing to special customer groups, product differentiation, ease of access and global reach, low cost advertising medium, increased efficiency and competitive advantage (Gonsalves et al., 1999; Nath et al., 1998).

Compatibility describes the degree to which an innovation is perceived as consistent with the existing values, past experiences and needs of a potential adopter (Rogers, 1983). For example, research shows that compatibility with existing systems is positively associated with adoption (e.g., Attewell, 1992; Duxbury & Corbett, 1996; Farhoomand et al., 2000). Compatibility also recognizes the extent to which a technology aligns with the firm's needs. The importance of aligning a firm's IT strategy with its business strategy cannot be overstated (King & Teo, 1996; Lederer et al., 1997). For example, research has shown that business strategy directly influences the adoption and integration of IT into the organization (Bhattacherjee & Hirschheim, 1997; Cunningham, 1999). Raghunathan and Madey (1999) noted that without a corporate e-commerce strategy for guidance, firms may adopt non-integrated information systems with conflicting goals. Business strategy should therefore determine technology selection and its uses.

Organizational Factors

Organizational factors address the resources that the organization has available to support the adoption (McGowan & Madey, 1998). This includes financial and technological resources, as well as support from top management and staff in respect of the adoption.

Financial Support refers to financial assistance from within or outside the firm's resources (e.g., loans and subsidies) that equip the firm to acquire e-commerce technology. Access to sufficient financial resources is generally acknowledged as a factor that enables adoption. For example, only businesses that have adequate financial resources are likely to adopt IT (Iacovou et al., 1995; Thong & Yap, 1995). Although the cost of technology adoption can vary widely, limited financial resources can inhibit uptake, especially for small firms (Cragg & King, 1993).

Technological Readiness considers internal IT sophistication and access to external IS support. *Internal IT sophistication* refers to the level of sophistication of IT usage, IT management and IT skill within the organization (Iacovou et al., 1995). Iacovou et al. (1995) found that organizations that are more IT sophisticated are more likely to adopt EDI. Attewell (1992) also suggested that internal technological knowledge and expertise could influence the technology diffusion process. For example, CEO knowledge of IT has been identified as a key factor influencing adoption (e.g., Thong & Yap, 1985) while lack of knowledge may inhibit uptake (e.g., AC Nielson, 2001).

External IS Support refers to IT-related assistance received from outside the firm (e.g. external consultants). Research suggests that strong support from external technical sources may lead to or accelerate technology adoption (e.g., Attewell, 1992; Raymond & Bergeron, 1996; Salehi, 1997). Small firms in particular often lack access to sufficient internal IT resources, and tend therefore to rely on external IS support to assist the uptake of technology (e.g., Cragg & King, 1993).

Top Management Support. Researchers argue that top management support of the innovation could lead to adoption or early adoption of an innovation (King & Teo, 1996; Rai & Bajwa, 1997; Raymond & Bergeron, 1996; Salehi, 1997). For example, research suggests that the support of a top management champion may have a positive impact on adoption (Angeles et al., 2001; Mehrtens et al., 2001). Characteristics of the CEO, including attitude toward an innovation, is also considered an important factor influencing IT adoption (Thong & Yap, 1985).

Staff Support is defined as the extent to which technology adoption is supported by staff, who are not members of top management, but who will be affected by the technology. Researchers argue the importance of staff support of the adoption of technology, especially when the technology could lead to changes in their routines (Cunningham, 1999; Headrick & Morgan, 1999; Salehi, 1997). Sheng et al. (1998) found that favourable staff attitudes toward the technology were important to implementation.

Environmental Factors

The environmental context embraces external pressures to adopt a technology. These may be exerted by competitors, clients, trading partners and characteristics of the marketplace. For example, Chircu and Kauffman (2000) suggested that the e-commerce adoption decision of traditional firms may be influenced by other e-commerce-able firms. Rogers' (1989) diffusion study suggested that interactive technology had zero utility until other individuals adopted the technology as well, so until a *critical mass* of adopters was achieved, the rate of adoption would be slow. As more customers begin to trade online and competitor adoption of Internet-based online trading increases, this can be expected to influence organizational adoption of a technology.

METHODOLOGY

In this study, case evidence was collected from six New Zealand-based brokerage firms that had adopted or were evaluating/trialing order-execution online share trading technology. Online share trading allows investors to buy or sell shares e-commerce through their own computer (or other electronic device) on the Internet, typically by way of an online trading website access provided by a brokerage firm. Online share trading provides a type of self-help environment, in which investors can access their trading history and investment research, place orders (in a secure environment) and receive electronic trade confirmation; payments are managed through a *call account* that is debited or credited accordingly. Online trading allows New Zealand investors to trade directly on the National Association of Securities Dealers Automated Quotation (NASDAQ) stock market, the New York Stock Exchange (NYSE) and other offshore exchanges that allow such access.

Increasing demand for online trading in the New Zealand market has spurred the New Zealand Stock Exchange (NZSE) to develop online trading systems to upgrade the current automated settlement and clearance system (Walker & Fox, 1999). When implemented, these systems will allow individual investors to place their orders directly to the exchange via the broker's Web-based interfaces (Walker & Fox, 1999). As a relatively new and emerging technology in the New Zealand financial market[1], the order-execution online trading context is particularly useful for studying e-commerce adoption.

The case data was gathered using semi-structured interviews with top management who were directly involved in the adoption decision. Where possible, internal documents, Web site information, and other publicly available documents were also reviewed.

The sample comprised three full-service[2] retail brokerage firms, two discount[3] retail brokerages and one firm that operated in the institutional market[4], as well as the retail market (see Table 1). Three of the firms were at the *evaluation* stage, one at the *trial* stage and three at the *adoption* stage (Rogers, 1983). Of these firms, five were NZSE-members, and were therefore able to trade directly on the exchange. NZSE-member firms have permanent connections with the NZSE, through *FASTER*, the automated settlement and clearance system. NZSE-member firms had developed Internet-based interfaces to facilitate trade between the investor, the broker's systems and NZSE systems using in-house expertise or external consultants. Alternatively, firms may subcontract the order-execution process to an online share trader on a shared commission basis. Sub-contracting the order-execution process is the most viable option for non-NZSE member firms who cannot trade directly on the exchange; only firm A1 was a non-NZSE member.

Table 1: Profile of cases

Case	A1	A2	A3	L1	L2	L3
Type of firm	Discount	Full service	Discount	Full service	Full service	Full service
Market focus	Retail	Institutional & Retail	Retail	Retail	Retail	Retail
Number of Staff	< 20	40 – 60	20 – 40	20 – 40	> 80	> 80
Assets (NZ$)	<$200,000	$5–7m	$10-20 m	> $2m	$400-600000	$5–7m
Type of Adopter	Early	Early	Early	Later	Later	Lagging
Adoption stage	Adopted	Adopted	Adopted	Evaluation	Trial	Evaluation

FINDINGS AND DISCUSSION

In this study, *early adopters* (Firms A1, A2, A3) were identified as firms that had already adopted the technology. These firms believed the technology provided strong support for firm business strategy (compatibility) and that the benefits outweighed the drawbacks. They had sufficient financial resources, technological resources and had the strong support of top management for the adoption. Although there was external pressure to adopt the technology, the main drivers to adopt were internal. These early adopters were found to be more proactive toward adoption rather than reactive to marketplace changes.

Later adopters (Firms L1, L2) were identified as firms that face some external pressure to adopt the technology. However, these firms did not believe that online share trading provided strong support for firm business strategy (compatibility). While they recognized some benefits through adoption (e.g., increased internal efficiencies), the disadvantages were a key concern of top management (e.g., weak profitability of online trading) fueling reluctance to adopt the technology. These firms had the necessary financial and technical resources to adopt the technology, but lacked the enthusiastic support of top management. Firm management preferred a "wait and see" approach to adoption. Later adopters were found to be more reactive (rather than proactive) to market changes. Relative to early adopters, the adoption process was more gradual.

Lagging adopters (Firm L3 only) face little direct external pressure to adopt the technology either from competitors or clients. Top management also did not believe the technology supports firm business strategy; they believed that the drawbacks outweighed the benefits, hence, little or no advantage. Although this firm had the necessary financial and technical resources to adopt the technology, they were not motivated to do so. However, as more firms go online and the dynamics of the financial market change (e.g. Poon & Swatman, 1999), these firms are likely to adopt online trading technology, but their adoption rate is expected to be slow.

Adoption Factors

Analysis of the case findings provided support for the usefulness of the three-class model adoption framework (innovation, organizational and environmental factors) for explaining the adoption of online trading technology by small firms and identifying individual factors that are significant to the adoption decision. Specifically, the case evidence suggests that perceived benefits, compatibility (specifically, support for firm business strategy), technical readiness, management support and competitor and client pressures influence e-commerce adoption. Of these factors, perceived benefits and compatibility appeared to be the most important drivers of the adoption, distinguishing early, later and lagging adopters.

Innovation Factors.

Compatibility. The findings showed that early adopters (A1, A2 and A3) believed online trading provided strong support for firm business strategy, enabling them to compete more effectively in the discount retail market. For example, firm A1 aims to be a one-stop shop for investors using the Internet as a key tool. Later adopters (L1, L2) believed online trading provided only partial support for firm business strategy. These firms emphasized client relationships as fundamental for firm business strategy and believed personalized service and quality of service provided competitive advantage; they did not believe the online trading technology supported such a strategy. Although there was some demand from clients for an online trading service, these firms (L1, L2) would prefer a technology that supported a wider range of advisory services and was therefore more compatible with business strategy.

Lagging adopter (L3) believed order-execution technology was not compatible with firm business strategy. Like L1 and L2, they believed online trading did not support the firm's focus on customer relationships and service quality. They did not face pressure from clients to adopt, believing their clients were neither interested in, nor had the time to execute their own transactions - firm L3 believed their wealthier clientele preferred the *personal (human) touch* and their advisors to manage portfolio transactions. Firm L3 was undecided about their approach to order-execution technology. Changes in the marketplace were likely to lead to the firm providing an online service but, in its current form, the technology did not align with business needs. L3 was therefore evaluating the potentials of the technology to determine an adoption strategy that would better support firm needs.

Full-service firms (A2, L1, L2, L3) held the view that order-execution technology was best suited to the discount market. Rather than being a technology that could support and enhance business strategy, it was seen as simply replacing the telephone or face-to-face meetings as the medium for executing transactions. This was considered efficient for dealing with small clients. However, for the wealthier client, the direct and personal advisor-client relationship was key and online trading was perceived as incompatible with creating, supporting and maintaining these relationships.

In summary, these findings are consistent with prior research that emphasises the importance of aligning business strategy and IT strategy (e.g. Kendall et al., 2001; King & Teo, 1996; Lederer et al., 1997). The case evidence suggests that compatibility was a key factor determining adoption and that the extent to which the technology was compatible with firm business needs may also determine rate of adoption. The findings also suggest that, while all the firms were not keen on adopting the order-execution technology as is, the changing dynamics of the industry were likely to lead to the adoption of similar and extended systems. This is consistent with previous research and critical mass theory; as more competitors go online, Internet commerce is likely to become a competitive necessity rather than a competitive advantage (Poon & Swatman, 1999).

Perceived benefits describes an awareness of direct and indirect benefits (Iacovou et al., 1995) that e-commerce can provide. The case findings suggest a positive relationship between perceived benefits and e-commerce adoption. Early adopters (A1, A2 and A3) expected to realize benefits such as reduced operating costs and increased market reach and market share by including new customers and broadening the existing customer base through online trading; these benefits were expected to impact the bottom-line and increase financial profitability. For early adopters, the benefits far outweighed the drawbacks, accelerating the adoption rate relative to later and lagging adopters.

Later adopters (L1, L2) recognized benefits similar to those of early adopters. For example, L1 believed the key advantage lay in increased efficiencies and savings in information distribution. However, these firms also recognized disadvantages, which slowed adoption. For example, although L1 had invested heavily in technology development, they believed many in the industry were losing money. They preferred to "hold off" and "learn from everyone else" before making a final decision on an adoption strategy. Firm L2 believed that, to achieve adequate cost savings, it was necessary to "totally eliminate the human component" from the order-execution process. However, NZSE restrictions meant that the human element was not totally eliminated, thereby slowing the order-execution process. This weakness inhibited adoption.

All of the firms believed online trading attracts the small independent retail client and is beneficial for firms targeting this market segment. However, lagging adopter L3 operated in a niche market segment (focusing on the wealthy client) which, like later adopters (L1 and L2) they believed would be better served by a technology that provided a range of services other than order-execution only. They believed there were few benefits for the firm and were therefore reluctant to adopt the current technology.

Later and lagging adopters identified weak profitability of internet ventures as a key inhibitor. For early adopters, low initial return was not an inhibitor, as they believed the market would eventually return a profit. For example, A1 acknowledged that their firm was operating at a loss, but believed that recent improvements in their revenue streams would lead to eventual profit.

The view of the benefits of the technology held by early adopters was more optimistic than that of later or lagging adopters. This suggests that if firms perceive the benefits to be significantly greater than the drawbacks, this may accelerate the adoption process, leading to the adoption or early adoption of a technology. These findings are consistent with prior research, suggesting these benefits are likely to accelerate to technology adoption. (e.g. Chwelos et al., 2001; Fink, 1998; Harrison et al., 1997). Premkumar and Roberts (1999) also found that relative advantage distinguishes adopters from non-adopters.

Organizational Readiness. Early adopters (A1, A2 and A3) and later adopters (L1, L2) possessed the necessary financial and technological resources, as well as management support, to adopt the technology. Although lagging adopter L3 possessed the financial and technological resources necessary for adoption, top management did not support the adoption, because they believed the technology was not aligned with firm business strategy.

Financial readiness. The results showed all firms were financially self-capable of offering an order-execution service to clients, whether by way of building the interfaces themselves or sub-contracting the order-execution process. While developing an interface to the basic system was not considered very expensive, developing fully integrated systems (which was a goal of the firms in this study) was costly – one firm estimated they had invested at least $40 million (NZ) throughout the past two years developing fully integrated systems. They believed that front-end online trading systems were inadequate for meeting the firm's needs; for the systems to work efficiently, integration was a necessary requisite. Clearly, this was an investment that could not be readily undertaken by the average small firm.

In this study, financial readiness did not inhibit adoption of a basic online trading technology, because cheaper technology acquisition options were available for firms with fewer resources. For example, of the firms, only A1 cited access to financial resources as a factor inhibiting technology uptake. Although the firm offered an online trading service by subcontracting the order-execution process, they lacked sufficient capital to integrate the order-execution process into backend systems. Hence, in the case of full integration (which would be necessary for firms to be able to realize the full potential of e-commerce), financial readiness could be a key constraint for firms without sufficient capital. Clearly, financial readiness was an enabler of technology adoption. This is consistent with prior research that suggests that only those firms with access to sufficient financial resources would be able to adopt the technology (Iacovou et al., 1995; Thong & Yap, 1995). Kuan and Chan (2001) also found that financial costs distinguished adopters from non-adopters, for small firms.

Technological readiness considered internal IT sophistication (i.e. IT usage, IT management and IT skills) and the firm's access to external IT support. All the adopting firms (A2, A3, L1 and L2) except A1 were found to be internally IT sophisticated and have access to sufficient external support for adopting the basic order-execution technology. However, technical concerns regarding technology development and integration highlight the importance of internal IT sophistication and external IT support for small firm adoption. While the firms were reasonably confident about their ability to adopt and support a basic online trading system, developing a more complex system (e.g., providing online client advisory services and access to firm research) and integrating it with back-end or legacy systems was more challenging. This contributed to the slower uptake of the technology by some firms (e.g. L1, L2). Access to limited technical resources is therefore likely to inhibit adoption and the firm's ability to develop integrated systems that enable them to realize the full potential of online trading for the firm.

Although lagging adopter L3 was not internally IT sophisticated, they had access to sufficient external support for later integration and did not anticipate difficulties. This may suggest that internal sophistication was not a strong adoption inhibitor, where firms were able to hire in the required expertise through external consultants. This confirms the importance of external expertise for small firm adoption (e.g. Cragg & King, 1993), especially where such firms lack requisite internal skills. However, where firms do not have access to sufficient IT expertise (whether by way of internal expertise or through external consultants), this is likely to inhibit adoption. For example, AC Nielson (2001) found that lack of access to IT expertise was a key barrier to Internet use for New Zealand firms.

The case findings are consistent with prior research suggesting the importance of IT support for adoption (e.g. Cragg & King, 1993; Fink, 1998). For example, access to sufficient internal IT support and/or external technical support may lead to or accelerate technology adoption (Attewell, 1992; Raymond & Bergeron, 1996; Salehi, 1997), while knowledge barriers may impede adoption (Attewell, 1992; Fichman & Kemerer, 1999). Sophisticated firms (e.g. A2, A3) seem less likely to feel intimidated by the technology (e.g. Pare & Raymond, 1991) and more willing to explore complex technology (Rai & Bajwa, 1997). This would seem to be the case for the firms that had adopted (or were evaluating options for) integrated systems. For an adoption to be successful, it is evident that firms must be reasonably IT sophisticated or have access to sufficient external IT expertise to address the technical issues that may inhibit adoption. Lack of access to sufficient technological resources is likely to slow adoption and inhibit the firm's ability to realize the full potential of the new technology.

Management Support. An organization's decision to adopt or not adopt an innovation is a function of individual decision-making (Zaltman et al., 1973). Hence, support of the decision-maker (or, alternatively, resistance by the decision-makers) can be expected to inhibit adoption. The findings showed that early and later adopters had the support of top management for adopting online trading. For early adopters, top management recognized benefits of adoption; for later adopters, while top management believed there were benefits to be gained, they had concerns such as weak investment profitability, which appeared to contribute to the slower rate of adoption. Lagging adopters did not recognize any relative advantage to adopting the technology; they therefore lacked the enthusiastic support of top management for the adoption. These findings are consistent with prior research that found that top management support could lead to adoption or early adoption of the innovation (e.g., King & Teo, 1996; Rai & Bajwa, 1997). For example, Premkumar and Roberts (1999) found that management support was an important determinant of firm adoption of communication technologies.

Although the interviewees discussed the support of non-management staff for e-commerce adoption, the case evidence did not provide strong support for the impact of staff support on adoption. Of the cases, only L1 identified a perceived pressure from staff as a key factor; they believed staff would leave if the firm did not keep abreast with the cutting edge. A1 had staff support, while A3 had the strong support of the younger staff only. Firms A2 and L2 did not believe the support of staff was relevant to the adoption decision. For lagging adopter L3, their staff viewed online trading as a *threat* to the advisor-client relationship – this factor encouraged the firm's more cautious approach to the technology.

Online trading is poised to change the nature of existing relationships between staff and clients, ushering in an era of disintermediation of the broker's role in order-execution (Fan et al., 2000). For full-service firms, the advisor-client relationship underpins firm success; building and maintaining such relationships in an internet-based environment will require staff (and the firm) to develop new strategies for supporting the advisor-client relationship. It is therefore important that firms do not overlook the role of staff in ensuring successful adoption. Staff support is likely to have a greater impact on successful integration (Headrick & Morgan, 1999; Sheng et al., 1998).

Environmental factors. External pressure to adopt the technology was defined as the influence that the external environment exerts on a firm to embrace a technology (Iacovou et al., 1995). In this study, external pressure to adopt e-commerce came from competitors, clients, and the changing face of the financial marketplace (e.g. from an automated settlement and clearance system for brokers to online trading systems for brokers and investors).

The early adopters (A1, A2 and A3) identified competitive pressure (exerted by other firms operating in the small retail client sector) and client demand for an online trading service as factors influencing their decision to adopt online trading. Later adopters identified client demand as an external pressure, but not as a strong factor; L2 also identified competitive pressure as a consideration, but not as a strong factor.

Lagging adopter L3 held the view that online trading was all about "discount brokering" and which firm can offer the cheaper deal. However, they believed their advantage lay in research, building personal relationships, trust and reassurance about investment decisions, not in offering "cheap deals." The firm believed their clients had little or no interest in the technology but, for a personal financial service, their clients were willing to pay a premium. Hence, they perceived little or no external pressure to adopt, either from competitors or clients.

The case evidence suggests that competitors and clients exerted some pressure on early and later adopters to embrace online trading. These findings are consistent with prior research. For example, Chircu and Kauffman (2000) suggested that other e-commerce-able firms might influence the e-commerce adoption decision of traditional firms. It is expected that, as more people gain access to the Internet and more brokerages offer online trading services, online trading will increase (e.g., Allgood, 2001). It is reasonable to expect that firms that have not yet adopted the technology will need to do so, in response to the changing nature of the marketplace (Poon & Swatman, 1999).

CONCLUSION

This chapter examined a number of factors influencing the adoption of e-commerce technology, namely innovation factors (i.e., perceived benefits and compatibility), organizational factors (i.e., technological readiness and management support) and environmental factors (i.e., external pressure from competitors and clients). The findings were consistent with prior studies in small firm adoption (e.g., Fink, 1998; Iacovou et al., 1995; Mehrtens et al., 2001). While financial readiness was shown to facilitate adoption (and its absence as likely to inhibit adoption), this factor did not allow distinctions to be drawn among adopting firms (e.g., between early and lagging adopters) in the study context. This outcome may be explained by the relative affordability of the order-execution technology for the firms included in this study. However, recent research found that cost of development was a key factor inhibiting Internet use within the finance sector (AC Nielson, 2001). Results regarding staff support were also inconclusive, with only one firm identifying staff pressure as a significant adoption factor.

The case evidence provided strong support for a priori expectations that there would be a positive relationship between support for firm business strategy (compatibility) and adoption (Lederer et al., 1997). The findings also suggested that *compatibility,* as well as *perceived benefits*, were key factors influencing the adoption decision and determining rate of adoption. These findings confirm a recent study that suggests compatibility and relative advantage may be the more important factors in explaining SMEs willingness to adopt e-commerce (Kendall et al., 2001). Recent study of IT in New Zealand also showed that lack of compatibility with business purpose was a key factor inhibiting firms from making more commercial use of their websites (AC Nielson, 2001). However, despite burgeoning interest in the use of the Internet to support firm business strategy, this is an area that still lacks research, providing opportunities for future research (Teo & Too, 2000).

Later and lagging adopters held the view that online trading was not an application that would enable the firm to gain major benefits or competitive advantage; rather, they viewed the technology as suitable for the discount market, but not for their niche market segments. This view may reflect a lack of understanding of the full potential of e-commerce for the firm, which, in turn, inhibits adoption and leads to a negative perception of benefits. For example, from the broker's perspective, the Internet has the potential to expand market reach, increase market share and reduce operating costs; the Internet is also ideal for the delivery of financial information (Allgood, 2001). Fear of disintermediation of the broker-relationship may be a factor inhibiting adoption of the technology, particularly for full-service firms. Firms that had adopted e-commerce did so because they were aware of the benefits of the technology and believed the technology was compatible with business needs. Assisting managers in overcoming their initial hesitancy about e-commerce technologies, through training and education, will be necessary to enable adoption and assist the firms in finding new and creative ways in which to use e-commerce technology if they are to remain competitive in a rapidly changing business environment.

Limitations and Future Directions

The research findings extend current understanding of the factors that influence e-commerce adoption among small brokerages in the financial services sector. However, the research methods and context in this study limit the conclusions that can be drawn. For example, a key weakness of the research methodology lies in the difficulty of generalizing the study findings. While this limitation is addressed, to some extent, by the use of multi-case studies, the singular nature of the study context (i.e., adoption of order-execution Internet-based technology by small on-line brokerages) limits generalizability and the applicability of the findings to other study contexts.

Another key limitation of this study concerns the relative significance of the key determinants vis-à-vis the adoption decision. The research method limits the extent to which conclusions can be drawn regarding cause and effect, as well as the significance of the proposed determinants on adoption. Further study is therefore needed to assess impacts, as well as the relative significance of these factors, to the adoption process. The proposed research model also did not take into account the impact of culture, management styles and environmental pressure from other players such as trading partners and the government, and other variables that may also impact adoption. These limitations provide opportunities for future researchers. In addition, future research should also consider the views of unsuccessful adopters to help develop a more holistic understanding of factors inhibiting adoption.

REFERENCES

AC Nielsen. (2001). *Electronic Commerce in New Zealand: A Survey of Electronic Traders*. (Report No. 1402282). New Zealand: Inland Revenue Department and Ministry of Economic Development. Retrieved July 2002 from the World Wide Web: http:// www.ecommerce.govt.nz/.

Allgood, B. (2001). Internet-based share dealing in the new global marketplace. *Journal of Global Information Management, 9* (1), 11-15.

Angeles, R., Corritore, C.L., Basu, S.C. & Nath, R. (2001). Success factors for domestic and international electronic data interchange (EDI) implementation for U.S. firms. *International Journal of Information Management, 21* (5), 329-347.

Applegate, L.M., Holsapple, C.W., Kalakota, R., Radermacher, F. J. & Whinston, A. B. (1996). Electronic commerce: Building blocks of new business opportunity. *Journal of Organizational Computing and Electronic Commerce, 6* (1), 1-10.

Attewell, P. (1992). Technology diffusion and organizational learning: The case of business computing. *Organizational Science, 3* (1), 1-19.

Bhattacherjee, A. (2000). Acceptance of e-commerce services: The case of electronic brokerages. *IEEE Transactions on Systems, Man and Cybernetics – Part A: Systems and Humans, 30* (4), 411-420.

Chau, P. K. & Tam, K.T. (1997). Factors affecting the adoption of open systems: An exploratory study. *MIS Quarterly, 21*(1), 1-24.

Chircu, A.M. & Kauffman, R.J. (2000). Re-intermediation strategies in business-to-business electronic commerce. *International Journal of Electronic Commerce, 4* (4), 7-42.

Chwelos, P., Benbasat, I. & Dexter, A. S. (2001). Research Report: Empirical test of an EDI adoption model. *Information Systems Research, 12* (3), 304-321.

Cragg, P. B. & King, M. (1993). Small-firm computing: Motivators and inhibitors *MIS Quarterly, 17* (1), 47-60.

Cunningham, M. (1999). The case for comprehensive business change. *Inform, 13* (4), 43-48.

Duxbury, L. & Corbett, N. (1996). Adoption of portable offices: An exploratory analysis. *Journal of Organizational Computing and Electronic Commerce, 6* (4), 345-363.

Fan, M., Stallaert, J. & Whinston, A. B. (2000). The Internet and the future of financial markets. *Communications of the ACM, 43* (11), 82-88.

Farhoomand, A. F., Tuunainen, V. K. & Yee, L. W. (2000). Barriers to global electronic commerce: A cross-country study of Hong Kong and Finland. *Journal of Organizational Computing and Electronic Commerce, 10* (1), 23-48.

Fichman, R.G. & Kemerer, C.F. (1999). The illusory diffusion of innovation: An examination of assimilation gaps. *Information Systems Research 10* (3), 255-275.

Fink, D. (1998). Guidelines for the successful adoption of information technology in small and medium enterprises. *Journal of Management Information Systems, 18* (4) 243-253.

Giaglis, G. M., Paul, R. J. & Doukidis, G. I. (1999). Dynamic modeling to assess the business value of electronic commerce. *International Journal of Electronic Commerce, 3* (3), 35-51.

Gonsalves, G.C., Lederer, A.L., Mahaney, R. C. & Newkirk, H. E. (1999). A customer resource life cycle interpretation of the impact of the World Wide Web on competitiveness: Expectations and achievements. *International Journal of Electronic Commerce, 4* (1), 103-120.

Han, K. S. & Noh, M. H. (1999). Critical failure factors that discourage the growth of electronic commerce. *International Journal of Electronic Commerce, 4* (2), 25-43.

Harrison, D.A., Mykytyn, P.P. & Riemenschneider. (1997). Executive decisions about adoption of information technology in small businesses: Theory and empirical tests. *Information Systems Research, 8* (2), 171-195.

Headrick, R.W., & Morgan, G.W. (1999). Measuring the impact of information systems on organizational behavior. *Journal of End User Computing, 11* (4), 16-21.

Iacovou, C.L., Benbasat, I. & Dexter, A.S. (1995). Electronic data interchange and small organizations: Adoption and impact of technology. *MIS Quarterly, 19* (4), 465-485.

Kendall, J. D., Tung., L.L., Chua, K.H., Ng, C.H. & Tan, S.M. (2001). Receptivity of Singapore's SMEs to electronic commerce adoption. *Journal of Strategic Information Systems, 10* (3), 223-242.

King, W.R. & Teo, T.S. (1996). Key dimensions of facilitators and inhibitors for the strategic use of information technology. *Journal of Management Information Systems, 12* (4), 35-53.

Kuan, K.Y. & Chau, P.Y.K. (2001). A perception-based model for EDI adoption in small businesses using a technology-organisation-environment framework. *Information and Management, 31* (8), 507-521.

Lederer, A.L., Mirchandani, D.A. & Sims, K. (1997). The link between information strategy and electronic commerce. *Journal of Organizational Computing and Electronic Commerce, 7* (1), 17-34.

McGowan, M. K. & Madey, G. R. (1998). Adoption and implementation of electronic data interchange. In T. J Larson and E. McGuire (Eds.), *Information Systems Innovation and Diffusion: Issues and Direction.* Hershey, PA: Idea Group Publishing.

Mehrtens, J., Cragg, P.B. & Mills, A.M. (2001). A model of Internet adoption by SMEs. *Information and Management, 39* (3), 165-176.

Nath R., Akmanligil, M., Hjelm, K., Sakaguchi, T. & Schultz, M. (1998). Electronic commerce and the Internet: Issues, problems, and perspectives. *International Journal of Information Management, 18* (2), 91-101.

Ngai, E. W. & Wat, F. K. T. (2002). A literature review and classification of electronic commerce research. *Information and Management, 39* (1), 415-429.

Pare, G. & Raymond, L. (1991). Measurement of information technology sophistication in SMEs. Niagara Falls, Ontario, Canada: *Proceedings of Administrative Sciences Association of Canada Conference,* (May, 90-101).

Poon, S. & Swatman, P. (1999). A longitudinal study of expectations in small business Internet commerce. *International Journal of Electronic Commerce, 3* (3), 21-33.

Prekumar, G. & Roberts, M. (1999). Adoption of new information technologies in rural small business. *OMEGA International Journal of Management Science, 27* (4), 467-484.

Raghunathan, M. & Madey, G. R. (1999). A firm-level framework for planning electronic commerce information systems infrastructure. *International Journal of Electronic Commerce, 4* (1), 125-145.

Rai, A. & Bajwa, D.S. (1997). An empirical investigation into factors relating to the adoption of executive information systems: An analysis of EIS for collaboration and decision support. *Decision Sciences, 28* (4), 939-975.

Raymond, L. & Bergeron, F. (1996). EDI success in small and medium-sized enterprises: A field study. *Journal of Organizational Computing and Electronic Commerce, 6* (2), 161-172.

Riggins, F. J. (1999). A framework for identifying Web-based electronic commerce opportunities. *Journal of Organizational Computing and Electronic Commerce, 9* (4), 297-310.

Rogers, E. M. (1983). *Diffusion of Innovations.* New York: Free Press.

Rogers, E. M. (1989). The critical mass in the diffusion of interactive technologies in organizations. In J.I. Cash, Jr., J.F. Nunamaker, Jr. & K.L. Kramer (Eds.), *The Information Systems Research Challenge: Survey Research Methods.* Boston: Harvard Business School, 245-271.

Salehi, S.E. (1997). Information technology as determinant of competitiveness. *Competitive Review, 7* (2), 52-58.

Sheng, O.R., Hu, J.H., Wei, C.P., Higa, K. & Au, G. (1998). Adoption and diffusion of telemedicine technology in health care organizations: A comparative case study in Hong Kong. *Journal of Organizational Computing and Electronic Commerce, 8* (4), 245-275.

Teo, T. S. & Too, B. L. (2000). Information systems orientation and business use of the Internet: An empirical study. *International Journal of Electronic Commerce, 4* (4),105-130.

Thong, J.Y. & Yap, C.S. (1995). CEO characteristics, organizational characteristics and information technology adoption in small businesses. *Omega, 23* (4), 429-442.

Turban, E., Lee, J. King, D. & Chung, H.M. (2001). *Electronic Commerce: A Managerial Perspective*. London: Prentice Hall.

Zaltman, G., Duncan, R. & Holbek, J. (1973). *Innovations and Organizations.* New York: John Wiley and Sons.

ENDNOTES

[1] For a brief history of the NZSE, see http://www.nzse.co.nz/exchange/about/brief_history

[2] Full service firms offer a range of financial services that include share-broking, financial planning, managed funds, portfolio management, research, investment service management, and investment advice.

[3] Discount firms receive and process share transactions, but do not normally offer advisory services.

[4] Institutional brokerage firms focus on institutional investors including fund managers and mutual fund companies. Such firms provide financial advisory, research, capital raising, and securities broking services to corporations and institutions.

Chapter IV

Conceptualizing the SMEs' Assimilation of Internet-Based Technologies

Pratyush Bharati
University of Massachusetts, USA

Abhijit Chaudhury
Bryant College, USA

ABSTRACT

This chapter conceptualizes a model for the assimilation of Internet-based technologies in small and medium enterprises (SMEs). The research examines the factors influencing the assimilation of Internet-based technologies and the penetration of these technologies in SMEs. Internet-based technologies are complex organizational technologies. The model uses the learning related scale, related knowledge and diversity, together with several control variables like host size, IT size, specialization and education. Several external factors, such as influence of customers, suppliers, vendors and competitors, that have been suggested in studies of SMEs have also been included to explain the assimilation and diffusion of innovation. The results of an exploratory survey are presented and future research is discussed.

INTRODUCTION

E-commerce is impacting the way the small and medium sized firms conduct business. They are increasingly being subjected to competition from firms located in different parts of the world. General Electric (GE) has had a big rise in bids from Chinese manufacturers to supply to U.S. plants. GE has, therefore, developed a new capacity to handle these suppliers (*The Economist*, 2000). Thus, e-commerce is exploiting the combined power of the Internet and information technology to fundamentally transform key business strategies and processes (Jones, 2000). This transformation has commenced. For example, the percentage of new Internet Business-to-Business (B2B) projects by small- and medium- sized firms in supply chain and procurement will rise from less than 25% in 1999 to more than 75% in 2003 (*The Economist*, 2000). In Denmark, Netherlands and Australia, Internet penetration is more than 50% among medium-sized businesses (*The Economist*, 2000). These facts show the speed at which change is taking place in small- and medium-sized firms. Therefore, it is pertinent to investigate the impact of e-commerce on small and medium firms.

The research project will examine the factors influencing the assimilation of Internet- based technologies and the penetration of these technologies in small-and medium-sized manufacturers.

Other questions that will be investigated are:

- How are Internet-based technologies impacting small-and medium-sized manufacturers?
- What elements or components of Internet-based technologies are they using and for what purpose?
- What factors have played a major role in their move to adopting these technologies?

BACKGROUND

The Internet Economy now directly supports 3.1 million workers. It grew by 62% in 1999 to $523.9 billion and by 58% in 2000 to an estimated $830 billion. The Internet-related revenue growth was 15 times the growth rate for the U.S. economy (Center for Research in Electronic Commerce, University of Texas Austin, 2001). Since 2001, the Internet Economy has slowed down but it still is an important part of the economy.

Small companies constitute a surprisingly high 90% of all U.S. exporters and account for about 30% of the total value of U.S. exports (US Alliance for Trade Expansion, Washington). There is considerable empirical evidence that the employment share of traditionally large-business-dominated industries is declining while that of traditionally small-business-dominated industries is increasing (Cordes, Hertzfeld

& Vonortas, 1999). Despite these facts, the implications of this new technology for small-sized enterprises have not received much attention to date.

Bill Clinton, former president of the United States, in his 1998 directive on e-commerce, urged the United States Government Working Group on Electronic Commerce to "facilitate small business participation in electronic commerce" (US DOC, 1998). The Working Group on Electronic Commerce, in its second annual report, recommended, "The Secretary of Commerce and the Administrator on the Small Business Administration shall develop strategies to help small businesses overcome barriers to the use of the Internet and electronic commerce," (US DOC, 1999). Some studies were launched by these governmental agencies, but no systematic data has been collected relating to small and medium firms.

A recent, though limited, survey by Inc (Winter, 2000) shows that about 47% of small businesses are using the Internet in some form or another for purchasing. A more elaborate empirical study is required to establish how the Internet-based technologies are being assimilated and used by small enterprises. We intend to conduct such a study among small-and medium-sized manufacturers in Massachusetts, with the possibility of expanding it nationwide.

RESEARCH MODEL

Internet-based technologies are complex organizational technologies. Complex organizational technologies impose substantial knowledge burden on would-be adopters (Attewell, 1992). Most theories of innovation incorporate communication of new information about innovations in one way or another. Attewell's work (1992) on organizational learning and innovation diffusion differentiates between the *kinds* of information involved and the *mechanisms* by which information is acquired and propagated. Attewell's research (1992) draws a clear distinction between the communication of "signaling" information about the existence and potential gains of the innovation, versus know-how and technical knowledge. It argues that the acquisition of this technical knowledge and know-how " ... plays a more important role in patterning the diffusion process of complex technologies than does signaling ...[and] should move to center stage in any theory of complex technology diffusion" (1992, pg. 5).

The absorptive capacity of the organization for an innovation is largely a result of the organization's pre-existing knowledge in areas related to the focal innovation and the diversity of knowledge in general (Cohen & Levinthal, 1990). These theories have been used to develop a model for the assimilation and diffusion of innovation (Fichman, 1995; Fichman, 2001). The model of assimilation and diffusion of innovations was validated using data from firms undergoing assimilation of software process innovations. The model uses the learning-related scale, related knowledge and diversity, together with several control variables like host size, IT size, specialization and education. Several external factors, such as influence of customers, suppliers, vendors and competitors, that have been suggested in studies of SMEs have also been included to explain the assimilation and diffusion of innovation (Chen, Justis & Chong, 2002; Mehrtens, Cragg & Mills, 2001; Premkumar & Roberts, 1999). We have used this model as the starting point, although, we have made changes so that the model is more relevant to Internet-based technologies and to small-and medium-sized manufacturers. This model will be used to investigate the assimilation and diffusion of Internet-based technologies in small-and medium-sized manufacturers in Massachusetts.

The hypotheses of our research are:

Hypothesis 1: Learning-Related Scale is positively related to E-Commerce Assimilation Stage.

Hypothesis 2: Related Knowledge is positively related to E-Commerce Assimilation Stage.

Hypothesis 3: Diversity is positively related to E-Commerce Assimilation Stage.

The research study will investigate what Internet technologies small-and medium-sized manufacturers are using. For the small-and medium-sized manufacturers who are utilizing these technologies, the research will investigate in what way and for what purpose are they using Internet technologies. The research plans to study the firms that are at the forefront of usage of these Internet-based technologies and how organizations can adopt these technologies to strategically benefit themselves. This will shed light on the practices that are becoming best practices in the industry. Thus, the research will develop an understanding of how Internet-based technologies are relevant to small-and medium-sized manufacturers, how does it impact them and what can they do about it.

METHODOLOGY ADOPTED

The model of assimilation and diffusion of Internet-based technologies was used to develop the research hypotheses. An exploratory survey instrument was designed to collect preliminary exploratory data. This exploratory survey was further modified into a survey instrument, which was an operationalization of the model. For the pre-test, the survey was sent to small—and medium-sized manufacturers in Massachusetts. The Greater Boston Manufacturing Partnership (GBMP), which is based in the College of Management, is helping in administering the survey.

The exploratory survey provided some interesting insights about the SMEs' usage of Internet—based technologies in Massachusetts. A majority of SMEs in Massachusetts have a Web site and use e-mail to conduct business. A substantial number of firms are using electronic data interchange (EDI) to communicate with other customers or suppliers. Quite a lot of SMEs are also using, deploying or exploring supply chain management (SCM), customer relationship management

(CRM) and enterprise resource planning (ERP). These firms already use information technology for support functions such as accounting and human resources (HR). Surprisingly, the SMEs are able to adopt these technologies with very small or outsourced information technology departments. These findings are preliminary and the follow-up survey would provide a clearer picture.

FUTURE RESEARCH

After the follow-up surveys have been analyzed, if need be, follow-up questions will be asked of the survey respondents. We are also currently exploring the possibility of conducting a few case studies in order to revise the model. After we have collected all the survey data, we will be using several statistical techniques to analyze the data. The statistical techniques that will be used are correlation analysis, multiple regression analysis, factor analysis and structural equation modeling. The statistical analysis will help in developing explanatory models that would shed more light on the relationships that are the subject of research.

Figure 1: Assimilation and diffusion model

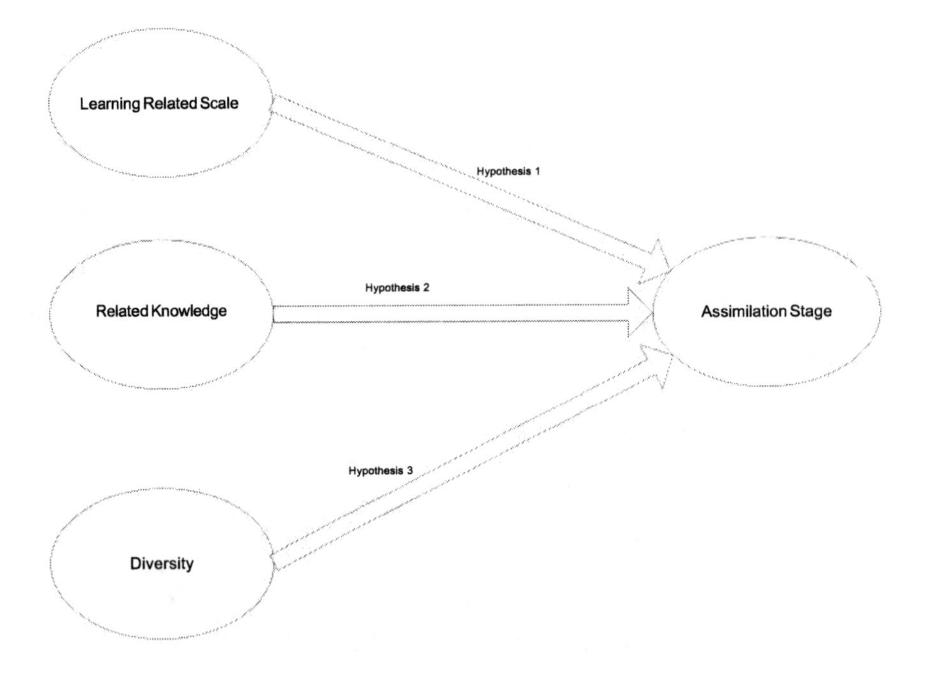

REFERENCES

Attewell, W. B. (1992). Technology diffusion and organizational learning: The case of business computing. *Organization Science 3* (1), pp. 1-19.

Bharati, P. & Chaudhury, A. (2002). Assimilation of Internet-based technologies in small and medium sized manufacturers. *Proceedings of 2002 IRMA International Conference.*

Center for Research in Electronic Commerce. (2001). Measuring the Internet economy. Texas: University of Texas at Austin. Retrieved from the World Wide Web in June 2001: http:// www.internetindicators.com).

Chen, Y., Justis, R. & Chong, P.P. (2002). Franchising and information technology: A framework. Burgess, S. (Ed.). In *Managing Information Technology in Small Business: Challenges and Solutions.* Hershey, PA: IRMA Group Publishing.

Cohen, W. M. & Levinthal, D.A. (1990). Absorptive capacity: A new perspective on learning and innovation. *Administrative Science Quarterly* 35, pp. 128-152.

Cordes, J. J., Hertzfeld, H.R. & Vonortas, N.S. (1999). A survey of high technology firms. U. S. Small Business Administration.

The Economist e-management survey. (2000, November 11). *The Economist.*

Fichman, R. G. (1995). The assimilation and diffusion of software process innovations. Doctoral Dissertation. Boston, MA: Massachusetts Institute of Technology.

Fichman, R. G. (2001). The role of aggregation in the measurement of IT-related organizational innovation. *MIS Quarterly 25* (4), pp. 427-455.

Jones, Frank (2000). E-business transformation in the manufacturing industry. Manufacturing Institute. National Association of Manufacturers.

Mehrtens, J., Cragg, P. B. & Mills, A.M. (2001). A model of Internet adoption by SMEs. *Information and Management*, 39, pp. 165-176.

National Association of Manufacturers M. (1999). Survey of small manufacturers.

Premkumar, G. & Roberts, M. (1999). Adoption of new information technologies by small businesses. *OMEGA. International Journal of Management Science,* 27, pp. 467-484.

U. S. Department of Commerce. (1998). The emerging digital economy report. Retrieved from the World Wide Web: http:// www.ecommerce.gov.

U. S. Department of Commerce. (1999). Second annual report of the working group on electronic commerce. Retrieved from the World Wide Web: http:// www.ecommerce.gov/ecommerce.pdf.

ENDNOTES

[1] Previous version of this paper was presented at IRMA, 2002. See reference Bharati and Chaudhury (2002).

Chapter V

Strategic Issues in Implementing Electronic-ID Services: Prescriptions for Managers

Bishwajit Choudhary
Norwegian Banks' Payments Central, Norway

ABSTRACT

During the past few years, e-security solutions (e.g., digital certificates, e-signatures, e-IDs) gained tremendous attention as they promised to plug security loopholes and create trusted electronic markets. Implementation of such critical, complex and costly security solutions demands thorough assessment at technical, as well as business levels. Based on the author's experience at one of Scandinavia's leading vendors of banking solutions and infrastructure, the paper develops basic concepts, discusses strategic (product, market and technical) concerns and, finally, summarizes the contemporary challenges facing the implementation of e-ID schemes.

INTRODUCTION

The diffusion of electronic services over 'open' (Internet and wireless) networks has accentuated concerns about privacy infringement, data corruption and false denial of services. This poses not merely business and legal questions, but also challenges the basic 'trustworthiness' of open networks as the potential motor of future e-commerce. Not surprisingly, the need for a robust e-security infrastructure has become essential to critical online support services (e.g., authentication, verification, authorization), value-added e-solutions (for banking, commerce, stock trading) and securing the legacy systems (like customer databases, transaction histories, archives, etc.). In brief, these issues summarize the backdrop for this paper.

In the first section, we introduce some basic concepts. The needs and roles of players (vendors of e-ID scheme, merchants and users) are discussed in the following section. Later, the implementation of e-ID schemes is explained using a so-called 'certificate value chain' and selected business and technical considerations. Finally, we summarize the contemporary challenges in implementing e-ID. Throughout the paper, we have tried to present simple methodologies that will help managers develop business and technology strategies. Our 'target' readers are the (product/project) managers in different stages of implementing e-ID schemes (planning-strategy, infrastructure establishment and e-ID 'enabling' of new services).

UNDERSTANDING THE BASICS

A Digital Certificate (or simply a 'certificate') is analogous to an electronic 'passport' and comprises a set of policies (or customers' rights) bound to a number of key-pairs besides user's Distinguished Name (DN), name of the certificate issuer (Certificate Authority or CA) and, sometimes, the user-profiles. An e-ID contains a digitally signed statement from the CA and provides an independent confirmation of the certificate. A certificate (usually) also contains three key pairs, one each for signing, encryption and authentication. Each key pair, in turn, comprises a Public Key (publicly available) and a Private Key (known only to the authorized user). This e-security technology is popularly known as 'Public Key Infrastructure' (PKI). Stated formally, *PKI is a collection of hardware, software, policy and human roles that successfully binds a subscriber's identity to a key pair (public and private) through the issuance and administration of digital certificates all through their 'life-cycle' (creation, maintenance, archival records and destruction).*

A certificate can be stored in a smart card or PC hard drive, diskette or server. It has a lifetime, after which it can be either suspended temporarily or terminated permanently (by the CA), if not renewed by the user. Depending on a CA's security policy, there can be different types of certificates:

- Identification Certificates: CA checks that the user-name corresponds to something in the non-digital world and binds this name to the certificate issued. CA identifies the client and confirms that the client is who s/he purports to be.

- Authorizing Certificates: In the medium-term, a CA is likely to begin certifying both the user-attributes and user-identity. Value-added online services, such as one-to-one marketing, loyalty programs, etc., then can be provided based on user-attributes. Such a certificate states the subject's address, age, relations to an organization, etc.
- Transactional Certificates: These attest that an observer witnessed some form of formality (e.g., lawyers confirming and authenticating their client's e-signatures in 'real time' from a remote location).

A digital certificate ensures user privacy, data-integrity, user's rights authentication and non-repudiation.

Given the centrality of a CA's role in running the infrastructure of an e-ID scheme, it is often referred to as a Trusted Third Party (TTP) because it secures the electronic communications between the recipients and senders (of payments, invoices, contracts, e-mails, etc.) using CA's certificates. *A TTP is an unbiased firm that contributes to and (often) takes liability for the lapses in security of electronic services.* In Figure 1, we have presented a simple business model needed to illustrate the issuance of a digital certificate. This would also help managers identify specific roles for their businesses and points of external partnerships.

Figure 1: Certificate value-chain[1]

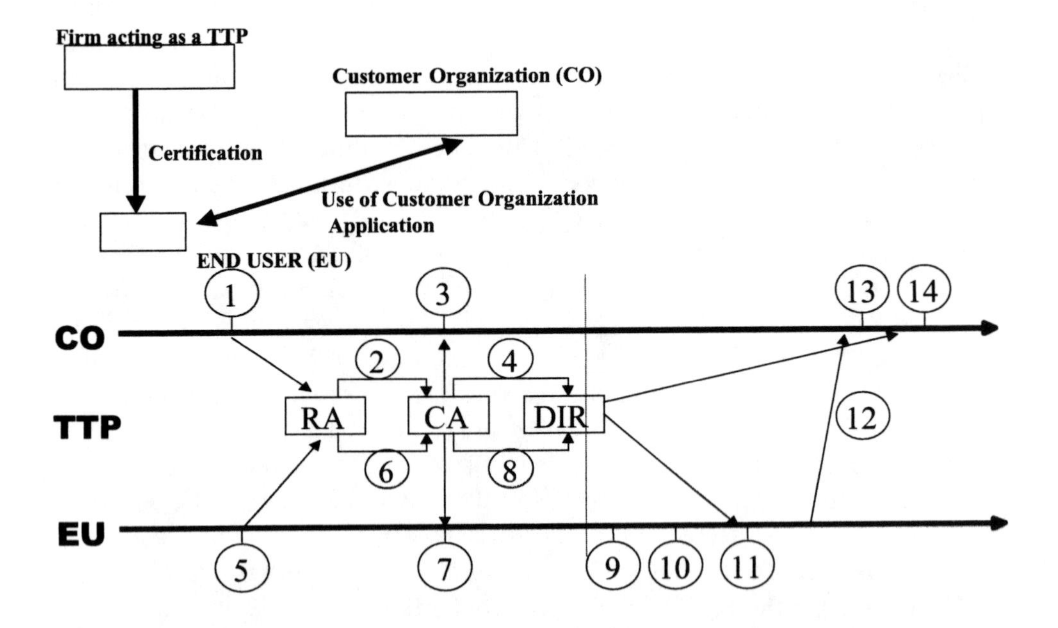

1. Customer Organization (CO) or End User (EU) applies for a certificate to an RA (Registration Authority), an office authorized (by a CA) to accept and verify certificate requests.
2. If RA approves the application, it sends a request to CA.
3. CA returns an acknowledgement to CO, together with the CO's private keys (if generated by CA).
4. CA publishes the CO's certificate(s) and certified public keys in the TTP's directory.
5. EU will follow the same sequence of events for steps 5 to 8 as followed above. All the events from 1 to 8 are 'once-only' events. After event step 8, the CO and the EU have certificates and are ready to do business. Events from 9 onward (right side of the dotted line) through the timeline show how a transaction is sent from an EU to the CO. These events may be repeated as many times as the end user wishes.
9. EU then runs a suitable application on his/ her system, perhaps a standard browser, a special application supplied by CO. This generates a message that is to be sent to the CO.
10. The message is digitally signed using the private signing key of the EU.
11. EU fetches the public encryption key of the CO from the directory; or, if s/he already has it, s/he checks that s/he has the correct key and that it is still valid (i.e., has not been revoked). The message is encrypted using the CO's public key.
12. Transaction message is sent to the customer organization.
13. The CO decrypts the message, using its own private decryption key. It discovers the identity of the end user, which claims to have sent the message.
14. The CO fetches or checks the end user's public signing key from the directory and then checks the signature on the message to ensure its origin from EU and integrity during transmission.

STRATEGIC MARKET ISSUES

The needs of merchants and end users define the value propositions that a CA's services must match. A CA may provide its own trusted electronic services, besides offering its trusted infrastructure (PKI) to the content providers. At CA's place, an elaborate PKI market development strategy usually comprises applications enabling priorities (list of 'killer' applications that need secure authentication and signing), besides partnership and communications strategies (to educate employees, partners and clients on e-IDs). Leading banks, post offices and telecommunication companies usually compete for the prized position of a TTP (which operates as the CA). TTPs can position themselves on the following criteria:

1. Geographical reach (national or regional or global);
2. Industry specialty (banking or telecom or government, etc.); and

Figure 2: Framework to compare the e-ID schemes and their respective vendors

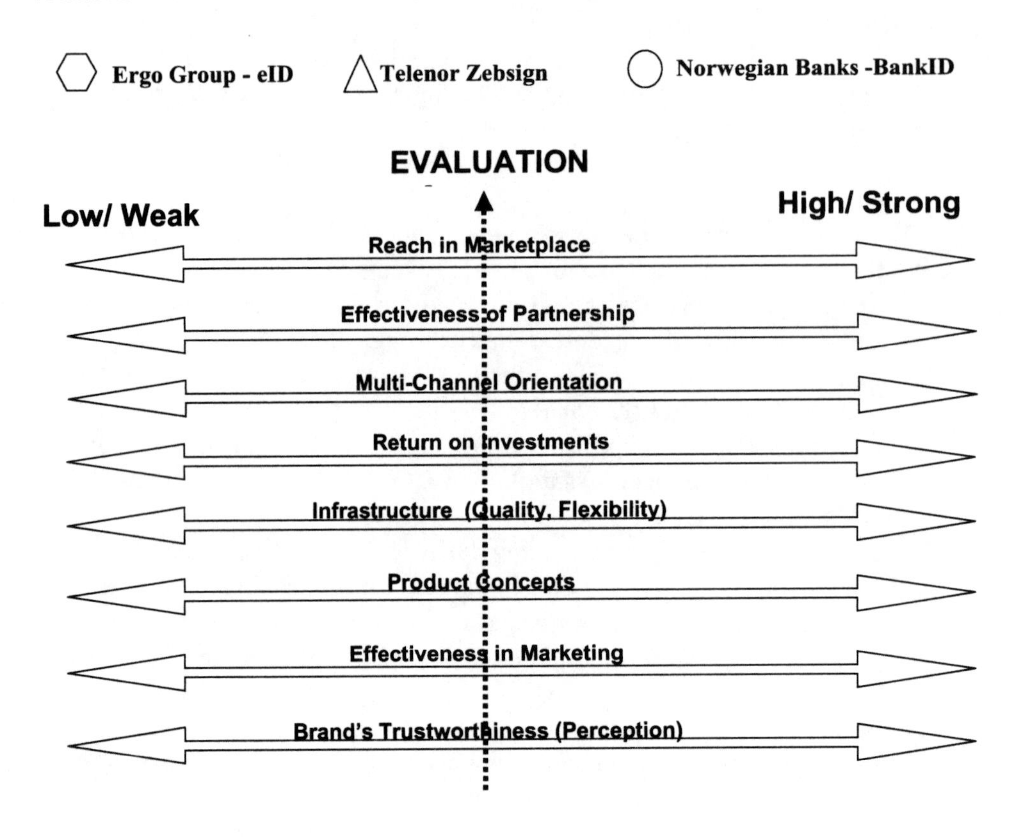

3. Specificity of certificate use with respect to:
   3.1 Segments: Business-to-Business, Person-to-Person or Person-to-Business
   3.2 Solutions: Banking or Entertainment or Gambling, etc.

We illustrate (Figure 2) a framework that was used in Norway to compare the then-e-ID schemes and their respective vendors (TTPs). Such a framework can be used to understand competition by firms aspiring to become TTPs.

Finally, a TTP must prioritize getting its PKI systems operationally stable and populating the users with digital certificates. Once certificate distribution has been initiated by the TTP, it is time to partner with the leading content providers and intermediaries (system integrators, consulting companies and solution developers).

Merchants' Needs

Some of the e-security needs of the merchants are similar to those of the end users. In this section, we describe some of the key security concerns of e-service providers.

- Authentication: The merchants want to confirm a buyer's identity before making the sale. A merchant may wish to build a database of customers and their buying profiles.
- Certification: The merchant may need proof that the buyer possesses an attribute required to authorize a sale. For example, some goods may be sold only to people older than 18 years.
- Confirmation: The merchant needs to be able to prove, to a third party involved in the transaction, (such as a credit card company) that the customer did indeed authorize the payment.
- Non-Repudiation: The merchant wants protection against the customer's unjustified denial on order placement or non-delivery of goods.
- Anonymity: The merchant may want to control the transaction information disclosed.

In addition to understanding merchants' desires, a TTP should align its business strategy with the needs of the market. This can be achieved if the TTP provides:

1. Updated roadmaps of services and functionality forthcoming in the future;
2. Education and training programs for the client organizations and partners; and
3. Discussion forums with merchants and partners (such a forum should comprise technical and business analysts from both sides).

In turn, the merchants must maximize their ability to reach the largest number of users through the rollout of 'killer (PKI-enabled) applications.' Yet another way for the merchant to maximize exposure to users is to accept more than one digital certificate (without compromising infrastructure security level or quality of support services).

End Users

Corporate or private users who place online orders or make payments have some general requirements of the e-security solution.

- Authentication: As discussed in the previous section.
- Integrity: As discussed in the previous section.
- Recourse: Comfort that there is an option if the seller fails to perform or deliver.
- Confirmation of order/payments through a receipt.
- Privacy/Anonymity: Control of the amount of information disclosed to merchant

Even as the users demand high levels of assurances of the e-security of their communications or transactions, one must never underestimate the need for 'friendly user interface' and 24/7 support. The market demands from e-security solutions can be summarized simply with the word 'PAIN' - Privacy, Authentication, Integrity (of data) and Non-Repudiation.

STRATEGIC TECHNICAL ISSUES

To complement the market-oriented discussions above, in this section, we identify a checklist of factors needed to establish a PKI, describe technology

Table 1

MANDATORY FACTORS & DESCRIPTIONS	
CA Establishment	PKI Solution must provide a capability to create a new Certification Authority (CA) that can issue, certify and manage certificates according to a pre-established policy.
Subscriber Initialization	PKI must provide a capability to initialize new subscribers through user- friendly interfaces. Usually banks do this through their Net-banks (for existing authenticated users), branch networks (for new subscribers) and, while doing so, TTPs are backed by a robust order distribution and certificate life cycle management system.
Registration and Certification	This is the capability to issue and certify (sign) certificates and includes: • **Initial request:** The capability for subscribers to initiate and submit a certificate request securely to the Registration Authority (RA) and/or CA in such a way that subscribers can generate key pairs. This helps RA/CA to validate subscriber's 'proof of possession'. • **Certificate issuance:** A capability for the RA/CA to issue a certificate containing information and provide a capability for the certificates to be signed by the CA(s) according to certificate policy.
Certificate Publication	A capability to publish certificates in a repository such that recipients (or 'relying entities') can verify their certificates, and entities who require a subscriber's public key to encrypt a message/ session can retrieve it from this repository.
Revocation Request Processing	A capability to process requests, from subscribers and administrators, that particular certificates be revoked.
Certificate Revocation List (CRL) Publication	PKI must provide capabilities to publish/ distribute the CRLs, such that certificate recipients can be informed of certificates that have been revoked.
Key Recovery	The PKI Solution should provide an optional capability to restore/recover a key pair that is lost/corrupted or whose protection mechanism (PIN, challenge phrase, physical token) is no longer known/possessed by the subscriber. Key recovery features are most useful in recovering encrypted data that would otherwise be inaccessible due to a lost or corrupted key.

management issues, develop a matrix comprising technology-business (decision-making) factors and, finally, identify some 'failure points' that may come up during the operational phase.

Checklist of Factors to Establish a PKI

We must mention that only the most important factors (from our experience) are summarized in the tables provided. In real-life implementation projects (and sub-projects), this 'check-list' is more detailed.

Table 2

INTEROPERABILITY FACTORS & DESCRIPTIONS	
Network Communications	Intercommunication among subscribers, managers and various PKI components must use the existing communications channels- TCP/IP, WANs, telephone networks, etc.
Pre-existing Software	PKI must be capable of managing certificates according to the formats and standards that exist in the current versions of the software already in use.
Pre-existing Standards	A capability to publish certificates to a directory structure using commonly used basis as Lightweight Directory Access Protocol (LDAP).
Interoperability with Legacy Systems	PKI should offer integration tool kits necessary to develop a complete security infrastructure; typically, firewalls, Virtual Private Network and Authorization systems.

Table 3

SCALABILITY FACTORS & DESCRIPTIONS	
Distributed human Administration	PKI solution support for distributed administration to multiple people operating at geographically distinct locations.
Policy Flexibility	PKI solution must be able to support a variety of certificates corresponding to different certificate policies that can differ due to type of e-service, segment or place of use.
Auditability	PKI must provide a capability to audit its main functions, include a running log of PKI activity, the capability to re-construct the state of specific certificates at some time in the past and log the activities of the PKI administrators. The logs must be tamper-proof and accessible only to authorized administrators.
Strength of Mechanism	PKI must issue and manage Public Key certificates based on strong cryptographic algorithms that provide protection equal to the strongest available algorithms.
Life Cycle Cost	The total cost of ownership of the PKI solution should be as low as possible. Such estimates should include the establishment costs, regular maintenance costs, costs incurred in operations (production stability and customer support).

Technology-Business Matrix

The checklist of factors mentioned previously will help project leaders manage the deliveries of specific system modules and functionality. However, this is *not* enough. Based on field experience, we have tabulated strategic decision making issues.

Table 4

Decision Making Issues	Description of Issues	Priority
Portability	• Ease of use • Ease of installation • Form factor (size-design)	
Mobility	• Ease of 'moving around' • Usage while 'on move' • Interoperability with devices	
Security Levels	• Degree of Access Control (read, write restrictions) • Number of 'failure' points	
Systems' Processing Capability	• Speed & Response time • Usage while 'on move' • Interoperability with devices	
Administration and Issuance	• Degree of administrative centrality • Degree of human intervention • Issuance process	
Costs	• Fixed costs • Operating costs • Upgrades/maintenance	
Business Model	• Number of partners needed • Revenues/liability sharing	
Scale of Implementation	• Ease of large scale roll out • Need for extra infrastructure in markets (reader units, key storage software, etc.) • Ease of 'enabling' new e-services	
Social Acceptance and Users' Perceptions	• Dialogues and Interfaces • Experience similarity	

E-ID schemes can be implemented in different key storage models:

- Local key storage (soft certificate), where the private keys are stored in the hard drive of a PC.
- Smart card key storage (hard certificate), where the private keys are stored in a smart card.
- Net-Centric key storage, where the private keys are stored in a tamper-proof and secure server.

Typically, the organizations in initial phases of implementing e-ID schemes can 'cross' the Technology-Business Matrix with the above-mentioned key storage models to decide the appropriate one for their case. This is the most fundamental issue to be sorted out during the initial phases. The decision variables mentioned in Technology-Business Matrix also can be used to decide the 'lower level' (functional) issues. In fact, we successfully used this matrix to identify the most 'optimal' access channel toward a Net Centric Certificate (storage) server.

Vulnerability Issues

Using CEN/ISSS workshop[2] as a reference, we have structured below some 'vulnerability' factors.

Type 1: Communication link between Signature Creation Application (SCA) and Secure Signature Creation Device (SSCD):

- Wrong signature through malfunction of physical interface;
- Eavesdropping or interfering at a wireless interface between SCA and SSCD;
- Wrong signature creation due to corruption of input-output communication link.

Type 2: Signer's Authentication Components (SAC):

- Unauthorized use of Signer's Device ;
- Observation of PIN/ Password and misuse of this data; and
- PIN / Password display.

Type 3: Signature Creation Application (SCA) & Trusted Path Components:

- Corruption of DTBS (Data-To-Be-Signed) components; and
- Breach of confidentiality of Signer's Authentication Data or DTBS components or DTBSF.

Type 4: Signature Creation Application (SCA) & Data-To-Be-Signed (DTBS):

- Generation of inappropriate signature;
- Ambiguity of the signer's certificate implied by the signature;
- Ambiguity of the Signature Policy or Commitment type; and
- Inappropriate presentation of the Signer's Document.

CONCLUDING REMARKS

Let us revisit the objectives we set initially for this paper: To help managers understand and assess e-ID services and make compelling business case, while gaining a fair understanding of the underlying implementation issues. We addressed these objectives by:

1. Describing the basic concepts in PKI and e-ID scheme;
2. Illustrating 'certificate value chain';
3. Discussing the roles and expectations of all players involved; and
4. Structuring key business and technical issues for the decision-makers.

At this stage, we know that the value of e-ID is enhanced when a CA (TTP) writes a flexible certificate policy, uses an open standards-based solution and ensures stable production systems. In addition, the value of e-ID will multiply if the certificates can be used with all popular access devices, smart cards, payment terminals and 'killer applications.' Addressing all these demands simultaneously is quite difficult, because many issues (especially on the design of access devices, technical standards and cross-border legal e-trade policies) are not under the control of a TTP. Again, 'channel independence' condition is easier said than fulfilled, since most of the existing electronic devices and Web browsers are not considered safe places to 'store' private keys. Add to all this fierce standards war and brand rivalries between content providers and security vendors, leading to complex revenue/ liability sharing models and the diffusion of PKI on a large scale is all the more difficult.

To summarize, the challenges in realizing a positive business case on e-IDs depend on a wide range of business, technology and legal issues. While the need for creating secure e-markets is well understood, the roadmap to realizing a global online trust community through use of e-ID schemes is less obvious.

ACKNOWLEDGMENT

I thank Mr. Øyvind Apelland, Senior Vice President (Trusted Services, Norwegian Banks' Payments Central, Oslo) for his support during my work in the BankID Project.

ENDNOTES

1 'PKI Report', Norwegian Banks' Payments Central, Oslo, 1999.

2 'Security Requirements for Signature Creation Systems' (v 3.9), CEN–ISSS, Berlin, 2001.

Chapter VI

Potential Roles for Business-to-Business Marketplace Providers in Service-Oriented Architectures

Markus Lenz and Markus Greunz
University of St. Gallen, Switzerland

ABSTRACT

The chapter introduces the concept of a service-oriented architecture and assesses its applicability to business-to-business (B2B) marketplaces. In order to provide substantial value to their members, marketplaces have to offer a comprehensive service offering which aims to support all phases of the transaction process. Building up such a service offering is not a one-time effort; electronic B2B marketplaces will have to evolve their service offerings continually and adapt them to the ever-changing needs of the companies they serve. Therefore, instead of trying to create such an extensive service offering on their own, we argue that B2B marketplaces have to make use of partnerships with specialized service providers.

They can make use of the speed and scale of such a collaboration of specialists only by partnering (Ernst et al., 2001). This view of B2B marketplaces as integrators of services from multiple parties puts some new demands on architectural issues of the marketplaces. A service-oriented architecture may help to cope with these new demands. To test this view, a survey among European B2B marketplaces has been conducted in order to match their service development and their needs with the characteristics of a service-oriented architecture and the potential roles that marketplace providers can play in such a concept.

INTRODUCTION

Currently, B2B marketplaces are expanding the scope of their services. While the core matching services that were at the center of attention in the beginning of the evolution of B2B marketplaces are still relevant, they have to be supplemented with services from such diverse areas as logistics, finance, collaborative design or quality assurance. A technological infrastructure that enables the necessary alignment of business needs and technological requirements will be critical for the success of B2B marketplaces (Alt & Fleisch, 1999). From this, we derive two main hypotheses. First, B2B marketplaces will rely heavily on specialized service providers, in order to be able to provide this wide array of services. In addition, they will have to redefine their own role. To examine this hypothesis, we surveyed a large number of European B2B marketplaces in mid- 2001, and have been asking them about their use of service providers in key service areas.

Second, a service-based architecture — built upon standards as envisioned by the Web services concept — is required for marketplaces to coordinate and integrate services from multiple service providers into a flexible and extensive service offering (Gottschalk et al., 2002). This allows marketplace providers to define their roles in a new way. This hypothesis has been tested by analyzing the current body of literature on service-based architectures and by examining its applicability for B2B marketplaces.

THEORETICAL FOUNDATIONS

According to Schmid (1999), electronic markets are media that foster market-based exchanges between agents in all transaction phases. He distinguishes information services, intention services, contracting services and settlement services. These service areas must be subsumed under the umbrella of a shared context, or "logical space," outlining the roles and protocols of the interacting parties. In a similar way, Bakos discerns three main functions of a market: matching buyers and suppliers, facilitation of transactions, and the institutional infrastructure, which contains the context-related services that have been mentioned above (Bakos, 1998). Dai and Kauffman (2001) extend this categorization by three additional tasks: the aggregation of product information, price discovery, and providing procurement and industry-specific expertise. These categorizations show the broad scope and diverse nature of the tasks associated with providing marketplace functionality. Also, the relevance of integrating the services offering into a wider concept of context and roles is emphasized by the authors mentioned above.

Marketplace providers can cope with this wide and diverse scope of services by defining their role as coordinators and integrators of specialized service providers — and their offerings — into the marketplace offering. By combining best-of-breed service providers (the current critical phase in the evolution of B2Bmarketplaces) to expand the service offering beyond basic transaction services, according to the needs of the members, can be facilitated. Thus, this paper is based on the assumption that it is highly critical — for the success of B2B marketplaces — that they not only see buyers and sellers as relevant stakeholders in their markets, but also take the interests and processes of service providers into account and find the best partners in the critical service areas for their marketplace (Nenninger & Lawrenz, 2001). However, by using specialized service providers, the ability to deliver diverse, high-quality services is traded for demanding integration tasks. Or, to put it differently: "interconnecting is the new competency" (Temkin, 1999) for B2B marketplace providers. In this view, market makers become aggregators of a multitude of individual services into customer-specific service offerings, and coordinators of processes among the parties involved.

THE USE OF SERVICE PARTNERS – EVIDENCE FROM EUROPE

Five hundred and seven electronic B2B marketplaces from throughout Europe had been identified in March 2001 through searching professional databases and the Internet. Because this is still a manageable number, it was decided to survey the complete population, instead of a random sample. Following, the marketplaces have been contacted by telephone in order to identify the relevant contact person for surveys on a strategic level. Thus, the survey took place from April to May, with mostly high-ranking members of the management team, from 248 marketplaces, participating.

The participating sample represents the overall population very well, with about 30 percent of the marketplaces being located in Germany, about 15 percent in the UK, slightly less than 10 percent in Sweden, and the remainder coming from 13 other European countries. An important characteristic of the sample is the distribution among horizontal and vertical marketplaces. Due to the complex transactions in most vertical markets, compared to horizontal markets, we expect our hypotheses will be true especially for vertical marketplaces. With more than 50 percent of the participating marketplaces defining themselves as purely vertical marketplaces, and another 25 percent stating that they perceive themselves as a hybrid vertical and horizontal marketplace, these marketplace pose the majority of participants.

Following, we will present the survey results from that impact the research question at hand. One major area of interest in this survey was the service offering strategy of the management, with regards to the current status and planned evolution, as well as the use of, specialized service providers in this strategy. Following Schmid and Bakos, we distinguished 13 services in three major service areas.

The results of the survey show that, currently, the service offering mainly consists of basic information services and core transaction services. However, the planned development of the surveyed marketplaces indicates that all services, but especially value-added services such as logistics, financial, backend integration and trust services, are being adopted widely by marketplaces in the near future.

From this evolution we derive two conclusions.

1. Marketplaces will expand their services to a far more comprehensive service offering, thereby increasing the complexity of the overall marketplace coordination task with respect to architecture, as well as processes.
2. This increase in overall complexity is reinforced by the more complex nature of the services that are added now, in the later phases of the initial set-up process. Logistics, financial, backend integration and business intelligence services intrinsically are more complex than basic information or transaction services.

Figure 1: Current and planned service offering (percentage of marketplaces that offer (left section) / plan to offer (right section) a service, n = 233 to 244)

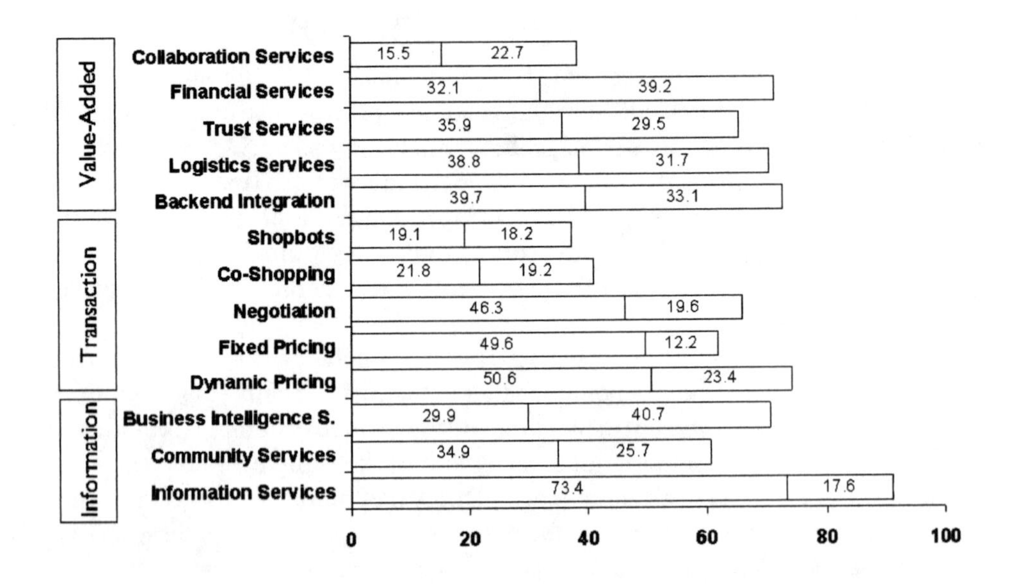

Figure 2: Use of service providers as partners (in percent, n = 49 to 175)

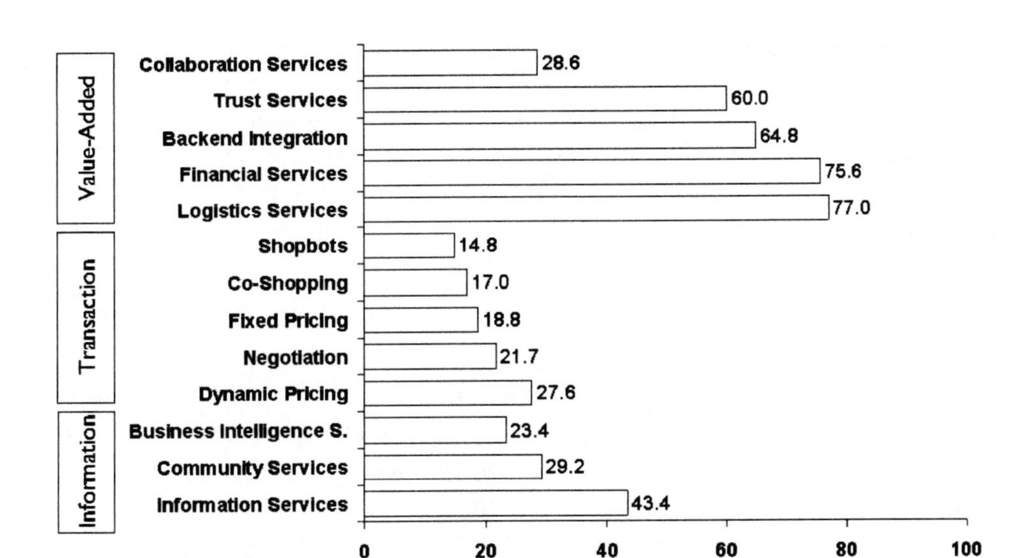

We already suggested that a suitable strategy, for the providers of B2B marketplaces to cope with this increasing complexity, is to make use of specialized service providers and to focus on the role of the service aggregator and coordinator.

The results from the survey support this view. The use of specialized service providers, especially in the more complex value-added services, is high among European B2B marketplaces.

The resulting rise in the significance of partnering (and thereby in the organizational structure of B2B marketplaces) will have profound effects on the technological architecture. Today, simple point-to-point integration dominates the landscape supported by EDI and VANs. With proprietary point-to-point integration based on EDI, it is difficult to manage more than 25 partners (Durchslag et al., 2001) and even harder to change partners. Therefore, according to our second hypothesis, we test the ability of the service-based architecture concept to align with such an organizational model.

SERVICE-BASED ARCHITECTURES AND THEIR IMPLICATIONS FOR MARKETPLACES

The current situation — and the further development (in the near future) — among B2B marketplaces toward an integration of specialized service providers into a comprehensive, customer-oriented service offering, clearly has to evolve jointly, with a different approach to technological architecture. This new approach has to enable B2B marketplaces to fulfill the demands of their customers, the buyers and sellers, on their marketplace more easily. At the same time, it must provide opportunities for the marketplace itself to generate revenue from its offerings. In our view, a service-based architecture may be a suitable solution for this alignment.

A service-based architecture basically features three distinct roles: a service provider, a service requestor and a service broker. In contrast to the three participating actors, Gisolfi (2001) distinguishes five business roles a company can play in a service-based architecture. Depending on the business goals, a specific company may play one or more of the roles. For a B2B marketplace, all five roles qualify and offer potential added value to the customers and to its partners.

Service requestor — Because the business activity of the marketplace includes the provision of services offered by its partners, it may take the role of a service requestor with respect to those partners. More specifically, the marketplace interacts with the partners in order to offer composite services to the customers, and therefore can be seen as an intermediary providing service aggregation. Hence, the marketplace adds value by solving a number of problems that, otherwise, customers would be left to address on their own. Intermediaries provide context-support missing from standards. Web services standards are powerful in their simplicity and openness. However, these standards do not provide for the context-oriented issues of the interaction among services. The marketplace can provide this context by offering a common interface (to the buyers and sellers) to access the partners' services (Truelove, 2001).

Service provider — A marketplace may provide services directly to its customers. Although the partners eventually implement these, the marketplace offers a standardized way of interacting with the required services. Interfaces are, therefore, more stable throughout time and minimize the customers' need for frequent and expensive adaptation. Therefore, the marketplace hides the heterogeneity of interacting with different service providers, which makes service usage easier for the customers and, at the same time, offers the marketplace the flexibility in choosing the most suitable provider. Services from multiple parties can be compounded to form comprehensive service bundles. This matches the aggregator and coordinator function that is described (above) for B2B marketplaces on a technical level.

Registry — Instead of directly providing services aggregated from partners, the marketplace also may collect and catalog data about the partners' services, in order to make them accessible to the customers. This catalog relates to a particular industry, or narrow range of related industries, the marketplace is serving. The marketplace hosts the registry, which is accessible to its customers only, providing contact information that allows the customers to interact directly with the service-providing partners. The partner, in turn, is offered access to the customers of the marketplace, and the listing in the registry indicates a certain level of trust, as well as, industry focus guaranteed by the marketplace.

Broker — In service-oriented architectures, service description and metadata play a central role in maintaining a loose coupling between service requestors and service providers. The role of the broker extends the registry, as it offers additional metadata about the partners' services and, based on this metadata, the functionality for searching and classification of services. In addition, industry-specific taxonomy data helps the customers find service providers and enables service providers to describe service offerings precisely.

Aggregator/Gateway — This extends the capabilities of the broker by the ability to describe actual policy, business processes and binding descriptions that form the standard way of operation on the marketplace, and which are fulfilled by service-providing partners. It would be the logical place to find standard Web service interface definitions for common business processes in the industry. Marketplace customers then cab use these as a reference to use the services. Services implementing the standard service interface definitions are offered directly by the marketplace (i.e., services intermediating among the customers and heterogeneous partner interfaces) or by the partners of the marketplace, who have the opportunity to acquire the standard processes used on the marketplace as a guide for service implementation. As marketplaces provide standard service interface definitions, services adoption will accelerate. This allows a marketplace to provide value-adds, like quality of service monitoring on the partner's Web services response times, and self-monitoring of its members' Web services business practices (Graham , 2001).

A marketplace can provide support for customers and partners in playing each of these five roles. Therefore, adopting a service-based architecture will be essential for business-to-business marketplaces in coordinating the multitude of contributions from service partners that a single transaction and, even more so, the management of complex relationships on B2B marketplaces will require. Therefore, the functionalities and services contained in the marketplace will have to be supplemented by Internet-based services, to a large extent (McCullough, 2000, p. 11).

The technological foundations for the implementation of flexible service-based architectures are evolving rapidly. These are available in the form of specifications only, as in the case of ebXML (Eisenberg & Nickull, 2001) or as specific technologies that have been deployed, i.e., the SOAP messaging protocol, the UDDI registry service, the WSDL service description language, etc. They support the marketplace in providing the roles discussed above. Registry services can be deployed by a marketplace in order to provide information about service providers, industry taxonomies and common process and interface definitions. These standardized processes and interface definitions can be described using WSDL, or they may be available as collaboration protocol profiles in an ebXML registry. In any case, the technology needs to provide an underlying common messaging protocol, a registry which administers the common knowledge available on the marketplace for integration of services and, preferably, a set of standardized components and services used by the participants of the marketplace, as well as a way to describe services and how to interact with them. Which specific technology will evolve as the dominant solution is hard to predict. , However, because of its association with OASIS and UN/CEFACT — and its wide support in the electronic B2B community – ebXML is perceived as an important element in the evolution toward a service-oriented architecture (Fitzgerald, 2001). For example, Covisint, the automotive B2B marketplace, chose ebXML as a cornerstone of its architecture (Covisint, 2001).

CONCLUSION

Partnering with specialized service providers is critical for the success of B2B marketplaces in creating a comprehensive service offering for their members. However, the complexity of integrating not only buyers and sellers but also service providers (dynamically) into the marketplace platform requires new architectures and new role definitions for marketplace providers. The introduction of service-oriented architectures has the potential to provide such flexible architectures and the corresponding role definitions.

Further research should be directed at how marketplaces and their members that require service providers find and bind those partners on a technical, as well as, a business and relationship level. Do they forge special and close relationships in order to create unique offers on their marketplace? Or, do they use market mechanisms to let the members decide which of a broad choice of service offerings they prefer? Also, future questions might include: How do they evaluate the performance of their service providing partners, and how do marketplace providers and service providers generate value and accordingly split revenues?

REFERENCES

Aberdeen Research. (2000). *The E-Business Marketplace: The Future of Competition.* Boston, MA: Executive White Paper.

Alt, R., & Fleisch, E. (1999). Key success factors in designing and implementing business networking systems. *Proceedings of the Twelfth Bled Electronic Commerce Conference. (Vol. 1, 219-235).* Bled, Slovenia.

Bakos, Y. (1998). The emerging role of electronic marketplaces on the Internet. In *Communications of the ACM, 41* (8), 35-42.

Bennett, K., Munro, M., Gold, N., Layzell, P., Budgen, D. & Brereton, P. (2001). An architectural model for service-based software with Ultra Rapid Evolution. In *Proceedings of the IEEE International Conference on Software Maintenance.* Florence, Italy.

Covisint. (2001). Covisint supports ebXML technology findings. Retrieved July 31, 2001 from the World Wide Web: http://www.covisint.com/info/pr/ebxml.shtml.

Doyle, T., & Melanson, J. (2001). B2B Web exchanges - Easier hyped than done. *Journal of Business Strategy, 10* (13).

Durchslag, S., Donat, C. & Hagel III, J. (2001). Web services: Enabling the collaborative enterprise. Grand Central Networks Inc. Retrieved October 2, 2001 from the World Wide Web: http://www.grandcentral.com. Access: 10/02/01.

Eisenberg, B., & Nickull, D. (2001). ebXML technical architecture specification v1.0.4. Retrieved April 5, 2001 from the World Wide Web: http://www.ebxml.org/specs/ebTA.pdf.

Ernst, D., Halevy, T., Monier, Jean-Hugues J. & Sarrazin, H. (2001). A future for e-alliances. *The McKinsey Quarterly 2001,* 2, 92-102.

Fitzgerald, M. (2001). *Building B2B Applications with XML.* New York: John Wiley & Sons.

Garicano, L. & Kaplan, S. N. (2000). *The Effects of Business-to-Business E-Commerce on Transaction Costs* (NBER Working Paper No. W8017). Cambridge, MA.

Gisolfi, D. (2001). Web services architect, Part 2: Models for dynamic e-business. Retrieved September 2, 2001 from the World Wide Web: http://www-106.ibm.com/developerworks/library/ws-arc2.html.

Gottschalk, K., Graham, S., Kreger, H. & Snell, J. (2002). Introduction to Web services. *IBM Systems Journal, 41* (2), 170-177.

Graham, S. (2001). The role of private UDDI nodes in Web services, Part 1: Six species of UDDI. Retrieved August 24, 2001 from the World Wide Web: http://www-106.ibm.com/developerworks/webservices/library/ws-rpu1.html?dwzone=webservices.

Kaplan, S., & Sawhney, M. (2000). E-Hubs: The new B2B marketplaces. *Harvard Business Review,* 78 (3), 97-103.

Lacoste, G., Pfitzmann, B., Steiner, M., & Waidner, M. (Eds.) (2000). *SEMPER – Secure Electronic Marketplace for Europe.* Berlin: Springer.

Lindemann, M., & Schmid, B. (1998). Elements of a reference model electronic markets. In *Proceedings of the 31st Annual Hawaii International Conference on Systems Science HICCS'98, Vol. IV* (pp. 193-201). Hawaii, HI.

McCullough, S. S. (2000). *eMarketplace Hype, Apps Realities* (Forrester Report). Forrester Research.

Nenninger, M. & Lawrenz, O. (2001). *B2B-Erfolg durch eMarkets.* Braunschweig/Wiesbaden, Germany: Vieweg.

Phillips, C. & Meeker, M. (2000). *The B2B Internet Report – Collaborative Commerce* (Report). Morgan Stanley Dean Witter Equity Research.

Schmid, B. F. (1999). Elektronische Märkte - Merkmale, Organisation und Potentiale. In A. Hermanns & M. Sauter (Eds.), *Management-Handbuch Electronic Commerce: Grundlagen, Strategien, Praxisbeispiele* (pp. 31-48). München, Germany: Vahlen.

Temkin, B. D. (1999). *Distribution Reconstructed* (Report). Forrester Research.
Truelove, K. (2001). Intermediaries that simplify inter-enterprise projects. Retrieved October 9, 2001 from the World Wide Web: http://www-106.ibm.com/developerworks/library/ws-netwrk.html.
Wald, E. & Stammers, E. (2001). Out of the Alligator Pool: A service-oriented approach to application development. *eAI Journal,* March 2001, 26-30.

Chapter VII

An Evaluation of Dynamic Electronic Catalog Models in Relational Database Systems

Kiryoong Kim, Dongkyu Kim, Jeuk Kim, Sang-uk Park,
Ighoon Lee, Sang-goo Lee and Jong-hoon Chun
Seoul National University, Korea

ABSTRACT

Electronic catalogs are electronic representations about products and services in the electronic commerce environment and require diverse and flexible schemas. Although relational database systems seem to be an obvious choice for their storage, traditional designs of relational schemas do not support electronic catalogs in the most effective ways. Therefore, new models for managing diverse and flexible schemas in relational databases are required for such systems. Proposed in this paper are several models for electronic catalogs using relational tables, and an experimental evaluation of their efficiency. The results of this study can be put to practical use and are, in fact, being applied in the design of a commercial software product.

INTRODUCTION

Electronic catalogs are electronic representations of information about the products and services of an organization (Segev et al., 1995). A typical catalog containing 100,000 products may contain thousands of different schemas (Jhingran, 2000). For example, a "TV" may have a 'voltage' attribute while a "pen" may not. Consequently, one of the biggest problems in electronic catalogs is diversity of schemas for products.

For this reason, XML seems to be a suitable alternative that meets the requirements of electronic catalogs. However, it is inefficient to store large numbers of catalog data as XML documents. Relational database systems are still the most practical choice for managing business data (Shanmugasundaram et al., 1999).

However, traditional relational databases are not geared toward managing several schemas at once or managing a universal table with many nulls. Consequently, careful application level design is required. For example, a frequently used model represents catalogs in the form of<id, attribute name, attribute value>. But this scheme requires multiple self-joins to retrieve information about a product and, thus, is inefficient in managing a large quantity of product data. We should therefore consider other models that support the view of thousands of tables efficiently from an application perspective, yet manage the database from a finite set of verticalized tables (Jhingran, 2000).

This paper suggests several models for electronic catalogs using relational databases, and verifies their efficiencies through experiments. The goal is to find the most efficient model, and to utilize it in a practical electronic catalog system.

Figure 1: Experimental process

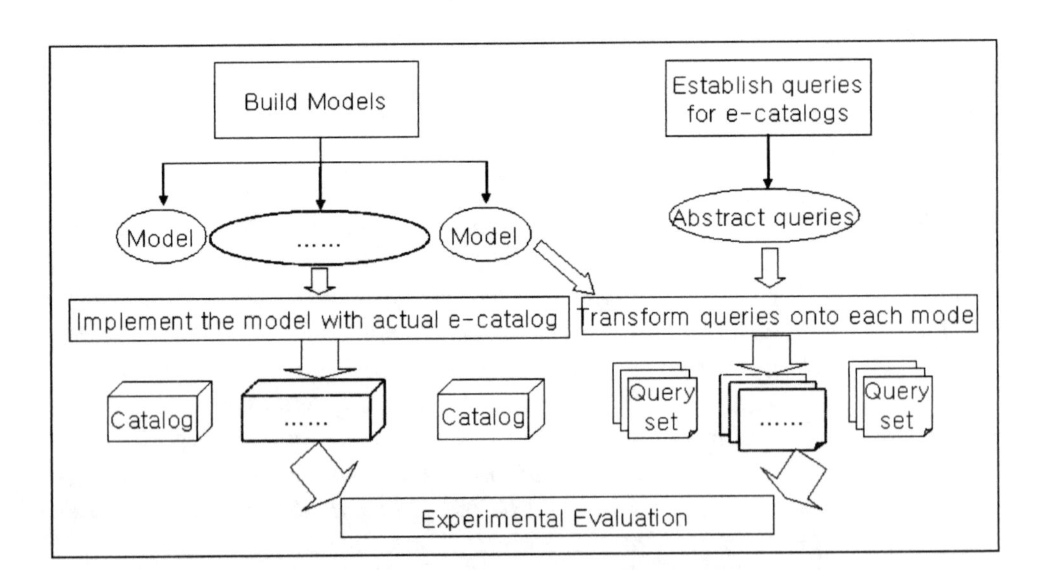

First, we suggest several relational database schemas for e-catalogs. The models are implemented and populated with actual e-catalog data. We then establish queries for e-catalogs considering practical usage, and transform them into queries for each model. The models are tested against these query sets. Figure 1 depicts the experimental process

RELATED WORK

This work is a part of studies on the Content Management System being developed jointly by the Intelligent Database Systems Laboratory, Seoul National University, and CoreLogX, Ltd., Korea. In a previous work, Jung et al. (2000) proposed a data model for electronic catalogs that offers effective searching and also management facilities to customize the electronic catalog system. They set a normalized structure for the content that can be shared among multiple catalogs. One of the contributions of their model is that it separated content from presentation and defined the content by itself to express the variety of the schema. The proposed model, EE-Cat, allows for the effective management of dynamic change of requirements on catalog systems. This paper shares fundamental notions of this previous work.

Anant Jhingran (2000) presents a clear insight on new requirements for database systems to support e-commerce processes. In his work, he introduces a concept of a verticalized schema that can support the view of thousands of different tables from an application perspective, yet manage the database from a single verticalized table. He argues studies on those schemas. There is much to be investigated for such requirement.

An interesting practical experiment on the electronic catalog model is presented by Rakesh Agrawal et al. (2001). They suggest two verticalized schemas – horizontal schema and vertical schema. The former is a schema of a universal table that contains all the attributes used in the catalog, while the latter is a schema in the form of <attribute name, attribute value> pair. The two are compared via experiments with varied factors, and the results show that a vertical schema is superior to a horizontal schema. The work is very similar work to ours, but the results are very different from ours. The difference is in the experimental environment and query models. Further discussion can be found in the later sections.

TERMINOLOGY

There may be some confusion on the terminology. As the terminologies for electronic catalogs are not yet established fully, we define important terms in order to eliminate unnecessary confusion.

Product group: A group of products that is treated as the 'same type' of products (Jung et al., 2000). We assume that products in a specific product group share the same set of attributes, and that a product belongs to only one product group only. Hierarchy of product groups is not considered here.

Common attributes: The attributes that are common to all the product groups (Jung et al., 2000). We represent the set of common attributes as $\boldsymbol{C}$:

$$\boldsymbol{C} = \{c_1, c_2, ..., c_n\},$$

where c_i is an attribute required for every product group in the database, such as product group id., and product id.

Dependent attribute: The attributes that are specific to a product group. Let $\boldsymbol{D}_i$ be a set of dependent attributes for the product group i, then:

$$\boldsymbol{D}_i = \{d_{i1}, d_{i2}, ..., d_{jm}\},$$

where d_{ij} is an attribute for product group i and not in $\boldsymbol{C}$, such as 'voltage' of refrigerators, 'diameter' of bearings, or 'frequency' of cellular phones.

Now, we can express a schema for product group i as $\boldsymbol{P}_i$.

$$\boldsymbol{P}_i = \boldsymbol{C} \cup \boldsymbol{D}_i$$

Product_group id: An attribute that identifies a product group such as the classification code of products.

Product_id: Attribute that identifies each individual product in the database such as the SKU#[1] or EAN/UCC[2] code. Each product has a unique product id.

PRIMITIVE MODELS

In this section, we present two primitive data models for constructing electronic catalogs in a relational database. The name of each model describes its representative table.

1) The Universal Table (UT) Model

The UT model stores product data in a table with a schema consisting of the union of common attributes and dependent attributes. The schema of this table, $\boldsymbol{T}_{\text{UNIVERSAL}}$ can be expressed follows:

$$\textbf{Schema}\ (\boldsymbol{T}_{\text{UNIVERSAL}}) = \boldsymbol{C} \cup (\cup \boldsymbol{D}_i)$$

Schema(T) denotes a schema of table ***T***. Each tuple in this table represents one product. This model requires another table to keep the meta-data identifying the attributes relevant for each product group. The schema for table $\boldsymbol{T}_{\text{UT\_META}}$ is defined as follows:

$$\textbf{Schema}\ (\boldsymbol{T}_{\text{UT\_META}}) = (\textit{product group id, attribute name})$$

Underlined attributes denote key attributes. Each instance of *attribute name* should be identical to the name of an attribute in $\boldsymbol{T}_{\text{UNIVERSAL}}$. To introduce a new

Figure 2: The UT model

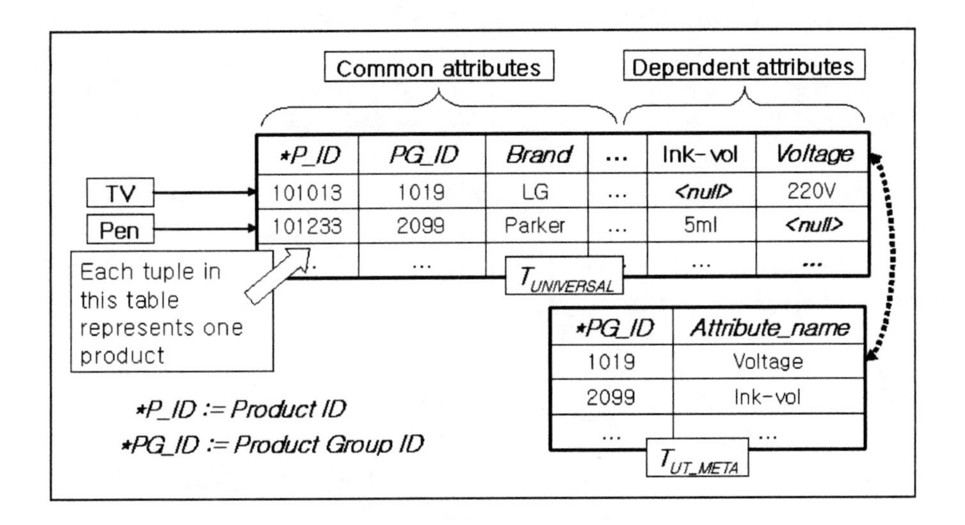

product group into the catalog, we should add new attributes to $\boldsymbol{T}_{\text{UNIVERSAL}}$ and insert corresponding meta-data for the product group into $\boldsymbol{T}_{\text{UT\_META}}$.

Now, we define the UT model as $\boldsymbol{M}_{\text{UT}}$:

$\boldsymbol{M}_{\text{UT}} = (\boldsymbol{T}_{\text{UNIVERSAL}}, \boldsymbol{T}_{\text{UT\_META}})$

Figure 2 shows an example of the UT model. The merit of this model is its simplicity. Not many tables are created. For example, since the table $\boldsymbol{T}_{\text{UNIVERSAL}}$ has both fields "Ink volume" for pens and "Voltage" for TVs, it can store product groups "TV" and "pen," So both product groups can be stored in one table. In addition, we can use the integrity constraints (e.g. domain constraints) provided by the DBMS because the model uses the relational schema directly. On the other hand, there may be too many nulls in the instance of $\boldsymbol{T}_{\text{UNIVERSAL}}$, because each tuple uses only some of the attributes in the table. However, because of its simplicity, UT is currently a popular model in actual electronic commerce applications (Jung et al., 2000).

2) The Name (PT) Model

The PT model stores product data in a table, $\boldsymbol{T}_{\text{PT}}$, whose schema is defined as follows:

Schema $(\boldsymbol{T}_{\text{PT}})$ = (*product id, product group id, attribute name, attribute value*)

Just like $\boldsymbol{T}_{\text{UT\_META}}$ in $\boldsymbol{M}_{\text{UT}}$, each instance of *attribute name* should be identical to the name of an attribute in $\boldsymbol{T}_{\text{UNIVERSAL}}$. As each attribute belongs to a specific product group, *product group id* is needed to express it. Table T_{PT}, which stores the product data, does not provide semantic scheme of each product data because this model cannot use RDB features directly. So we need additional information to express constraint conditions for the attributes. Thus, we must keep meta-data about the attributes each product group uses. Let T_{PT_META} be a table for these meta-data:

$$\textit{Schema}\ (T_{PT\_META}) = (\textit{product group id, attribute name}, c_1, c_2, \ldots, c_n)$$

c_k is a constraint for each attribute. For example, its data type, whether it allows null, or whether it is a unique value.

Now, we can define M_{PT} as follows:

$$M_{PT} = (T_{PT}, T_{PT\_META})$$

Figure 3 shows an example of the PT model with two product data that are identical to Figure 2. This is the most flexible model currently used with relational databases, as new attributes can be added without changing the schema of the table. To introduce a new product group, we simply insert tuples describing meta-data into T_{PT_META}. For this reason, this model is also popular.

On the other hand, this model cannot use the integrity constraints provided by relational database systems. For example, we should translate all values into character strings or some other common data types, and check each type of data by managing the meta-data. Besides, all the data must be stored in fixed-length records.

Another drawback is that this model uses too many tuples, because it requires one tuple for each attribute, not each product.

Figure 3: The PT model

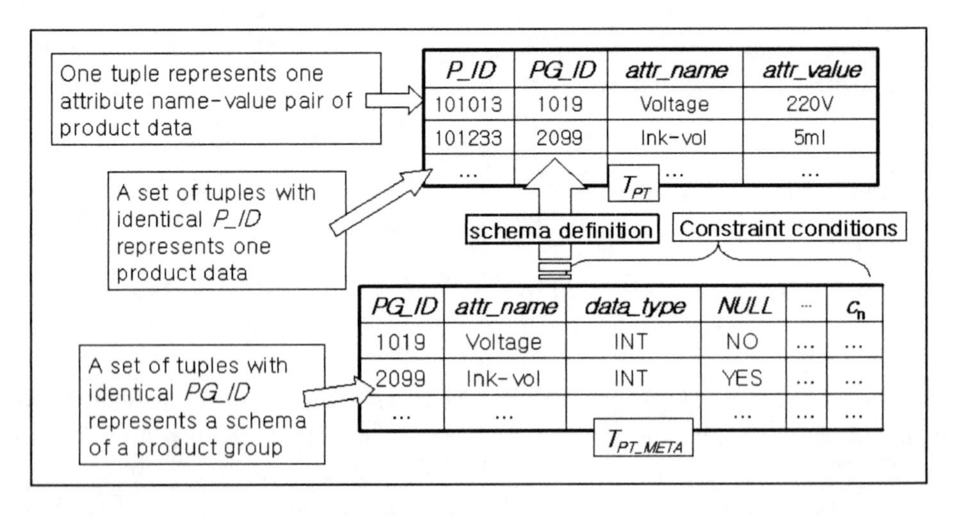

HYBRID MODELS

Each of the primitive models described in the previous section deals with all the attributes in the same way. However, it might be more efficient to handle common attributes and dependent attributes differently, because the former are fixed for all product groups, while the latter vary with the product groups. In this section, we present hybrid models, combining the primitive models.

Common attributes do not vary with product groups, so it is reasonable to manage common attributes using a single table, as done in $\boldsymbol{T}_{UT}$. Thus, we define a schema $\boldsymbol{T}_{COMMON}$ that is a projection of common attributes from $\boldsymbol{T}_{UT}$.

$$\boldsymbol{T}_{COMMON} = \text{“}_{c}(\boldsymbol{T}_{UT})$$

1) The HYBRID_1 Model

HYBRID_1 combines UT and PT. It is composed of three tables.

$$\boldsymbol{M}_{HYBRID\_1} = (\boldsymbol{T}_{COMMON}, \boldsymbol{T}_{PT}, \boldsymbol{T}_{PT\_META})$$

$\boldsymbol{T}_{COMMON}$ contains common attributes of all the products in the same way as UT, whereas $\boldsymbol{T}_{PT}$ and $\boldsymbol{T}_{PT\_META}$ store dependent attributes in the same way as PT. In contrast with PT, instances of $\boldsymbol{T}_{PT}$ and $\boldsymbol{T}_{PT\_META}$ in this model do not contain tuples for common attributes because they are included in instances of $\boldsymbol{T}_{COMMON}$. An example instance of the HYBID_1 model is shown in Figure 4.

Figure 4: The HYBRID_1 model

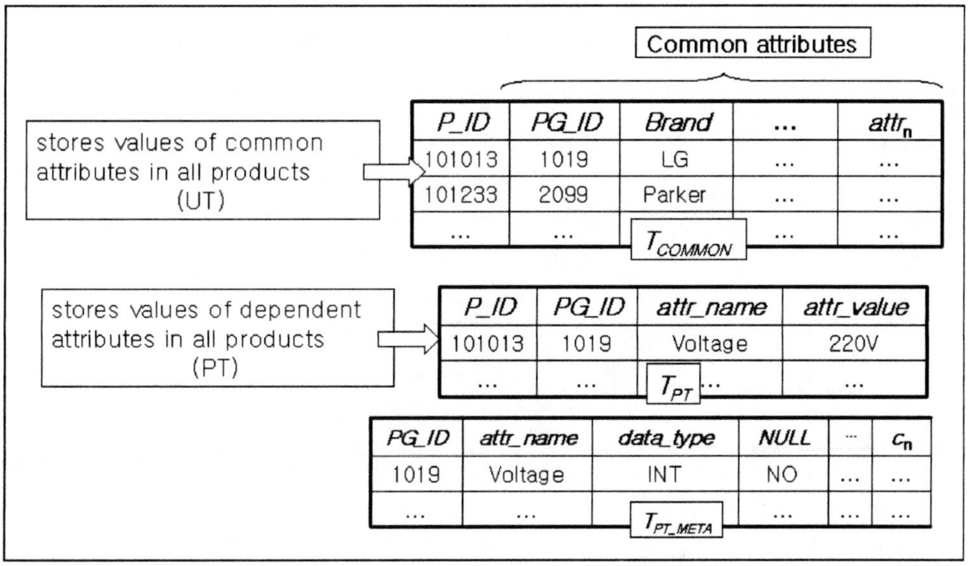

Common attributes

stores values of common attributes in all products (UT)

P_ID	PG_ID	Brand	...	attr$_n$
101013	1019	LG	...	...
101233	2099	Parker	...	...
...	...	T_{COMMON}	...	...

stores values of dependent attributes in all products (PT)

P_ID	PG_ID	attr_name	attr_value
101013	1019	Voltage	220V
...	...	T_{PT} ...	...

PG_ID	attr_name	data_type	NULL	...	c$_n$
1019	Voltage	INT	NO	...	...
...	...	T_{PT_META}	...	...	...

2) The HYBRID_2 Model

The main problem with PT and HYBRID_1 is that they require multiple tuples for a product data. To solve this problem, we introduce T_{OPTION} as follows:

Schema (T_{OPTION})
= (*product id, product group id, optional field*$_1$, *optional field*$_2$, ..., *optional field*$_n$)

Each product group can use *optional fields* in T_{OPTION} for its attributes. If the number of *optional fields* is at least the maximum of the number of elements in P_i, only one tuple would be sufficient for one product. To use this table, we require another table that manages its meta-data:

Schema ($T_{\text{OPTION\_META}}$)
= (*product group id, optional field #, attribute name,* c_1, c_2, ..., c_n),

where c_k is a constraint for the attribute.

$T_{\text{OPTION\_META}}$ is similar to $T_{\text{PT\_META}}$ except that it contains an attribute that indicates a specific optional field in T_{OPTION}. Combining (joining) T_{COMMON} and T_{OPTION}, we define $T_{\text{COMMON\_OPTION}}$ as a table containing the common attributes and optional fields ($T_{\text{COMMON}} \bowtie T_{\text{OPTION}}$).

Figure 5: The HYBIRD_2 model

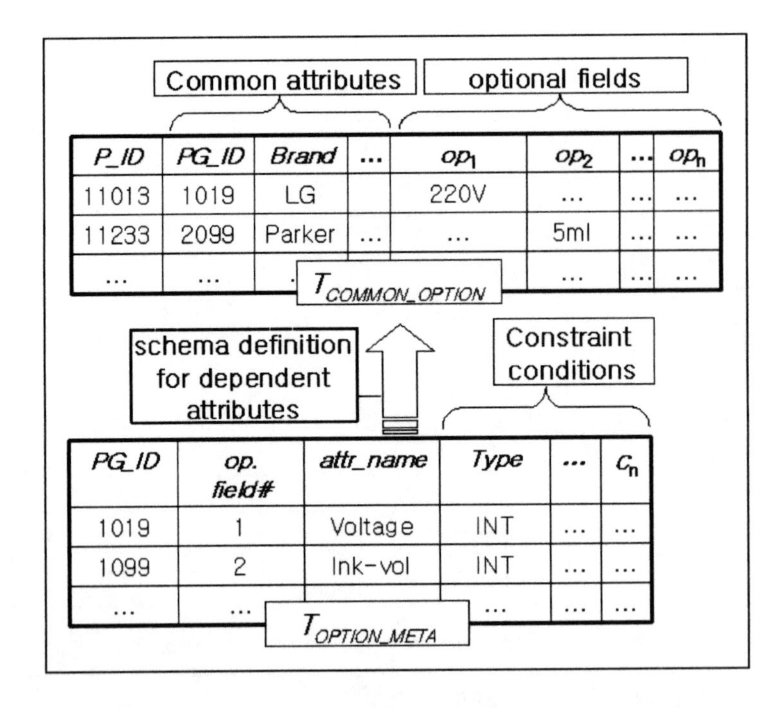

Figure 6: The HYBRID_3 model

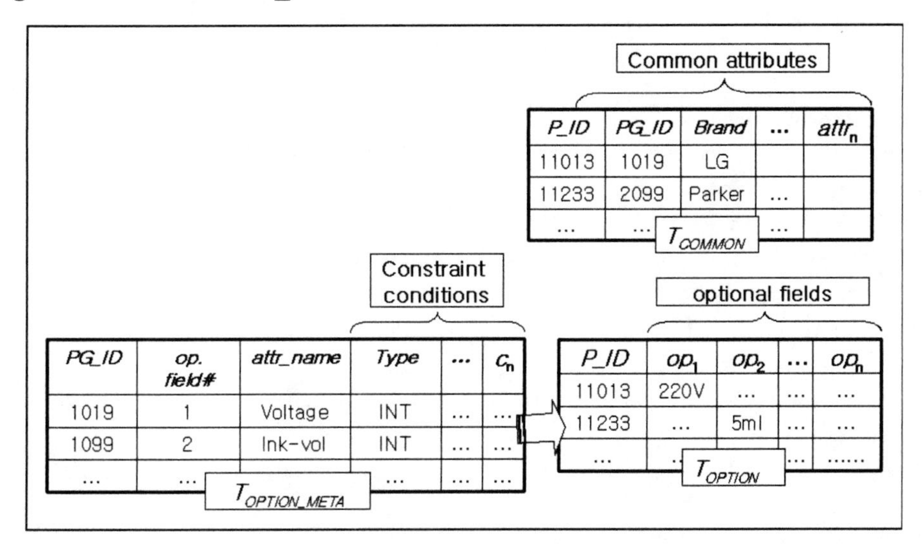

HYBRID_2 is a model using these tables.
$\boldsymbol{M}_{\text{HYBRID\_2}} = (\boldsymbol{T}_{\text{COMMON\_OPTION}}, \boldsymbol{T}_{\text{OPTION\_META}})$

Values of dependent attributes are stored in *option fields* of $\boldsymbol{T}_{\text{COMMON\_OPTION}}$, with $\boldsymbol{T}_{\text{OPTION\_META}}$ storing their meta-data. In this case, the number of *option fields* must be large enough so that it is at least the maximum of the number of elements in $\boldsymbol{D}_i$, for all i.

3) The HYBRID_3 Model
HYBRID_3 is composed of three tables.
$\boldsymbol{M}_{\text{HYBRID\_3}} = (\boldsymbol{T}_{\text{COMMON}}, \boldsymbol{T}_{\text{OPTION}}, \boldsymbol{T}_{\text{OPTION\_META}})$

HYBRID_3 is similar to HYBRID_2 except that it separates $\boldsymbol{T}_{\text{OPTION}}$ from $\boldsymbol{T}_{\text{COMMON}}$, so we can expect better performance for queries with conditions only on common attributes or only on dependent attributes, but worse performance is expected for queries with conditions on both common attributes and dependent attributes, because of join operations.

ANTICIPATED QUERIES FOR ELECTRONIC CATALOGS

When we manipulate electronic catalogs, our requirements are expressed as relational queries. There are several classes of these queries. We define these classes in this section.

1) Queries that retrieve desired data from the electronic catalogs:
These queries are most frequently used because the main purpose of an electronic catalog is to provide product information to buyers.

Query group 1.1: Retrieval of key with exact conditions for common attributes.
Query group 1.2: Retrieval of key with exact conditions for dependent attributes.
Query group 1.3: Retrieval of key with exact conditions for common and dependent attributes.
Query group 2.1: Retrieval of key with range conditions for common attributes.
Query group 2.2: Retrieval of key with range conditions for dependent attributes.
Query group 2.3: Retrieval of key with range conditions for common and dependent attributes.
Query group 2.4: Retrieval of key by pattern matching for common and dependent attributes.
Query group 3: Retrieval of an entire set of product data.

2) Queries that manipulate data in the electronic catalogs:
These queries are necessary to insert, delete, or update product data in the electronic catalogs.

Query group 4.1: Insert new product data.
Query group 4.2: Update product data.
Query group 4.3: Delete product data.

3) Queries that manipulate schemas of the electronic catalogs:
Electronic catalogs change throughout time. For example, a new product group might be introduced, or new attributes might be required for an existing product group. These queries allow changes to schemas for product groups in electronic catalogs.

Query group 5.1: Add a new product group.
Query group 5.2: Modify product group information.
Query group 5.3: Drop a product group.

EXPERIMENT

1) *Environment*
Experiments were null on a Solaris Ultra Sparc II machine with 332 MHz CPU and 512 MB RAM. The tests were carried out on two different DBMSs because some factors associated with a particular DBMS could affect the results. The

DBMSs were Oracle 8i[3] and MySQL.[4]

Experimental data were provided by the Public Procurement Service (PPS)[5] Korea, EAN Korea[6], and LOTTE.com Inc.[7] The combined catalog contained data on 51,766 individual products in 100 product groups. The product groups shared 18 common attributes, and the number of dependent attributes for a product group ranged from 2 to 15. Product groups contained from 4 to 986 products. There were 100 different attributes in the catalog. All product data were transformed into the tables presented in the previous sections.

2) *Module for Measuring Time*

We implemented a measuring module with the Java programming language and JDBC. This module reads queries in script files written in a specifically defined format we defined, and measures the time to process them.

The script file consists of groups of query sections. Each section is composed of queries required to complete one transaction, and each group corresponds to the query group described in the previous section. The module processes each group of query sections and takes the average of the times to process each of them. Each script file was executed 10 times, and the results were recorded.

3) *Index*

Indexes increase the speed of retrieval, but decrease the speed of inserting or updating data. In an electronic catalog system, we usually use retrieve queries more frequently than insert or update queries. Moreover, the models we mentioned above manage data in one or two tables only, so the number of tuples would become prohibitively large to process without indexes.

We experimented with indexes with each model with following criteria.

UT: A table on UT has so many fields that it is not easy to decide which fields should be indexed. In the experiments, we used indexes on *product group id* and some 'frequently referenced attributes,' that are used more frequently in predicates of the test queries. These 'frequently referenced attributes' are key attributes of electronic catalogs. For example, product name, price, and manufacturer are some of 'frequently referenced attributes.'

PT: In PT, there are only a few fields in the tables, so indexes were created for all the fields.

HYBRID_1: HYBRID_1 has features of both UT and PT, so methods of both models are applied. Indexes were created for all the frequently referenced common attributes and all attributes in $\boldsymbol{T}_{PT}$.

HYBRID_2: Indexes are created for *product group id* and some fields that are frequently referenced. Indexes for some optional fields are also created.

HYBRID_3: Two different groups of indexes for HYBRID_3 were created. One was identical to those of HYBRID_2, and the other was a group of composite indexes on all the *option fields* and frequently referenced common attributes.

RESULTS

1) *Oracle 8i without Indexes*

Table 1 shows the results of experiments using Oracle 8i in milliseconds. A bolded cell indicates the best result in each row. As a result of caching, values of first run results were too high for all cases and increased the variance of the results so they are excluded.

Performance of UT is not inferior to other models. The width of a table has little detrimental effect upon performance in this case.

Model PT performs worst, on average. In the experiment, instances of T_{PT} in M_{PT} include 1,354,493 tuples, and it requires too many self-joins because each attribute is expressed as a separate tuple. While PT shows relatively good perfor-

Table 1: Result on Oracle 8i without indexes

Query #	PT	UT	Hybrid-1	Hybrid-2	Hybrid-3
1.1	5673.04	379.42	**290.93**	696.31	363.46
1.2	2878.40	398.27	1053.71	368.60	**246.68**
1.3	3465.28	358.18	882.38	295.62	**272.62**
2.1	3156.44	6.82	6.74	7.25	**6.40**
2.2	2527.03	277.11	809.23	200.67	**129.96**
2.3	7971.07	290.45	1547.74	**196.01**	1323.25
2.4	22234.48	**122.37**	2824.89	238.11	338.78
3	2850.89	**15.70**	876.40	372.88	36834.59
4.1	57.89	3.67	14.18	**2.37**	4.33
4.2	7008.78	**260.22**	1021.07	294.66	411.44
4.3	148815.70	17523.26	64446.33	**10521.81**	22216.67
5.1	50.11	8.56	36.78	8.67	**8.33**
5.2	3911.56	2.67	1.67	1.78	**1.33**
5.3	3016.67	1845.89	2292.33	1320.89	**433.33**

mance on **query group 3**, as self-joins are not required in those queries, it seemed to be an inefficient model overall. For the same reason, HYBRID_1 produces to be an inefficient model among the hybrid models, because it stores dependent attributes in $\boldsymbol{T}_{PT}$.

HYBRID_3 outperforms the other models, on average. However, it is very inefficient for queries accessing both common attributes and dependent attributes, because it manages them in separate tables. On the other hand, HYBRID_2 is uniformly efficient for most queries.

2) *Oracle 8i with Indexes*

Table 2 shows the results with indexes, using Oracle. PT also shows the worst performance on average, but shows the best performance on **Query group 3**, which does not require self-joins. Results for the other models are similar to those without indexes except that processing time generally decreases and HYBRID_1 shows relatively good features for many queries, except for Query group 3, although it stores dependent attributes in $\boldsymbol{T}_{PT}$. This implies that a hybrid model can utilize good features of two models. Because the HYBRID_1 model can reduce width model by using name-value pair schema for dependent attributes, excessive indexes are not created. In addition, it has less tuples than the PT model, so the index for $\boldsymbol{T}_{PT}$ in the HYBRID_1 model can work more properly than that in the PT model.

Table 2: Result on Oracle 8i with indexes

Query#	PT	UT	HYBRID-1	HYBRID-2	HYBRID-3	H-3-ext
1.1	4332.02	301.51	299.18	621.62	389.20	**8.50**
1.2	1921.96	255.71	34.38	365.51	303.60	**5.88**
1.3	2939.60	293.60	17.84	234.20	14.31	**13.18**
2.1	2995.19	7.41	6.48	7.15	7.22	**4.90**
2.2	2468.89	244.22	**5.15**	244.74	158.89	10.57
2.3	6496.85	279.30	60.30	196.56	**5.48**	8.07
2.4	17116.26	116.81	7.07	44.78	6.19	**5.93**
3	**4.89**	16.07	**1876.52**	349.41	**1234.70**	15.73
4.1	45.11	**2.52**	28.67	5.41	14.89	34.90
4.2	8708.11	253.26	**151.89**	374.44	253.11	623.43
4.3	149544.10	**22214.37**	151330.40	28107.93	30706.96	31239.94
5.1	162.56	**9.89**	67.00	16.33	28.11	171.23
5.2	9200.33	1.89	21.56	**1.44**	29.67	8703.41
5.3	7504.22	1410.00	**119.44**	1155.33	633.89	3219.81

= milliseconds

We performed another experiment on HYBRID_3 with additional indexes, to verify how much this model could be improved by indexes. These were created on all the fields of $\boldsymbol{T}_{\text{OPTION}}$ and each frequently referred field of $\boldsymbol{T}_{\text{COMMON}}$ in $\boldsymbol{M}_{\text{HYBRID\_3}}$. A column, **HYBRID_3_ext,** in Table 2 shows the result for this experiment. As shown in the figure, indexes greatly improve the performance.

It was difficult to create indexes on the table in UT because of the large number of attributes.

3) *MySQL*

Because some factors associated with a specific DBMS can affect the efficiency of a model, the same experiments were made on MySQL. Table 3 shows experimental results for UT, HYBRID_2 and HYBRID_3 on MySQL. We didn't test PT and HYBRID_1 because they showed significantly worse performance in the previous experiment on Oracle. We also didn't test with indexes on MySQL because we didn't have enough time for the experiment, but we believe that indexes cannot change the order of results on each model, judging from the results on Oracle.

Because MySQL is a far simpler DBMS than Oracle 8i, the results on MySQL have somewhat different features in comparison with those on Oracle 8i. UT is far worse than the others on almost all queries, and HYBRID_3 shows relatively good performance on **Query group 3**. These results indicate that the efficiency of each

Table 3: Result on MySQL without indexes

Query#	UT	HYBRID-2	HYBRID-3
1.1	2066.22	861.01	**592.38**
1.2	2068.33	863.69	**551.47**
1.3	2052.96	**858.98**	1001.98
2.1	4349.96	3539.67	**3193.26**
2.2	2136.93	992.19	**691.04**
2.3	2113.74	**962.67**	20251.15
2.4	2729.19	1701.74	**1407.33**
3	2081.00	**843.43**	1132.82
4.1	**1.70**	1.78	2.52
4.2	3874.59	**2821.15**	4342.56
5.1	**5.00**	5.89	5.33
5.2	3.00	**1.11**	1.13

= milliseconds

Table 4: Space analysis

	PT	UT	HYBRID-1	HYBRID-2	HYBRID-3	HYBRID-3-ext
Tables on Oracle	126.67	39.85	62.59	**35.51**	50.14	50.14
Indexes on Oracle	112.47	**2.28**	57.66	11.23	12.6	51.02
Tables on MySQL	72.29	24.68	41.20	**23.53**	33.28	-

= megabytes

model could vary with the DBMS, because of differences in storage management, query processing, buffer management, transaction management, and so on. However, hybrid models shows:

4) *Space Analysis*

Table 4 shows total amounts of storage used to store catalogs in each model on Oracle 8i and MySQL. Each value was measured in megabytes. We presumed that UT would have many null values and incur a good amount of wasted space. However, the result shows that UT did not overuse storage. On the contrary, PT had the worst space utilization, because each attribute name was stored in each tuple in T_{PT}. HYBIRD_3_ext, which has extended indexes, required additional storage for indexes, more than for the tables themselves.

DISCUSSION

A similar work by Agrawal, et al. (2001) shows a slightly different result. They made experiments with the 'vertical schema' and the 'horizontal schema' that are homologous to the 'PT model' and the 'UT model.' The experiments were made with varied factors, such as number of attributes and number of product data. The results show the 'vertical schema' is superior to the 'horizontal schema' in the performance. This contradicts our results.

The main difference is in the experimental data. We used electronic catalogs that are used in e-commerce organizations, while the other work used virtual catalogs with randomly generated data. Also, their experiments were made with various parameters for some key factors such as number of attributes and number of product data.

Efficiency of the models can be varied with environments and various factors. Our work has limitation with this respect. Existing e-catalogs have less than a few hundred attributes, and e-catalogs must carefully be designed not to have excessive attributes, because many attributes are shared among product data. This is not only for performance reasons, but also for interoperation and standardization. Thus, experiments with extreme parameters are not necessary. Our work is based on real e-catalogs, so the parameters for the experiments are reasonable.

In addition, even if we accept the result of the other work — that PT model (vertical schema) is superior to the UT model (horizontal schema) — in some cases, it is also inefficient to use. Both the PT model and the UT model have problems in utilizing existing relational database features efficiently. The UT model has an excessive number of attributes, thus too many indexes are needed. Moreover, the result on MySQL shows that some DBMSs cannot manage large numbers of attributes at once. The PT model also has problems because it cannot express a product data with one tuple.

On the other hand, proposed models – HYBRID_2 and HYBRID_3 are more efficient than those primitive models, because they can keep the width of tables from excessive expanding and store a product data in a tuple.

Table 5: Comparison of proposed model

Model	Pros.	Cons.
UT	•Good performance in some DBMS •Low storage usage	•Poor performance in some DBMS(MySQL) •Hard to mange indexes
PT	•Flexibility	•Poor performance •High storage usage •Hard to manage indexes
HYBRID_1	•Flexibility	•Poor performance •High storage usage •Careful application design
HYBRID_2	•Good performance	•Careful application design
HYBRID_3	•Good performance (esp. with proper indexes) •Separate common attributes and dependent attribute	•Careful application design

CONCLUSION

It is believed that the PT mode; is a proper model for managing diverse and flexible catalogs in relational databases. However, our experimental results show that this model is inefficient when applied in practical cases, despite its flexibility. Moreover, applications must work too hard to maintain integrity constraints, as DBMSs cannot guarantee them. Despite its popularity, we conclude that the PT model is not a suitable model for current relational database catalog systems.

UT, which is another popular model, also cannot be regarded as a good model because of the fact that as the number of attributes increases, it becomes harder to manage. It may also show bad performance in the DBMSs that deal poorly with nulls. Meanwhile, the UT model requires the least storage of all the models in the experiments. This is believed to be the result of efficient null value management in current relational database systems.

We made an attempt to combine the good features of the primitive models by using hybrid models. In the results, HYBRID_2 and HYBRID_3 are good in both performance and space complexity, so we can expect them to be applied to practical electronic commerce systems. Moreover, these models can be improved by using indexes because storage currently is a far cheaper resource than time.

ACKNOWLEDGMENTS

We would like to thank Kyungsuk Kim and Hyunyoung Song for their helpful suggestions and efforts in the implementation. We are also grateful to PPS Korea, LOTTE.com Inc. and EAN Korea for offering their catalog data for our experiments. This work was supported by Brain Korea 21 Project and facilities supplied by The Research Institute of Advanced Computer Technology.

REFERENCES

Agrawal, R., Somani, A., & Xu, Y. (2001, September). Storage and querying of e-commerce data. *Proceedings of 27th International Conference on Very Large Data Bases,* 149-158.

Jhingran, A. (2000). Moving up the food chain: Supporting e-commerce applications on databases. *Association For Computing Machinery Special Interest Group on Management of Data. 29*(4). Industry Perspectives, 50-54.

Jung, J.H., Kim, D.K., Lee, S.G., Wu, C.S., & Kim, K.S. (2000, May). EE-Cat: Extended electronic catalog for dynamic and flexible electronic commerce. *Proceedings of the Information Resources Management Association International Conference,* 303-307.

Segev, A., Wan, D., & Beam, C. (1995, November). Electronic catalogs: A technology overview and survey results. *Proceedings of the 4th International Conference on Information and Knowledge Management,* 11-18.

Shanmugasundaram, J., Tufte, K., Zhang, C., He, G., DeWitt, D.J., & Naughton, J.F. (1999). Relational databases for querying XML documents: Limitations and opportunities. *Proceedings of 25th International Conference on Very Large Data Bases,* 302-314.

Stonebraker, M. & Hellerstein, J.M. (2001, May). Content integration for e-business. *Electronic Proceedings of Association for Computing Machinery Special Interest Group on Management of Data Conference.* Catalog Integration.

ENDNOTES

[1] Stock Keeping Unit number

[2] Universal Code Council / European Article Numbering

[3] Oracle 8i Database Release 2 (Version 8.1.6), Oracle Corporation, Redwood Shores, California, U.S.A.

[4] MySQL 3.22.20a, The MySQL AB Company, www.mysql.org. The software is distributed under the GNU General Public License.

[5] An administrative agency of the Republic of Korea which is responsible for procuring commodities and related services, for procuring works for major government projects, for stockpile management, and for government property management.

[6] A member organization of EAN International which offers system of identification and communication for products.

[7] One of largest B2C companies in Korea.

Chapter VIII

Extending Client-Server Infrastructure Using Middleware Components

Qiyang Chen and John Wang
Montclair State University, USA

ABSTRACT

Embracing inapt infrastructure technology is a major threat in developing extensive and efficient Web-based systems. The architectural strength of all business models demands an effective integration of various technological components. Middleware, the center of all applications, becomes the driver—everything works if middleware does. In the recent times, the client/server environment has experienced sweeping transformation and led to the notion of the "Object Web." Web browser is viewed as a universal client that is capable of shifting flawlessly and effortlessly between various applications on the Internet. This paper attempts to investigate middleware and the facilitating technologies, and point toward the latest developments, taking into account the functional potential of the on-market middleware solutions, as well as their technical strengths and weaknesses. The paper would describe various types of middleware, including database middleware, Remote Procedure Call (RPC), application server middleware, message-oriented middleware (MOM), Object Request Broker (ORB), transaction-processing monitors and Web middleware, etc., with on-market technologies.

INTRODUCTION

Evolution of Internet-based computing from local area networks (LANs), after transitioning from unconnected computers to networks, is the hallmark of all business models today. The technological backbone of this evolution is the middleware. First connecting, then communicating, and, finally, seamlessly integrating the distributed systems to external sites (customers, suppliers, and trading partners across the world) is the real challenge for the business world. It doesn't stop there. Also required is the talking between client and server across heterogeneous networks, systems architectures, databases, and other operating environment. All this is facilitated by the middleware technologies that offer undercover functions to integrate various applications with information seamlessly and instantly make it accessible across diverse architectures, protocols, and networks. Automation of back-end and front-end operations of business is also affected by the middleware. Middleware binds discrete applications, such as Web-based applications and older mainframe-based systems, to allow companies to hook up with the latest systems and developments that drive new applications without making their investments in legacy systems unyielding.

The chances of huge returns expected due to enabling middleware technology are, however, controlled—and often diminished—by the fact that the consequence of unpredictability or improper configuration of the middleware technology is extremely severe. Web browser war has given way to the middleware war. Numerous vendors offer various middleware product families—"the operating system of the Web"—with an estimated growth of about 65% for Object Request Broker, 50% for Messaging and 15% for Transaction Processing Monitor for the year 2002 (Slater, 2002).

FUNCTIONS OF MIDDLEWARE

Middleware functions are generally classified into:

- application-specific functions to deliver services for different classes of applications, such as distributed-database services, distributed-data/object-transaction processing, and specialized services for mobile computing and multimedia;
- information-exchange functions to manage the flow of information across a network for tasks like transferring data, issuing commands, receiving responses, checking status and resolving standoffs; and
- management and support functions to locate resources, communicate with servers, handle security and failures, and monitor performance.

Database, Web and legacy application middleware are three basic middleware application types. Database middleware is the major application in most systems and

Web middleware is spreading fast and extensively. SQL mainly provides communications between clients and servers in database middleware. Vender-dependent standard, such as ODBC, is developed to keep the client application from database server implementations and can be reached easily from most application programming environments. Non-relational data now can be accessed through Universal Data Access. Remote procedure calls, messaging middleware, transaction process monitors and object-oriented middleware are four basic middleware communication categories (Leibmann, 2000).

Because middleware is a relatively immature technology, the competing standards and fast-changing technologies factor into the selection process, making it a very complex decision-making area that requires a meticulous insight and analysis of the correlation of and interaction between the middleware and applications. The two major issues that concern and influence this process include (Edward, 1999):

- the types of applications to be carried out by the middleware technology, based on the current application environment, such as distributed relational database project or non-relational file servers dealing with VSAM and ISAM data; and
- the application development environment that is based on whether a client application will be written in a traditional procedural language such as COBOL, an event-driven language like Visual BASIC, or an object-oriented environment such as C++ or Java.

MAJOR TYPES OF MIDDLEWARE

The selection of middleware technology is determined by what information is required to be communicated. For example, database middleware is the choice if database is the main requirement. However, the following are the major categories of middleware:

- Database Middleware,
- Remote Procedure Calls (RPC),
- Object Request Broker (ORB),
- Application Server Middleware,
- Message Oriented Middleware (MOM),
- Transaction Processing Monitor (TP),
- Object Transaction Monitor, and
- Web Application Servers.

The most widely used, easy-to-install and relatively economical middleware, *database middleware usually is chosen* to complement other types of middleware. It facilitates communication among applications and local or remote databases, but cannot transfer calls or objects. However, database middleware does not allow the two-way communications between servers and clients. SQL type command is generally subjected to the middleware gateway, which would convey the command to the end database to collect and send back the reply of the SQL query. Synchronous point-to-point type of communications is the characteristic of database middleware, and can pose problems when multiple demands from multiple users produce huge traffic and congestion. Database middleware is the most mature middleware technology.

Remote Procedure Calls (RPC) permits a client program to call procedures located on a remote server program. RPC is not isolated as distinct middleware level, and is entrenched in the application with calls embedded into the client portion of the client/server application program. Stubs are developed for both the client and the server to call up synchronously when the client makes a call to the server. The intricacies of distributed processing are reduced by RPC by maintaining the semantics of a remote call, no matter whether the client and server are located on the same system. The synchronous nature of the RPC makes it most appropriate for smaller applications, where all communications are one-to-one and not asynchronous.

Object Request Brokers (ORB) are language-independent, object-oriented, synchronous RPC in which an affiliate function of an object can be brought into play remotely by means of the same essential notation. Asynchronous communication suitable to large applications can be made possible by extending the main standards, as in CORBA and DCOM, the main competing standards. Java's Remote Method Invocation (RMI) is also like an ORB, but is not language-independent. Interfaces among objects can be defined by an IDL (interface definition language). The codes can be re-compiled to avoid troubles associated with changing these interfaces dynamically. ORB technologies are based on the reliability of the transport layer, which is required for the functioning. The application programmer is secured from the details of the client/server approach by using IDL interfaces that allow the application code to call a remote object, as if it were supported locally. Thus, the maintainability is improved as the object communication details are concealed from the application and isolated on the ORB. Hence, ORB-based middleware applications are becoming standard for the multi-tier model. CORBA, enterprise Java Beans and Active X/COM are among the major technologies. The main limitation of ORB technology is that it goes well only with CORBA, JAVA, or COM-oriented applications in the network. However, it can be used with TP Monitor, which supports the client/server communication management. CORBA with TP Monitor tenders a good communication middleware in running Web-based business transactions.

Message Oriented Middleware (MOM), or enterprise message technology (EMT), provides asynchronous message delivery. The messages are lined up, just as objects, permitting the application that sends messages to complete other tasks without getting blocked until it receives the response. Generally located at a higher level than that of RPC, MOM assembly provides more than simply passing information. MOM also offers provisions for translating data, security, broadcasting data to multiple programs, error recovery, and prioritization of messages and requests. MOM enhances flexibility by allowing applications to switch messages without the requirement of knowing on which platform or processor the other application is located. MOM facilitates communications across a range of messaging systems, such as request-response, prolonged conversation, application queues, publishing and subscribing, and broadcasting. Messages are processed asynchronously with appropriate priority levels. Examples of messaging middleware are IBM's MQSeries, BEA's Message Q and Microsoft's MSMQ.

Transaction Processing Monitor (TP) is more than 25-year-old technology that controls interactions among a requesting client and databases. It is a database-independent technology. TP provides a three-tier client/server model and ensures an appropriate updating of the databases. This technology facilitates and controls the transport of data among numerous terminals and the application programs serving them. It can provide services to thousands of clients in distributed client/server environments by multiplexing client transaction requests, by type, onto a controlled number of processing routines that support particular services. TP monitor technology can also capture the application transitions logic from the clients. TP monitors are not employed for general-purpose program-to-program communications, but offer an environment for transaction type applications that utilize a database. On initiation of transactions by clients, the TP monitor transports the transaction relating to the database as required, and the response is sent back to the client. By using TP monitors, the developers can define segments of the application as transactions with clear start and stop points. TP monitors also offer failure protection, and ensure atomic transactions by turning back a transaction if it is not completed successfully. TP monitors often are used synchronously to ensure that failures are detected before further transactions take place. This is accomplished by breaking down complex applications into pieces of code called services. Three types of jobs are performed through TP monitors: process management, transaction management and client/server communication management. Examples of TP monitors include IBM's CICS and BEA's Tuxedo. BEA's Tuxedo is used by some companies as the middle-tier to allow an application developer to write applications as a set of services, each executing a single business function, such as add customer, process orders, and bill customers. Tuxedo supports LAN and Internet-based clients through its Tuxedo IWS and BEA Jolt. C, Cobol and Java can do programming in Tuxedo applications. Standard Tuxedo communications integrate the business logic proficiently with legacy applications. Tuxedo also supports many security levels, such as the application password, end-user authentication, access control and mandatory access control.

Object Transaction Monitor (OTM) is a coherent form of middleware consolidated and integrated from discrete technologies available, such as TP Monitors, Message Q and ORB. For instance, the Object Transaction Monitor (OTM) combines the functionality of MOM, TPM and ORB into one product (Boucher & Katz, 1999). BEA's Tuxedo integrates with ORB in one single OTM product known as M3. In addition, Microsoft's MTS, Iona's OrbixOTM, and Inprise's Visibroker ITS also offer OTMs. MTS, COM and MSMQ are integrated in COM+, and IBM integrates these technologies into the family of WebSphere products. The latest trends in middleware solutions support the integration of various technologies to increase returns by reducing costs of employing experts (in different technologies and applications) in the design and functioning of e-business models.

MIDDLEWARE AND THE WEB APPLICATION SERVERS

With the domination of the Internet over all business models, middleware technology has grown remarkably in recent years, and is expected to grow even more. For instance, MOM, the middleware technologies candidate with the most potential, is expected to grow at a compound annual growth rate of about 20%. The mission-critical obligation of systems engages far-reaching use of distributed computing to share information across expanding heterogeneous networks. Pushing required technology across the Internet requires the significant, consistent diffusion of information across numerous applications on a wide range of machines and embedding it in the range, from the reproduction of a database to the extraordinary push technologies. By and large, the information traverses throughout the network in the form of messages and the infrastructure entails software development.

HTML pages are served through the application servers' interaction with back-end databases and applications. However, application servers differ by operation speed, support for component models, programming standards and database interfaces. Application server software permits originating applications as a set of software components, such as Enterprise JavaBeans or ActiveX controls, and loading these applications on the application server. The components use the features, or the services, of the application server, such as accessing back-end databases, controlling transactions to those databases, interacting with the front-end Web server and even balancing the workload over several copies of the application server (Cox, 2001).

Sophisticated and easy-to-use Web development tools are being developed by the Web application server vendors. Simplification of coding, and integration of distinct tasks, allows the next-generation application servers to be developed with much-reduced development loads for a successful Web operation, particularly in terms of time. However, one must be careful about increased dependence on the vendor's technology due to several factors: proprietary application components, sizable investments in software and product-specific training, and extensive use of the vendor's own development and consulting services, (Liebmann, 2000).

High vendor-specificity of the application server architecture, infrastructure, and connectivity, as well as increasing business logics being stored in a vendor's application server, lead to increased dependence of a company on a vendor's product line. There are two ways to deal with the problem. One way is to create and design applications in a manner that separates each application function into its own discrete components, so that if any particular feature of the application server becomes outdated, it can be substituted with another technology without breaking down the whole system. The second solution to the problem can be aimed at by integrating standard programming environments such as C++ and Java (Rosencrance, 2002).

With pervading Web-technology over all business models, the use of application servers as critical components of the infrastructure has become more fundamental than even operating systems and databases. A fast-paced growth of the Net also requires the Web-based applications to be much more scalable, reliable and transaction-supportive. The standard application deployment platform of these applications with the Java 2 Enterprise Edition (J2EE) supports their use as application integrators with XML capabilities. BEA, IBM and iPlanet/Sun/Netscape/AOL are among the leading vendors of the Web-application servers on-the-market. These Web-application servers are based on Java and Enterprise Java Beans (EJB) component model and J2EE standards. Each claims leadership in some or the other dimension. BEA, for example, with its WebLogic Server, is the unit volume leader. IBM leads in revenue, a claim supported by its associated implementations of AS/400 and S/390 in traditional enterprises, whereas iPlanet is the leader in the high-growth/high-value Internet sector (CSIRO, 2001).

BEA's WebLogic Server offers an easy-to-deploy J2EE platform that marks the mainstream market. WebLogic Enterprise builds on this capability, but requires two programming environments to accomplish all of its functions. iPlanet is consistent in almost all areas. IBM's strength in high-end transaction services, application-development tool integration, as well as its leadership in XML and UDDI, make it a powerful product. IBM's product also requires two programming environments to accomplish all of its capability, bridging CORBA and legacy application systems. BEA WebLogic Enterprise offers strong J2EE conformity, Web-presentation services, object services, security, reliability and scalability. WebLogic is as good as IBM's offering in management functions, and as good as iPlanet Application Server 6.0 in Java Messaging Services (JMS). WebLogic's execution of J2EE is considered cleaner, or easier to install. On the other hand, iPlanet and IBM offer integrated implementations of XML. However, WebLogic 6 integrates XML into the package.

The Sun-Netscape-AOL alliance's iPlanet Application Server offers strong capability in J2EE. The first web-application server to be certified J2EE compliant, iPlanet 6.0 has a unique strength in simplified application development at the enterprise level, based on its Unified Integration Framework that offers single programming environment to build enterprise applications. By contrast, BEA WebLogic requires the user to dub a separate server, called E-link. IBM's WebSphere EE offers both the Advanced Edition container and CORBA-based Component Broker. However, iPlanet Application Server 6.0 lingers in management and transaction functions. IBM's WebSphere EE 3.5 leads the pack in transaction services and development-tool integration. It also is a leader in driving and propagating the XML and UDDI application-integration standards. WebSphere EE includes WebSphere AE as well as IBM's Component Broker technology. However, WebSphere lags in J2EE certification (Moorehead, 2000).

The factors that influence and drive this high-growth market segment — characterized by furious competition — come from the ever-expanding demand for application integration, growth of services, scalability and reliability, ease of implementation, and security control.

MAJOR PRODUCT FAMILIES OF MIDDLEWARE TECHNOLOGY IN WEB APPLICATION SERVERS

BEA's WebLogic integrates EJB with other Java-standard services required for business-critical applications and provides database integration, event management, remote object access, and directory service integration. Additional technologies sustaining the latest version include WML, XML and SNMP, with the add-on software kit offering support for the latest draft of the EJB 2.0 specifications. WebLogic provides see-through replication for automatic load balancing and fail over; strong RSA-based security; rich, graphical management of Enterprise JavaBeans; comprehensive support for the Java Enterprise Standards; and full Java types as arguments to Enterprise JavaBeans.

The WebLogic EJB Deployer tool provides the controls for managing multiple EJB.jar files and for configuring WebLogic Server deployment properties and resources. Deployer tool supports two levels of EJB deployment validation by automatically checking properties and references to make sure they contain the appropriate values, and by verifying that key EJB required classes are compliant with the EJB 1.1 specification. WebLogic Zero Administration Client (ZAC) Publish Wizard, a graphic utility, allows creating, publishing and managing packages containing an application, applet, or library of Java code. ZAC permits building the client-side Java applications and packaging them for distribution.

IBM's WebSphere can be used for e-business applications in multi-tier client/server environments. The WebSphere software platform consists of three layers:

- WebSphere Foundation, which consists of Application Server and MQ series, integrates the business processes and applications and transfers them to the Web;
- Foundation Extension, which helps the business develop, present, and deploy Web applications, and extend the back-end systems to the Web; and
- Application Accelerators, which help the business develop customized solutions for collaboration and B2B integration.

WebSphere supports every phase of e-business development, from the start to the end. The WebSphere provides standards-based Java environment, XML and XSL support, and support for applications built in Java Servlet specifications, transaction monitor and message queuing software. It also provides site analysis tools and native language support (Ling, Rihao, David, & Chou, 2001).

WebSphere Application server includes an HTTP server, user interface logic, the data to construct user interface of the e-business application, the business logic, and the connectors (Conner, 1999). The connectors support the communications with external applications, data and services. The user interface logic is separated from the business logic and follows the user interface style and mechanisms. The client browser communicates with WebSphere Application server through HTTP, IIOP and MQ Series. The Application Server integrates e-business applications with data server, existing enterprise resources and business partners. WebSphere is capable of providing connectivity to many different types of databases and offers high-speed database access using JDBC.

The WebSphere application family includes many products that maintain the existing enterprise systems, like enterprise resource planning (ERP), and aid the integration with business partners. It also makes possible the access of the existing enterprise applications (currently available to the customers, employees, and suppliers) through Intranets and Extranets by making these applications securely available on the Web. WebSphere Host Integration Solution makes possible the seamless integration of legacy applications and critical enterprise data with the Web. The Transcoding Publisher and WebSphere B2B Integrator enhance B2B data transaction. IBM's MQ Series can be used for flexible and rapid integration of core business processes, data and applications, and B2B transactions. IBM also presents MQ Series Integrator, which integrates applications originally developed for different systems with supply chain management and customer relationship management. WebSphere's VisualAge for Java offers patterns for expanding ERP systems (SAP R/3) to e-business (Rosencrance, 2000). IBM offers products ranging from simple Web applications to the more complex, such as the assorted Transcoding Publisher. This technology automatically translates and transforms Web-content (HTML) to XML or WML, the markup languages that are currently used in creating content and applications for pervasive devices such as wireless application protocol (WAP) devices, personal digital assistants (PDAs), and so on.

Sun-Netscape's iPlanet Application Server is the result of efforts to remake Netscape's Application Server, combined with NetDynamics technology, to take its place as a foundational piece of the iPlanet Internet Service Deployment Platform. iPlanet is expected to provide an extra level of support. iAS 6.0 meets the requirements of the latest J2EE specification, and has passed the J2EE Certification Test Suite. Basic core services of iAS include a transaction monitor, multiple load-balancing options, full clustering and failover support, an integrated XML parser and Extensible Stylesheet Language Transformations engine, and full internationalization support. iPlanet products, including Directory Server, Web Server, and other add-ons for Enterprise Application Integration, are well integrated. The iPlanet Application Deployment tool is a Java-based program that leads through the process of deploying an application. Several well-documented sample applications present an overview of creating and organizing real-world applications. iAS offers a separate product, iPlanet Application Builder, that integrates with such third-party tools as WebGain Studio, Inspire JBuilder, IBM Visual Age, and Sun Forte for Java Enterprise Edition. iPlanet offers a good e-commerce platform with applications and performance required to successfully deploy today's cutting-edge Web applications (Astley, Sturman, Daniel, & Agha, 2001).

iPlanet provides the enterprises with a scalable, high performance platform on which to develop and deploy their large-scale, transactional Web applications. iPlanet provides advanced availability features, such as failure recovery and session management, and scalability features, including dynamic load balancing. Developers can tool their applications in Java or C++, in addition to the offered support to the Client Independent Programming Model (CIPM) for the rapid development of applications for differing client types. iPlanet accomplishes high performance through its fully multi-threaded, multi-process architecture and advanced connection caching and pooling. Scalability is endowed with dynamic load balancing and point-and-click application partitioning, allowing applications to scale animatedly to support numerous users. New enhancements include a standards-based application model supporting Enterprise JavaBeans, JSP, JDBC, and the Java Servlet API. In addition, XML-based, distributed transaction support enables high performance and centralized management, while "fail over" functionality provides enhanced application availability.

Table 1 shows a comparison of several products of Web application server.

CONCLUSION

The traditional two-tier client server system no longer can manage very large amounts of data and transactions, because of a wide range of expanding and diverse operating systems and applications in the Internet environment. What is required now is a translating middle layer that can make these systems, applications, and environments communicate seamlessly at the same time. There are many middleware technologies developed to streamline these efforts. An imminent solution is that businesses adopt a multi-tier client/server architecture and infrastructure supported by several middleware technologies.

The choice of technology will be based on the specific needs of the business situation. It has been established now that middleware solutions offered by IBM's WebSphere, BEA's WebLogic, and Sun's iPlanet, etc. lent a hand to the businesses by integrating legacy systems into various applications on diverse platforms. These middleware solutions proved to be proficient and cost-effective in constructing and supporting complete e-business models. No organization will be able to compete and survive today's fast-changing, Net-ized, e-business environment, characterized by huge data flow, unless it adapts and adopts a multi-tier client/server architecture. The middleware solutions, such as WebSphere, WebLogic, or iPlanet not only integrate the Web, client/server and legacy systems, but also support integration of various middleware technologies to secure investment in existing technologies and skills, and incorporate with ERP and business partners.

Table 1

Product	Operating Systems	Support and Integration	Legacy Integration	DB Integration	Format Integration	Content Management
BEA Systems WebLogic Server 5.1	Windows NT/2000, Solaris, HP-UX, Compaq Tru64, SCO Unixware, Siemens MIPS with Reliant Unix, SGI Irix, Dynx, AIX, Linux, AS 400, OS/390	J2EE implementation with certified APIs, native HTML support, integrated HTML, JSP, EJB, O/R mapping development tools, integrates with Macromedia Dreamweaver, Visual Age for Java, Inprise Jbuilder, supports XML, COM, CORBA	More than 30 packaged plug-ins including SAP, Clarify, other mainframe, XML and MQSeries, adaptor developers kit for custom application integration	Oracle, MS-SQL Server, Informix, Sybase, any via JDBC-compliant driver	JMS (API+ SMTP), FTP, TCP/IP, POP, IMAP, SNMP	Workflow and personalization product plus native integration with Documentum, Interwoven and other content managers
IBM WebSphere Application Server 3.5	Windows NT/2000, AIX/6000, Solaris, Red Hat Linux, HP-UX, IBM OS/400, IBM OS/390, IBM OS/2	Supports third-party IDEs, integration with IBM VisualAge for Java and IBM WebSphere Studio IDEs, support for Java and non-Java applications, XML, JSP, EJB, COM, CORBA	IBM DB2, Oracle, Sybase, Informix, SQL Server, Lotus Domino, others, via JDBC, Versant EnJin, eXcelon, IMS, MQ-enabled Systems, MQSeries, CICS, TXSeries/Encina, SAP R/3	Oracle, DB2, SQL Server, Informix, Sybase,Versant, eXcelon, Lotus Domino	SMTP, FTP, LDAP, Telnet, TN3270, TN5250, TCP/IP, JMS, JDBC, JTS, JTA, JNI, XML/XLS, EJB, JNDI, J-IDL, CORBA, RMI, IIOP, RMI/IIOP	Workflow, personalization, rules-based engines, proxy, caching, serving, portal serving
iPlanet E-Commerce Solutions iPlanet Application Server 6.0	Windows NT, Solaris, IBM AIX, HP-UX	Includes an integrated J2EE development tool, O/R mapping, integrates with Sun Forte for Java tools and third-party IDEs including WebGain VisualCafe, supports XML, JSP, EJB, Servlets, COM, CORBA	CICS, SAP R/3, PeopleSoft, Tuxedo, MQSeries, Sun JMQ, includes an SDK for developing Enterprise Connectors for other Enterprise Information Systems	Oracle, DB2, Informix, Sybase, SQL Server	JMS, JNDI, XML, RMI, IIOP, RMI/IIOP, JTA, JTS, JDBC, LDAP, NSAPI, ISAPI, SMTP, SNMP, HTTP, XML, JNDI, FTP, TCP/IP, EDI	Native J2EE Workflow development tool and runtime, personalization, content management

REFERENCES

Astley, M., Sturman, D., Agha, C. & Gul, A. (2001). Customizable middleware for modular distributed software—Simplifying the development and maintenance of complex distributed software. *Communications of the ACM, 44*, (5).

Boucher, K. & Katz, F. (1999) *Essential Guide to Object Monitors*. New York: John Wiley & Sons, Inc.

Conner, M. H. (1999). Structuring e-business applications. (IBM, Library Papers). Retrieved from the World Wide Web: http:// www.IBM.Com/Software/ Developer/Library.

Cox, J. (2001). Web app servers bulk up on new features. Retrieved from the World Wide Web: http:// www.nwfusion.com.

CSIRO. (2001). Helping to choose the right middleware. Retrieved from the World Wide Web: http://www.cmis.csiro.au/adsat/publications.htm.

Edwards, J. (1999). *3-Tier Client Server At Work*. New York: John Wiley & Sons, Inc.

Liebmann, L. (2000). Web application servers-Are you ready for the commitment? Retrieved June 28, 2000 from the World Wide Web: http:// www.internetweek.com.

Ling, R., Rihao, Y., David, C. & Chou, D. (2001). From database to Web browser: The solutions to data access. *Journal of Computer Information Systems*, Winter 2000-2001.

Moorehead, K. (2000). Database middleware technologies. Retrieved September 4, 2000 from the World Wide Web: http://scis.acast.nova.edu.

Rosencrance, L. (2002). Middleware. *Computerworld, 34* (41), p. 84.

Slater, D. (2002). Middleware demystified. Retrieved from the World Wide Web: http://www2.cio.com/archive/051500_middle_content.html.

Chapter IX

E-Business Experiences with Online Auctions

Bernhard Rumpe
Munich University of Technology, Germany

ABSTRACT

Online auctions are among the most influential e-business applications. Their impact on trading for businesses, as well as consumers, is both remarkable and inevitable. There have been considerable efforts in setting up market places, but, with respects to market volume, online trading is still in its early stages. This chapter discusses the benefits of the concept of Internet marketplaces, with the highest impact on pricing strategies, namely, the conduction of online business auctions. We discuss their benefits, problems and possible solutions. In addition, we sketch actions for suppliers to achieve a better strategic position in the upcoming Internet market places.

INTRODUCTION

Electronic commerce will be the enabling technology for the forthcoming revolution in local and global trading. Virtual, Internet-based markets allow for entirely different forms of trading (Höller et al., 1998) than are known so far. Local and, therefore, sometimes monopolistic markets become global and more competitive.

Expectations are that, in the forthcoming decade, the Internet will be a market-enabler of unforeseen possibilities. Just in the last few years, it became apparent that doing business on the Internet can simplify marketing and purchasing considerably. Figure 1 shows estimates of the worldwide trading volume on the Internet (Forrester Research, 1999) that remarkably still hold. The part of e-business that is based on online auctions is growing equally. Therefore, an increasing number of online marketplaces have come into existence. Online marketplaces simplify the establishment of new alternative purchasing and selling partnerships. In the C2C and B2C areas, this fact has been widely recognized, and numerous new B2B marketplaces have been emerging in the last two years.

In most market places, sellers may advertise their offers, and consumers and industrial purchasers can distribute their demands via the Internet. Whereas these forms of establishing connections are important, it is the use of online auction systems that has an effective impact on the pricing structure. Auctioning is among the most efficient and fastest concepts available to achieve fair and competitive prices and identify the optimal business partner.

The chart in Figure 2 shows the core issues of electronic sourcing, which can be separated in an economic and a technical layer. The technical layer includes exchangeable documents based on the EDI standard, e.g., EDIFACT (ISO, 1993) or (UN/EDIFACT, 1993), or XML as a technical infrastructure (W3 Consortium, 2000); protocols for their safe transmission; and electronic payment systems, etc. The technical layer strongly supports, and is driven by, the economic layer. The economic layer focuses on the introduction of new strategies and techniques to let

Figure 1: Worldwide trading volume on the Internet (Forrester Research, 1999)

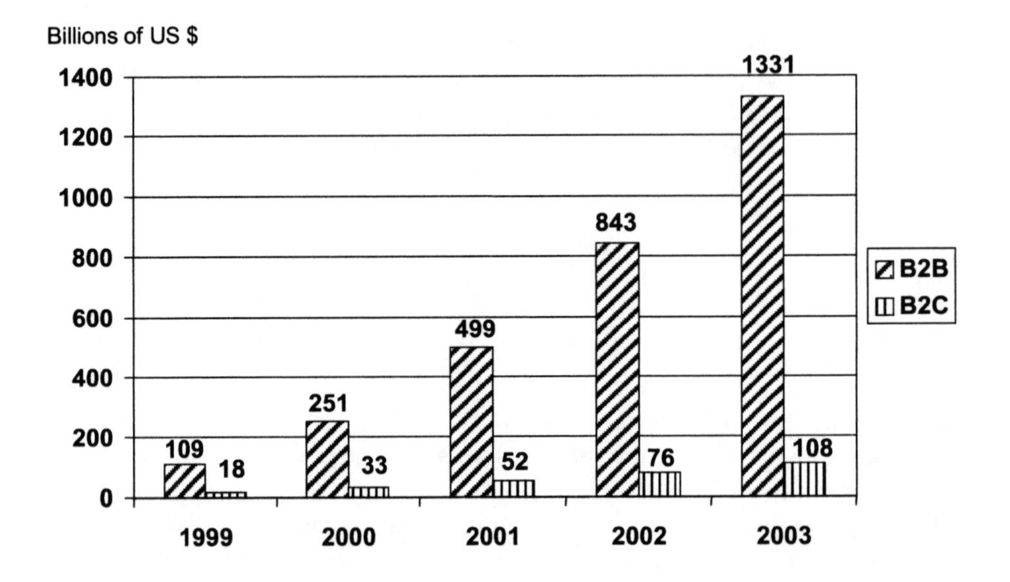

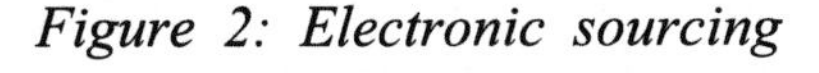

Figure 2: Electronic sourcing

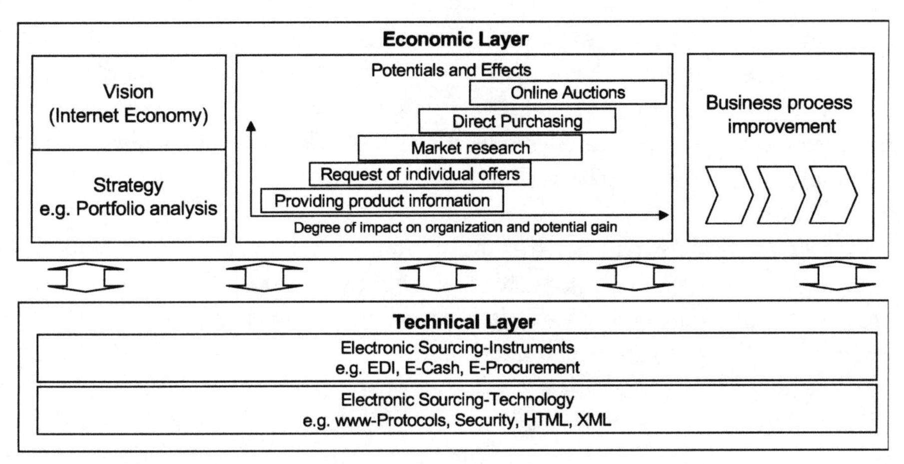

the vision of the Internet economy come true. This includes new interfaces to other companies, as well as restructuring of the company internal business processes, with the goal of overall improvement and cost reduction. The chart also shows that, among the existing concepts, the potentials and effects, together with the impact on the organization and, most importantly, the potential gain, grows at its best when using Internet auctions. Whereas the other concepts mainly concentrate on improving the overall sales and purchasing process, the use of Internet auctions has an impact on the pricing structure of traded goods.

Many marketplaces focus on e-procurement elements when trying to make the purchasing process more efficient. Contact with professional purchasing departments of large and middle-sized European companies shows that, by far, the largest gain in time and money arises from a proper identification of materials that qualify for Internet auctions, a set of possible suppliers and an appropriate auction format.

AUCTION FORMATS

The real world already has defined the primary auction formats that are possible and useful. Standard (also called English) auctions are used to offer one product to several potential buyers. On the contrary, reverse auctions (also called purchaser's auctions) allow one purchaser to auction a demand among several potential suppliers. In reverse auctions, the bidders submit decreasing bids.

If a supplier has more than one item to sell (e.g., a fixed number of tulip bulbs), he may use the Dutch auction format. The price for one product constantly decreases during a period of time. Whenever a bidder submits a bid, he gets one item for the valid price at that time. The auction is finished when all items are sold.

The use of electronic resources invites bidders throughout the world to participate in the auction simultaneously and provides for more complex and, depending on the particular situation, more specific auction formats. Long-term auctions may last up to four weeks, throughout which the bidders repeatedly review the current situation. However, it appears that such auctions are of interest mainly during the closing phase. Short-term auctions concentrate on these last few hours immediately. They may even be as short as 30 minutes, provided that the participants are invited beforehand.

Further, auction formats, such as multi-round biddings, have been defined. In a multi-round bidding, each bidder is forced to submit exactly one bid per round. All bidders are informed of their competitor's bids only after the round has ended, and the next round begins. Multi-phase auctions are quite similar; After each phase, only a subset of the earlier bidders is admitted to proceed to the next phase.

Recently, experience has been gained in conducting multi-dimensional auctions. Here, the price is left open and negotiated through the auction, and several other variables are determined, as well. For example, the price may be combined from the supplier and the logistics entrepreneur (Prince, 1999).

In running Internet auctions, it became increasingly apparent that the auctioneer must be independent to ensure both supplier and buyer have enough confidence in the fairness of the auction process. The experiences with other marketplaces have shown, in an apparent way, that an auction marketplace is not very successful when operated by the buyers or sellers themselves.

Auctions also can be much more complicated. An auction can, for example, exist of multiple slots that allow the slicing of materials and goods in order to auction them among several suppliers. This setting is important to prevent dependency from a single supplier – a critical issue when strategic goods are involved. These slots also may depend on each other. For example, if two competing types of materials can replace each other, and the price difference between both determines the purchased quantities. Another interesting question focuses on determining at what time certain information should be revealed to specific participants. This should prevent illegal price agreements between bidders and, at the same time, ensure confidence in a fair auction. For example, should competitors know each other? An interesting variant reveals identities of bidders at the beginning of a very short-term auction.

These examples show that, today, the power of Internet auctions has been by no means explored. Moreover, there are still a number of unforeseen variations to come.

Having conducted quite a number of online auctions, it became obvious that identifying auctionable goods and materials is not an easy task. Therefore, it is a common way to rely on the assistance of a consultant to define the actual auction set-up, starting with the identification of the demands and possible suppliers. On the other hand, it also turned out that it is rather irrelevant to have large supplier lists at hand, because companies that buy material in industrial sizes usually know their probable suppliers beforehand. They keep watching the suppliers' situation and they want certifications that prove the suppliers' capability of delivering high quality material in time.

FEATURES OF AN ONLINE AUCTION SYSTEM

In several seminars with major industrial purchasers, it soon became apparent that complete e-Procurement strategies are a time-consuming task to define and implement. On the one hand, aligning a whole purchasing department with electronic procurement is tiresome and brings an extra load of work. On the other hand, the desired effects of more efficient purchasing processes pay off only in the long run. However, it also became apparent that identifying the commodities with high purchase volume and buying them via auctions could significantly reduce the prices spent for these commodities and, what is most important, leads to a high and immediate return on investment. From that point of view, an initial set of requirements for online auctions was defined: It was evolved and enhanced during the following period. In the following, the most important of these features have been described. Also, other technical issues of interest, such as flexibility, ease of use, security, performance, robustness, and compatibility to existing systems, both on the buyer's as well as the supplier's side have been discussed.

1. The software is capable of online auctions, both the normal English format and the reverse format, thus allowing the auction of goods for buyers and demands for suppliers.
2. An intuitive graphical user interface is accessible through the Web, without any installation necessary. Being able to support access to online auctions without any installation proved a great success factor, because it often is difficult to install new software within company networks.
3. Auctions are running in real-time. This means that clients always have current information visible. This is especially important for short time (approximately 30-minute) auctions, where the frequency of bids is relatively high.
4. An auction may consist of several slots, allowing the buyer to split the material desired among several suppliers. This prevents dependency on a single supplier, and enables the auction to split the material for different delivery points. Later, we will discuss a mechanism to extend the auction time automatically, in order to ensure competitors are able to react to incoming bids. However, this extension time will get shorter as the auction proceeds. We had auctions where the bids arrived within seconds.
5. Different auctions may depend on each other. For example, depending on the results of simultaneous auctions, the buyer purchases percentages of competing materials. The auction system must reflect this dependency, e.g., with additional messages that describe which of the competing materials will be bought.

6. Persons may participate in an auction in different roles. The auctioneer, the bidders, the originator of the auction (buyer in reverse auctions, seller in the normal auctions) and guests shall be admitted. In particular, the role of guest is useful to show potential—and not fully convinced—participants how online auctions work, without revealing any information on auction details (neither currency, nor value, nor the buyer, nor the kind of traded goods).
7. Different roles receive different available information. Only the auctioneer can co-relate the bids to their bidders during the auction. Bidders appear to each other anonymously, but know how many competitors exist. Furthermore, bidders see their ranking. External observers following the auction see percentage values, instead of real currency.
8. Reverse auctions may have a historic and a target value. The historic value describes what the buyer paid for the auction goods so far, whereas the target value describes what he would like to pay this time. If the auction result hits the target value, then the buyer is obliged to sign the contract. If the target prize is not hit, the buyer is free to choose.
9. The auction times may vary. Very short auctions may have an auction time as short as 15 minutes. Typical auction times are between one and three hours, consisting of a main part and an extension part.
10. The auction time is extended whenever a bid arrives shortly before the auction's end. This allows all other bidders to react. The provided reaction time may vary, for example, starting from three minutes as an initial extension down to a few seconds at the very end.
11. A login mechanism is imperative. Passwords are distributed through safe channels, among them PGP-encrypted e-mails.
12. A report on the auction results is provided for all participants. This report allows the participants to view the auction results. The winner has evidence of his success. The other bidders have evidence that they have been out-bidden and perhaps should think about the pricing structure of their products.

Although the field of e-commerce is highly innovative, to our current knowledge (August, 2000), no existing online auction system provides all the functionality characterized above (Glänzer & Schäfers, 2000; Grebe & Samwer, 2000; Wahrenberg, 2000). Therefore, we created our own auction system, which is now online since March 2000. After two phases of intensive elaboration, a strongly increasing number of online auctions has been conducted. The rest of this paper discusses the lessons learned from these online auctions, with this system, for major industrial companies in Europe.

The Emporias' auction engine systems can cope with the above described auction formats. It provides standard and extension phases, historical and target prices, multiple slots, visibility constraints for multiple participants (bidder, guests with different access rights), and a ranking of bidders and other participants. Based on the

Figure 3: Screenshot of an Emporias online auction

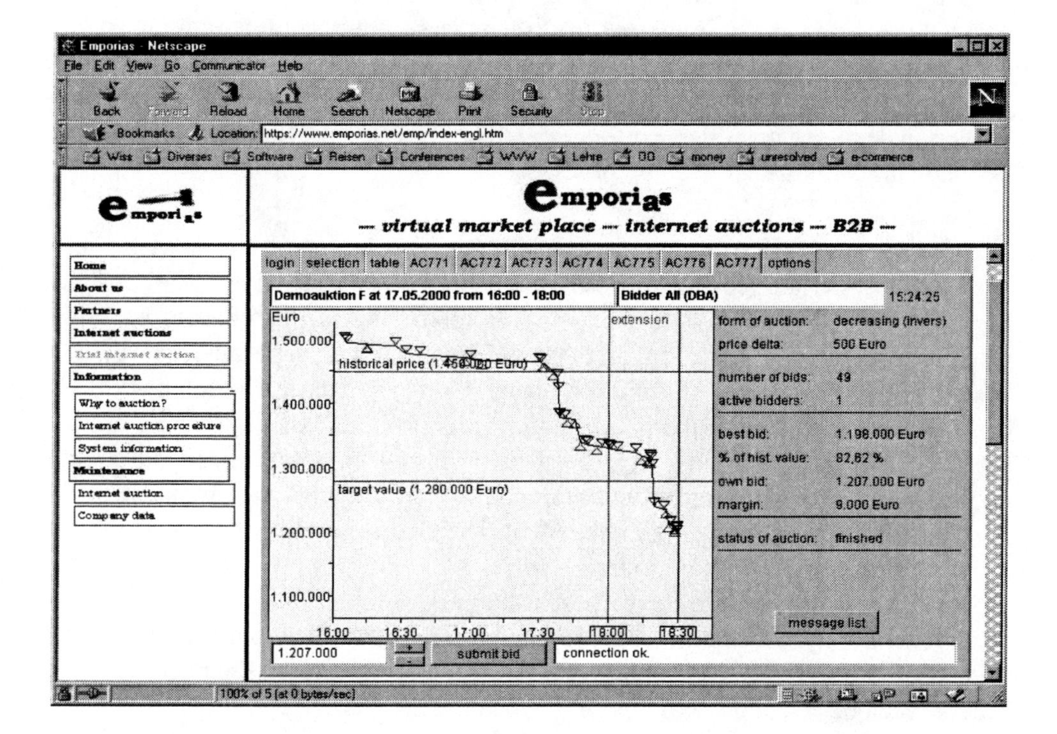

Java technology used for implementation and, therefore, its emerging Internet-realtime features, it is capable of active monitoring of short-time auctions, as well as bidding agents.

LESSONS LEARNED

So far, we have conducted online auctions for a volume of almost $100 million (U.S.). These auctions have occurred in different formats, settings and, of course, different market types. In the following, we concentrate on the results from reverse auctions, because most of the conducted auctions were this type. We have learned a lot about the differentiation between various markets, problems that can be encountered and the solutions to master these problems. Therefore, we concentrate on the most common problems encountered and sketch solutions for them.

Emotional Refusal of the Internet

Although — in the C2C and the B2B market — the number of online customers is increasing exponentially, knowledge of the Internet in the area of B2B purchasing is still somewhat limited. In particular, many managers do not trust the Internet's security and ability to conduct safe trading online. That emotional barrier can be reduced only by demonstrating – to such a manager — how an online auction works. For that purpose, the guest role for online auctions has been implemented, enabling us to invite foreign guests to participate in an auction, without giving them knowledge of auction details (e.g., price). Furthermore, it helps if, at least for the first auctions, a consultant is assisting the setup process.

Political Resistance

As always, when people collaborate, different interests may give rise to conflict. In particular, some people are interested in improving the purchasing process, whereas others would like to stay with the old, approved paths, because they do not wish to revise their habits or opinions. Therefore, it is a political process within a company to conclude that the purchasing process will, at least partly, use online auctions. This works best in companies with a centralized purchasing department. It may, therefore, be of interest to reorganize the purchasing structure, along with introducing electronic sourcing strategies. However, this should not be conducted in a "big bang" manner. Instead, a start in small online auction projects with specific auctionable goods seems to be the best strategy.

Occasionally, IT departments are one particular obstacle that we encounter. Very restrictive IT-departments are reluctant to allow their colleagues any Web-access. Such departments will not allow even the installation of Web browsers.

Another, more severe reason, is that IT-development departments tend to focus on overall solutions. IT-personnel usually favor strategic and expensive solutions. They would, for example, favor a complete restructuring of the procurement process, starting from a single pencil up to gigawatts of electrical power, through the same e-Procurement system. This kind of solution will be costly and will require a longer project for its implementation and, therefore, have no immediate effect. For some e-Procurement strategies, it is still unclear whether there will be a benefit in the long run, at all.

Online auctions, instead, have proven their immediate effect. It is possible to reduce the price paid for commodities, as well as for strategic materials and goods, within hours, after only a few days of preparation. Most importantly, online auctions tackle the major part of the total cost of ownership, namely, material costs. Standard e-Procurement techniques focus more on internal process costs, merely 7% of the total costs. Figure 4 shows the separation of the TCO into material and internal process costs.

Identifying Auctionable Goods

Surprisingly, many purchasing departments do not have a detailed overview of the materials and goods they buy. Of course, they have the numbers available, which tell what percentage of the budget is spent on each good. However, to have a competitive advantage (starting with the purchasing process) means to know the structure of potential suppliers. How many competitors exist globally? What is the competitive market price? Does one or do few companies have the monopoly on selling a particular material? How easy is it for the company to switch to a competing material? Must all the material (or product) be from the same supplier? How reliable is the supplier? Is it useful to split strategic goods among several suppliers?

Purchasers often have no quantitative comparison available that would allow an easy identification of the most promising auctionable goods.

Therefore, it is often useful to analyze the material structure before identifying online auctions. The technique of portfolio analysis proved useful.

How to Set Up an Auction

Sometimes, it is very simple to set up an auction. In a reverse auction, the buyer usually knows quite well what he wants to buy. Therefore, a number of potential suppliers have to be identified and invited to participate in the auction. Apart from the current supplier, a number of potential suppliers usually is well-known already. Additionally and, if desired, a web-search can help to identify new suppliers. The online auction contract is settled. After that, the auction is conducted, and the purchasing contract is closed.

When auctioning strategic goods, however, the situation often is more complicated. For example, if the buyer does not want to depend on one supplier only, he can split the material into several slots, with the side condition that each supplier will get

Figure 4: Separation of total costs into material and process costs

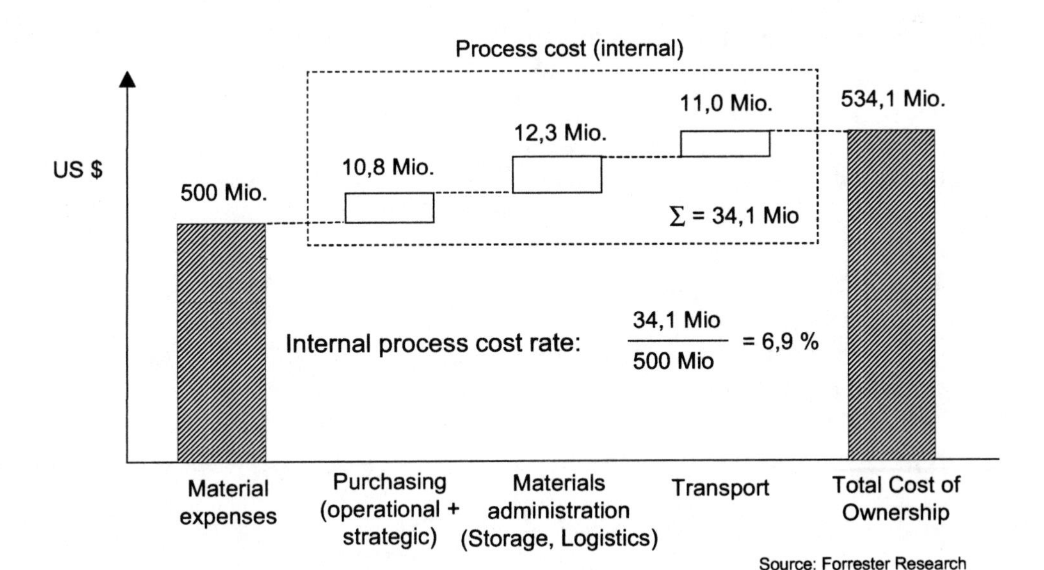

a contract for only one slot. Splitting material in slots is also of interest when the points of delivery are partitioned among time or space (e.g., in different countries).

To increase the competition, it is also possible to run two auctions on competing goods simultaneously, and define the contract in such a way that only the most competitive goods will be bought in the end. For example, if one material is of better quality but also more expensive, then the ratio of which goods to buy to what extent can depend on their price differences.

The concept of target price can be used to decrease barriers to the buyer conducting an online auction. The buyer is allowed to define a target price, which must be hit so that the buyer is obliged to sign the purchasing contract. If the target price is not hit, the buyer is free to sign. The definition of a good target price is not an easy task. In some occasions, and some markets, the definition of a competing target price had very good results. In other markets, the target price was missed completely. The problem occurred when the target price was set too competitively, which prevented the suppliers from starting to bid seriously at all. The most impressive results have been achieved by talking to the potential suppliers before the auction was conducted, and determining how they react to a more or less ambitious target price.

In most auctions, the bidders did not know each other, but there are markets where bidders can guess the other competitors easily. To decrease the risk of making special arrangements among bidders, it is useful to identify at least one or two additional bidders in the emerging global market. In addition, bidders can see, during the auction, how many competitors are online. Another strategy is to reveal the anonymity of the bidders with the start of the auction. To prevent arrangements between bidders, the auction is then conducted in 10-15 minutes only. This helps, for example, in markets where the competitors know and dislike each other.

The final problem encountered was the onlooker problem. Some suppliers were not interested in committing bids, but wanted to get an overview of the market prices of their competitors. In the markets where this problem can occur, multi-phase auctions are particularly useful. For example, the first phase of the auction is run as a normal auction, whereby a. first target price must be hit, before the participant is admitted to the second phase. In the second phase, the target price is usually lower, but not visible to onlookers, who are no longer allowed to cast bids. Two or three phases usually are enough.

To summarize, based on our experiences, different markets need customized variants of auctions. Some kinds of auctions are pretty straightforward. In other situations, markets are less competitive and, therefore, need additional techniques to ensure the finding of a fair price in an online auction.

SUPPLIER REACTIONS

The Emporias' auction engine was able to reduce the costs for certain materials up to 46%. This is a great gain for the purchasers, but also a loss of profits for suppliers. Therefore, suppliers have started to realign their purchasing structures. Marketplaces that mainly focus on establishing contact between buyers and suppliers are not as critical. Online auctions are the critical part of the new E-business economy. They drastically increase competition in the global market place. The supplier problems can be narrowed to the following three questions:

- How can I be prevented from participating in an online auction?
- What is the best strategy when I am forced to participate in an online auction?
- How do I deal with the consequences arising from online auctions?

Suppliers have a great deal of possible actions at hand, ranging from strategic, long-lasting contracts with their purchasers to dramatic cost reductions by using online auctions in their own purchasing processes. Depending on the goods, the market peculiarities, and the geographic and contractual situations of the supplier, a number of additional actions can be taken to deal with that new situation. Aggressive suppliers do not fear participating in online market places, but try to use them as a chance to establish contact with new purchasers. As always: the early bird catches the worm.

To optimize suppliers' strategies in various markets, it is useful to conduct workshops with the goal of realigning organizational structures, defining reaction variants for each type of material, redefining pricing strategies and, finally, leading to an improved composition of the companies sales portfolio. See Figure 5 for a structuring of such a workshop.

Although it is possible to define general strategies for suppliers that have to react in online auctions, the results and, therefore, the behavior of competing suppliers greatly depends on the auction format. Therefore, auction formats have to be taken into consideration when defining the auction's individual bidding strategy. There are many possible choices of action, actions such as preventing an online auction, or breaking up a buying syndicate, but today, it seems, one of the most promising strategies for a supplier is to reduce material costs himself in online auctions.

SUMMARY

In this chapter, we have introduced a number of auction formats and their impact on the conducting of competitive online auctions in business-to-business trading. Online auctions are among the Internet techniques that will become inevitable in the B2B purchasing area. They are rather easy to implement, and have a high, and immediate, gain of results in return.

Figure 5: Workshop structure for suppliers' strategic realignment

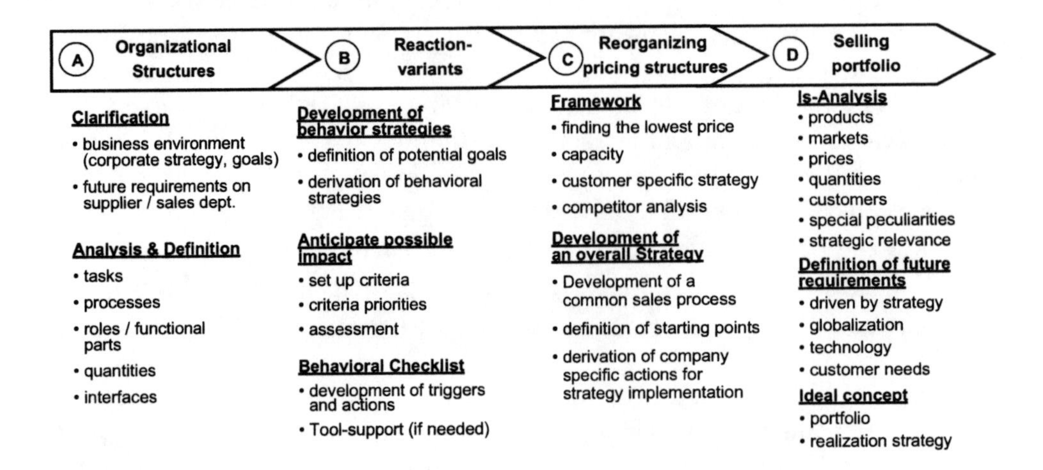

Current numbers and forecasts indicate the B2B E-business will explode even more in the coming years. However, the techniques that the new economy offers have to face the traditional fears and problems that the old economy—and the people in that economy—are still holding. Online auctions are one of the electronic sourcing techniques that will find their terra firma in the E-sourcing portfolio. However, the way to arrive there is not as easy and simple as the people had in mind in the early days of E-commerce.

ACKNOWLEDGMENTS

This work was supported by the Bayerisches Staatsministerium für Wissenschaft, Forschung und Kunst through the Bavarian Habilitation Fellowship and the German Bundesministerium für Bildung und Forschung through the Virtual Softwaereengineering Competence Center (ViSEK).

REFERENCES

Glänzer, S. & Schäfers, B. (2000). Auctainment statt nur auktionen. In: *Events and E-Commerce*, Peter F. Stephan (Ed.), Berlin: Springer.

Grebe, T. & Samwer, A. (2000). Die dynamische entwicklung einer Internet auktionsplattform. In: *Events and E-Commerce*, Peter F. Stephan (Ed.), Berlin: Springer.

Höller, J., Pils M. & Zlabinger, R. (1998). Internet und Intranet. *Auf dem Weg zum Elektronic Business*. Berlin: Springer.

Prince, D. (1999). *Auction this. Your Complete Guide to the World of Online Auctions*. Rocklin, California: Prima Publishing.

Wahrenberg, M. (2000). Die fußball WM-börse: Konzeption und durchführung des weltweit größten börsenexperiments. In: *Events and E-Commerce*. Peter F. Stephan (Ed.), Berlin: Springer.

XML 1.0 Specification (2nd Edition). Retrieved from the World Wide Web: http://www.w3.org/XML.2000.

United Nations Directories. (1993). *Electronic Data Interchange for Administration, Commerce and Transport*. EDIFACT Syntax Rules (ISO 9735) and EDIFACT Data Element Directory (ISO 7372).

Chapter X

Implementing Privacy Dimensions within an Electronic Storefront

Chang Liu, Jack Marchewka and Brian Mackie
Northern Illinois University, USA

ABSTRACT

Many electronic businesses will attempt to distinguish themselves from their competition and gain a competitive advantage by customizing their Web sites, in order to build a strong relationship with their customers. This will require the collection and use of personal information and data concerning the customer's online activities. Although new technologies provide an opportunity for enhanced collection, storage, use, and analysis of this data, concerns about privacy may create a barrier for many electronic businesses. For example, studies suggest that many people have yet to shop or provide personal information online due to a lack of trust. Moreover, many others tend to fabricate personal information. To this end, many electronic businesses have attempted to ease customers' concerns about privacy by posting privacy policies or statements, or by complying with a particular seal program.

Recently, the Federal Trade Commission has proposed four privacy dimensions that promote fair information practices. These dimensions include: (1) notice/awareness, (2) access/participation, (3) choice/consent, and (4) security/integrity. An electronic storefront was developed to include these privacy dimensions as part of a study to learn how privacy influences trust and, in turn, how trust influences behavioral intentions to purchase online. The empirical evidence from this study strongly suggests that electronic businesses can benefit by including these privacy dimensions in their Web sites. This chapter will focus on how these dimensions can be implemented within an electronic storefront.

INTRODUCTION

Even after the rise and fall of the dot-coms, e-commerce activities on the Internet are still thriving, as both small and large organizations try to reach potential customers online. For example, IDC forecasts more than $5 trillion throughout the next four years will be invested in developing more efficient modes of conducting e-business worldwide (IDC, 2001).

Many e-businesses today are attempting to distinguish themselves from their competition by strengthening the relationship with their customers. This means getting to know their customers' buying habits and preferences, so that an e-company can customize the online buying experience to better meet its customers' needs. Subsequently, the online daily receipts and banner ad click-throughs provide a base of data that is becoming a valuable asset for many e-businesses. The pronouncement is, "get inside your customer's heads (and their wallets too)" (VanScoy, 2000).

However, many people have yet to shop, or even provide personal information, online, due to a fundamental lack of trust. According to Hoffman and Novak (1999), almost 95% of Web users decline to provide personal information to Web sites at one time or another, when asked. Moreover, 40% of those who do provide personal information online go to the trouble of fabricating it. It appears that many customers simply do not trust most Web sites enough to engage in activities that involve the exchange of their personal information.

In addition, opinion polls have shown an increasing level of concern about privacy (Smith et al., 1996). In fact, a recent study by *Purchasing* magazine showed that, on a scale of 1-10, the level of concern about online privacy returned a weighted average rating of 7.6, with 67% of customers putting their concern level between 8 and 10 on the scale (Porter, 2000). Obviously, this concern will become heightened even more as more customers engage in e-commerce activities that collect personal information. Even though customers may benefit substantially from online information gathering, concerns about privacy are on the rise and create barriers for successful e-business (Mendel, 1999; Stepanek, 1999; Kleinbard, 2000; Green et al., 2000).

The focus of this chapter will be to provide an overview of the issues and challenges of online customer privacy, and to suggest ways of implementing privacy dimensions into an e-business storefront, in order to mitigate privacy concerns. Many e-business Web sites will continue to gather personal customer information. However, it seems, that if a site is going to collect customer information, it also should address customer privacy concerns about that information.

BACKGROUND

In November 1999, the *New York Times* reported that RealNetworks Inc. — whose products deliver audio and video content on the Web — used its downloadable RealJukebox CD player to secretly collect various data on its customers' listening habits. RealNetworks then automatically sent the data back to Web servers at its corporate offices (Robinson, 1999). RealNetworks claimed that they were collecting the information to customize services for customers and to offer music selections based on what RealNetworks knew about customer listening habits (Bradner, 1999).

Facing public outcry against privacy invasion, and lawsuits, RealNetworks quickly announced that it had disabled the collection features and posted an online policy to address privacy concerns (Harrison, 1999). Although the RealNetworks' case is an example of the risks companies face when they use online technology to collect and process customers' personal information, it is not an isolated incident.

For example, in August 2000, the Marketing News reported that the toy retailer Toys R Us, a Fortune 500 firm, and its baby sites, Babies R Us, Lucy.com and Fusion.com, forwarded customer identifiable information to the marketing company Coremetrics. Coremetrics then analyzed the data to build demographic information and to determine which Web pages and promotions were the most popular. Not only did Coremetrics learn customer names and addresses, but it also determined what pages customers visited on a site and what products they browsed. Moreover, the company was able to track customer activities between site visits (Hopper, 2000).

These examples show that advances in e-commerce provide many opportunities to collect and use customer information. However, it may be difficult for businesses to continue to pursue the opportunities, without risking public backlash, if the applications do not reflect a shared understanding about privacy. Indeed, 57% of U.S. respondents believe, "the government should pass laws now for how personal information can be collected and used on the Internet" (Garfinkel, 2000).

The issue of privacy protection has arisen not only in the U.S., but also in the international marketplace. The Internet and telecommunications technology have made national terrestrial distances insignificant. In recognition of the international marketplace, many vendors have begun to customize their Web sites for use outside of their native language audience. In fact, Forrester Research expects a growth rate of 50% a year for those companies who are customizing sites by providing multilingual versions and other tools (Engler, 1999).

However, customized language alone is not enough. E-business also must consider political, cultural, social, and legal differences. For instance, China has lax rules regarding the use of company data (D'Amico, 2001). Yet in other countries, laws might be broken if business firms transfer customer data internationally. For example, the European Union (EU) Directive on the protection of personal data and the free movement of such data prohibits the transfer of personal data to non-EU nations that do not meet the adequate protection standard set forth in European data privacy laws (Goldstein et al., 2001).

In order to ease customers' concerns about online privacy, public Web sites have begun to post privacy policies or statements regarding online collection, use and dissemination of personal information. In addition, several seal programs, such as TRUSTe, BBBOnLine (Better Business Bureaus Online Seal), and ESRB (Entertainment Software Rating Board Seal) have been developed. Seal programs require their licensees to abide by posted privacy policies, in accordance with fair information practices, and to submit to various types of compliance monitoring in order to display a privacy seal on their Web sites (Benassi, 1999). Privacy seals offer an easily identifiable way to reassure customers. The use of privacy seals represents an industry-wide effort to use self-enforcement mechanisms to make the Internet a safe, comfortable forum for customers to exchange accurate information and conduct transactions, and for Web commerce to reach its full potential.

Moreover, the Federal Trade Commission (FTC) has proposed and advocated that fair information practices include four dimensions (FTC Congress Report, 2000):

- **Notice/Awareness** - where appropriate, prior to collection of data;
- **Access/Participation** - allowing people access to the data collected about them;
- **Choice/Consent** - providing people a choice to share or use their information; and
- **Security/Integrity** - keeping the data secure both internally and externally.

The FTC promotes adherence to these principles to ensure effective privacy protection. However, a recent study conducted by Liu and Arnett (2002), suggests that only slightly more than 50% of large business Web sites provide privacy policies or appropriate links to policies on their home pages. In addition, comprehensive privacy policies that address the four privacy dimensions (notice/awareness, choice/consent, access/participation and integrity/security) are less common.

The results of this study are summarized in Figure 1, and show the distribution of the four privacy policy dimensions for Fortune 500 Web sites with privacy polices. It appears that less than half of the Fortune 500 with privacy policies comply with the *choice/consent* and *security/integrity* components of the fair information practice. Overall, the low percentages in the *choice/consent*, *access/participations* and *security/integrity* dimensions strongly suggest that many Fortune 500 Web sites fail to cover all four privacy dimensions.

Figure 1: Percentage of Fortune 500 privacy policies that address the FTC privacy dimensions (source: Liu & Arnett, 2002)

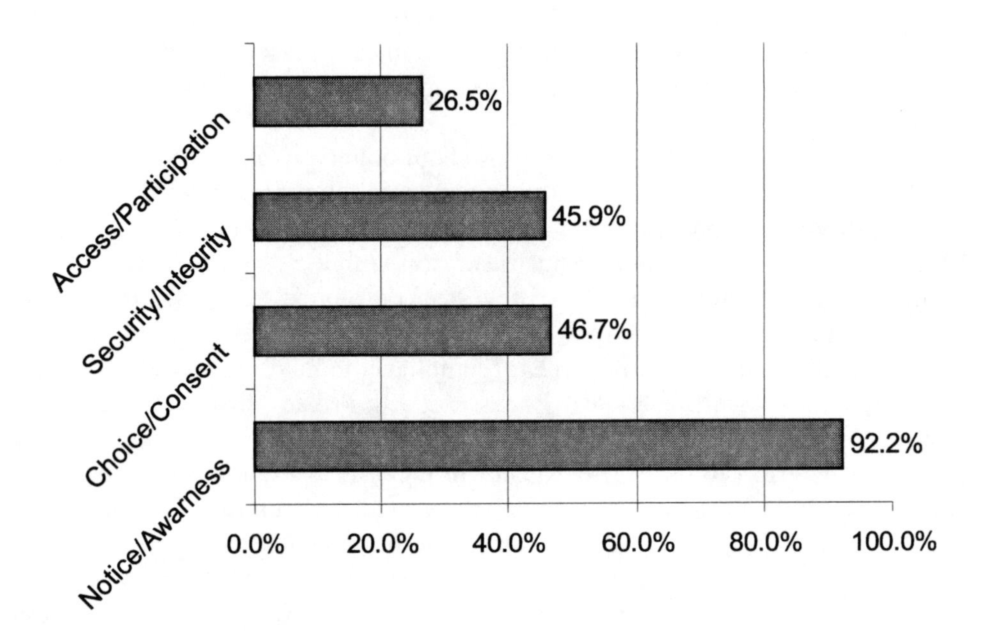

A recent study by Liu, Marchewka, and Ku (2002) examined the relationships among the three constructs of privacy concerns, trust and behavioral intentions from a global perspective. In this study, they proposed and tested a theoretical model, and found that online privacy policies have a strong influence on whether customers trust a commercial Web site. In turn, this level of trust influences their behavioral intentions to purchase from or visit the site again, whether they will have positive things to say about the business and whether they will recommend the site to others.

INFLUENCES OF E-COMMERCE ON PRIVACY CONCERNS

Privacy has long been defined as the right of an individual to be left alone and to be able to control the flow of information about him or herself (Warren & Brandeis, 1890). Although concerns for personal privacy are not new, today's information technologies, with enhanced capabilities for the collection, storage, use and communication of personal information, present both opportunities and challenges (Webster, 1998; Milberg et al., 1995; Culnan, 1993; Clark, 1988).

E-commerce brings considerable product-related information to the shopping environment, and customers can make better-informed purchasing decisions. On the other hand, it collects vast amounts of personal information about customers. Customer information can be collected both explicitly, through registration forms, order forms, online contests, and/or survey forms and, implicitly, by using tracking software and/or cookies. The collection of personal data allows businesses to follow customers' online activities and gather information about personal interests and preferences. This data has proven to be extremely valuable to online companies, because it not only enables them to sell products/services that are tailored to customers' demands, but also permits companies to boost their revenues by selling advertising space on their Web sites (Gilbert, 1999).

Intra-organizational applications focus on using Web technology to collect and disseminate information throughout a single business. Intra-organizational applications enhance existing relationships among parties within the business by promoting the efficient exchange of information and internal collaboration. Internal personal data also needs to be protected against misuse, and here, too, privacy is an issue. Consider, for example, the case where internal data might be collected without the knowledge of the employee, or might be collected for one purpose, but used for another.

Moreover, e-commerce has nurtured the emergence of various virtual business relationships that include virtual business-supplier relationships, virtual strategic alliance relationships, virtual business-client relationships and virtual business-to-end-consumer relationships (Speier et al., 1998). Until recently, the focus on privacy has been primarily in the B2C (business-to-consumer) area, where businesses collect customers' data directly. The privacy implications of B2B (business-to-business) transfers of personal, identifiable information have been neglected (Goodman, 2000).

Although businesses can benefit substantially from B2B applications, there are strong concerns about information disclosure to third parties and external secondary uses of customer personal information without customer consent. Based upon the privacy dimensions outlined by the FTC, customers should have the right to know what information businesses disclose to third parties. In addition, customers also should have the right to control their own information and opt to decline external secondary data use. In order to protect privacy in B2B applications, it is necessary to address the responsibilities of both the business that collects personal information and the business that receives the customer information secondarily.

IMPLEMENTING PRIVACY INGREDIENTS

Although there have been studies to explore privacy issues, examples of how to integrate privacy dimensions within an electronic storefront, and the lessons learned, could not be found in a literature review of information systems and marketing. Therefore, a research project was undertaken to learn more about the integration of the four privacy dimensions.

PROJECT OVERVIEW

A research project was designed to include a hypothetical virtual bookstore called Husky Virtual Bookstore (HVBS) at a mid-western university. Students (customers) could visit the HVBS electronic storefront, browse the site based on their academic departments, view textbook requirements and descriptions, and purchase books. The emphasis was to evaluate the four dimensions of privacy as defined by the FTC. To this end, we developed two identical electronic storefronts. However, one of the storefronts was modified to include the four privacy dimensions outlined by the FTC:

(1) *HVBS Privacy Statement (Figure 2)*: This page provided a detailed explanation of the information that would be collected and how it would be used by the HVBS bookstore. The intent of this page was to satisfy and support the *notice* dimension.

(2) *Main Page with Privacy Seal (Figure 3)*: This page contained general information about the HVBS bookstore. This page also included a prominent privacy seal that provided a means of telling the high privacy subject treatment group that certain standards were in place, in order to increase customers' trust perception. In addition, a scrolling message was displayed at the bottom of the

Figure 2: Privacy statement page at HVBS

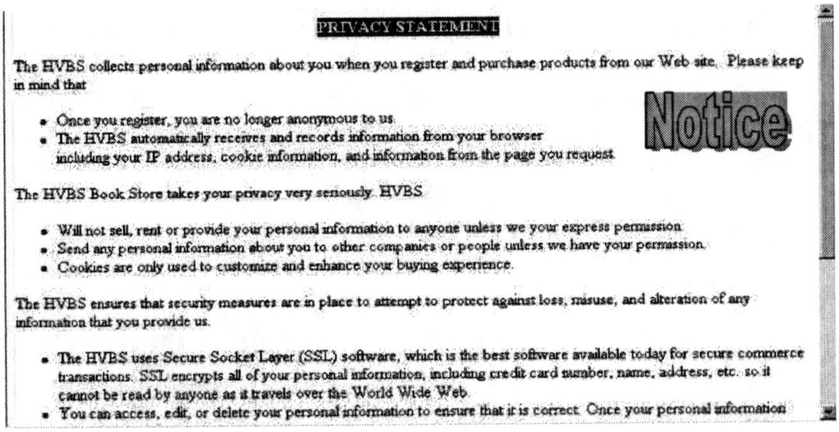

PRIVACY STATEMENT

The HVBS collects personal information about you when you register and purchase products from our Web site. Please keep in mind that

- Once you register, you are no longer anonymous to us.
- The HVBS automatically receives and records information from your browser including your IP address, cookie information, and information from the page you request.

Notice

The HVBS Book Store takes your privacy very seriously. HVBS

- Will not sell, rent or provide your personal information to anyone unless we your express permission.
- Send any personal information about you to other companies or people unless we have your permission.
- Cookies are only used to customize and enhance your buying experience.

The HVBS ensures that security measures are in place to attempt to protect against loss, misuse, and alteration of any information that you provide us.

- The HVBS uses Secure Socket Layer (SSL) software, which is the best software available today for secure commerce transactions. SSL encrypts all of your personal information, including credit card number, name, address, etc. so it cannot be read by anyone as it travels over the World Wide Web.
- You can access, edit, or delete your personal information to ensure that it is correct. Once your personal information

screen that assured the subject that the HVBS had taken reasonable measures to ensure data security and privacy protection.

(3) *Textbook Category Page (Figure 4)*: This page contained available textbooks that included Database Applications, Project Management, Operating Systems, and System Analysis and Design. When a customer visited this page, a pop-up window informed them that an electronic cookie was being downloaded to their computer. The intent of the pop-up window was to satisfy and support the *notice* dimension.

Figure 3: Main page with privacy seal at HVBS

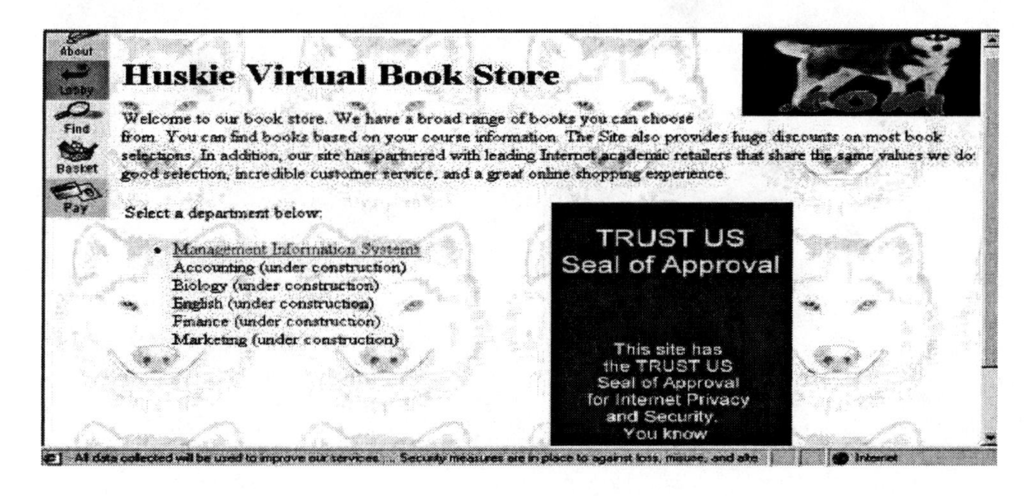

Figure 4: Textbook category page at HVBS

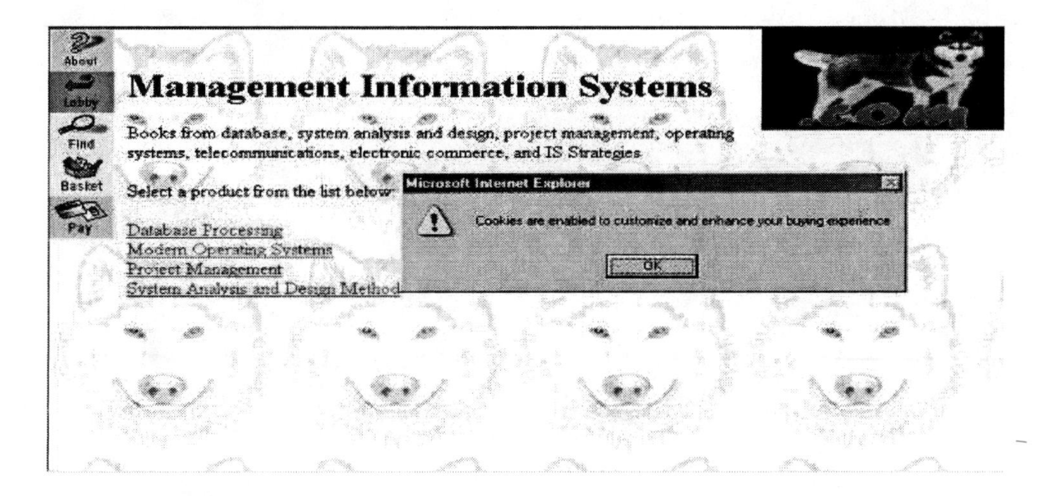

(4) *Textbook Description Page (Figure 5)*: This page contained information specifically related to the book selected by the subject, and included the book's description and price. After viewing this information, a customer could then add the book to his or her electronic shopping cart. In addition, a scrolling message at the bottom of the screen explained and enforced the perception of how *data security* was ensured at the HVBS site. The intent was to satisfy and support the *data security* element of the privacy principles.

Figure 5: Textbook description page at HVBS

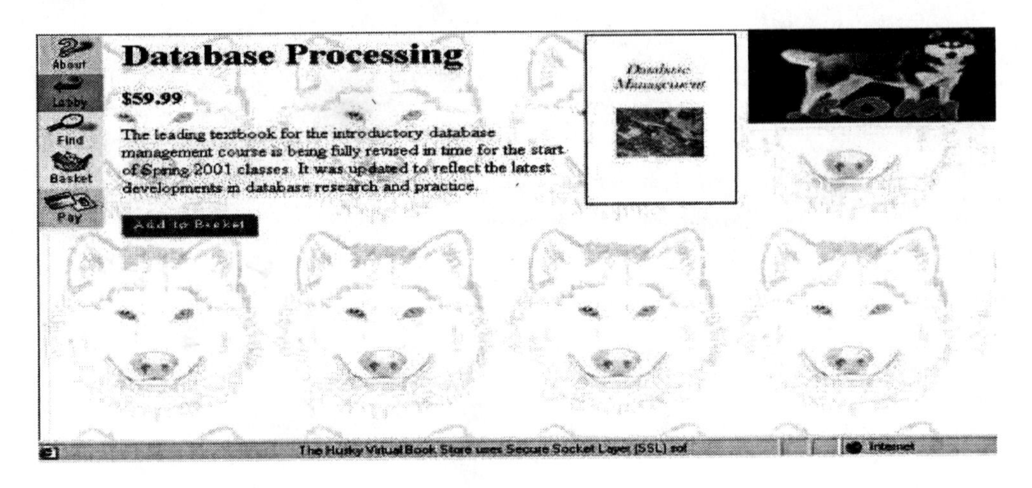

Figure 6: Shopping cart page at HVBS

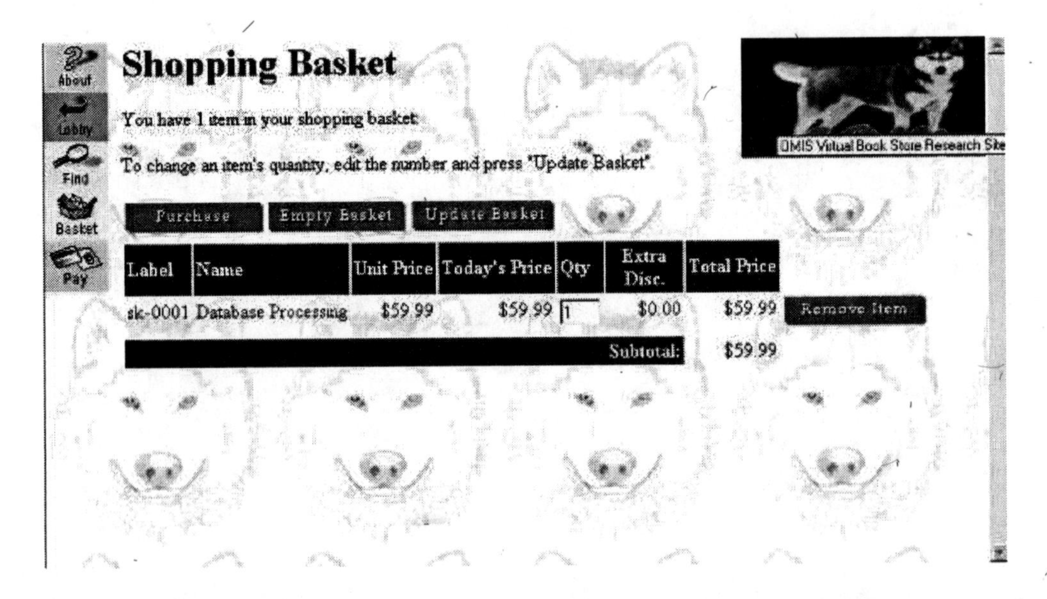

(5) *Shopping Cart Page (Figure 6)*: This page contained a list of books that were selected by the subject for purchase. The subject could alter the quantity or cancel the order and return to make another selection.

(6) *Order Registration Page (Figure 7)*: This page collected personal information, such as: name, address, telephone number, e-mail, and so forth. It also displayed a scrolling message stating that only authorized personnel would have access to the customer information. The intent was to satisfy and support the *data security* element of the privacy principles.

(7) *Verification Page (Figure 8)*: This page was used by the HVBS site to allow customers to view and update their personal information. Because the site was integrated with a database management system, any changes to personal information would be updated immediately. The intention of this page was to satisfy and support the *access* dimension.

(8) *Order Processing Page (Figure 9)*: This page supported the subject's perception that the HVBS site was secure. The subject was able to view a message that the HVBS Web site uses Secure Socket Layer technology, a firewall, and stringent auditing procedures.

Figure 7: Order registration page at HVBS

Enter your shipping address etc.

Enter a Ship-To Name:

Use the form below to tell us where you want your order to be shipped. Required fields are indicated with an asterisk (*).

Title:
* First Name:
Middle Name*
* Last Name:
* Address (line 1):
Address (line 2):
Address (line 3):
* City:
* State/Province: -- Select a State --
* Zip/Postal Code:
* Country: UNITED STATES
Daytime Phone Number: In case we need to contact you about your order.

Security

Only certain employees are authorized to access to the information you

Internet

Figure 8: Verification page at HVBS

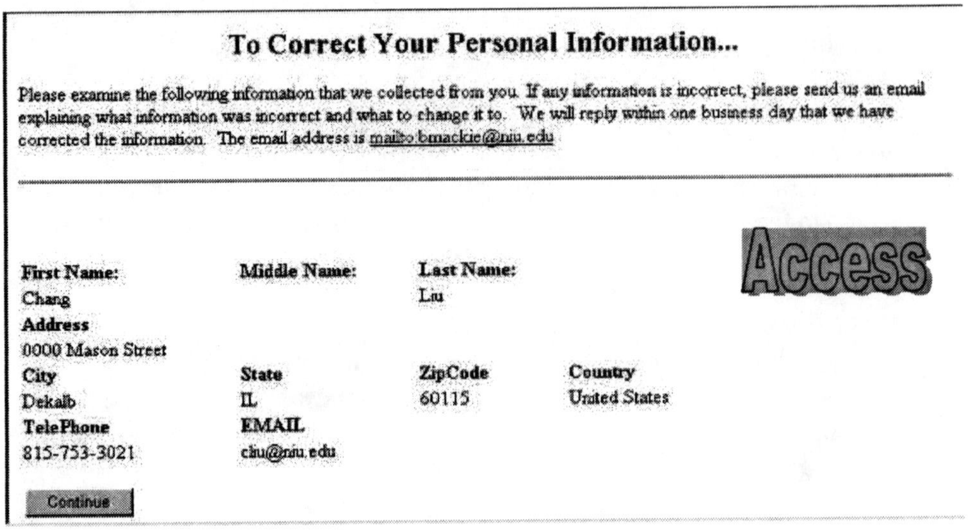

To Correct Your Personal Information...
Please examine the following information that we collected from you. If any information is incorrect, please send us an email explaining what information was incorrect and what to change it to. We will reply within one business day that we have corrected the information. The email address is mailto:bmackie@niu.edu
Access
First Name:
Chang
Middle Name:
Last Name:
Liu
Address
0000 Mason Street
City
Dekalb
State
IL
ZipCode
60115
Country
United States
TelePhone
815-753-3021
EMAIL
cliu@niu.edu
Continue

Figure 9: Order process page at HVBS

Dekalb, IL 60115
United States
cliu@niu.edu
Book SKU
Book Description
Unit Price
Quantity
Extended Price
sk-0001
Database Processing
$59.99
1
$59.99
Subtotal:
$59.99
Shipping:
$3.00
Total Cost:
$62.99
Visa
American Express
Master Card
Discover
HVBS Beleives your Credit information must be secure and gurranties that security
*Credit Card Number:
*Expiration Date:
Month: 11
Year: 2000
*Cardholder's Name:
Enter the name exactly as it appears on the credit card.
Security
The Husky Virtual Bookstore will provide you with means to ensure that your personal information is accurate
Internet

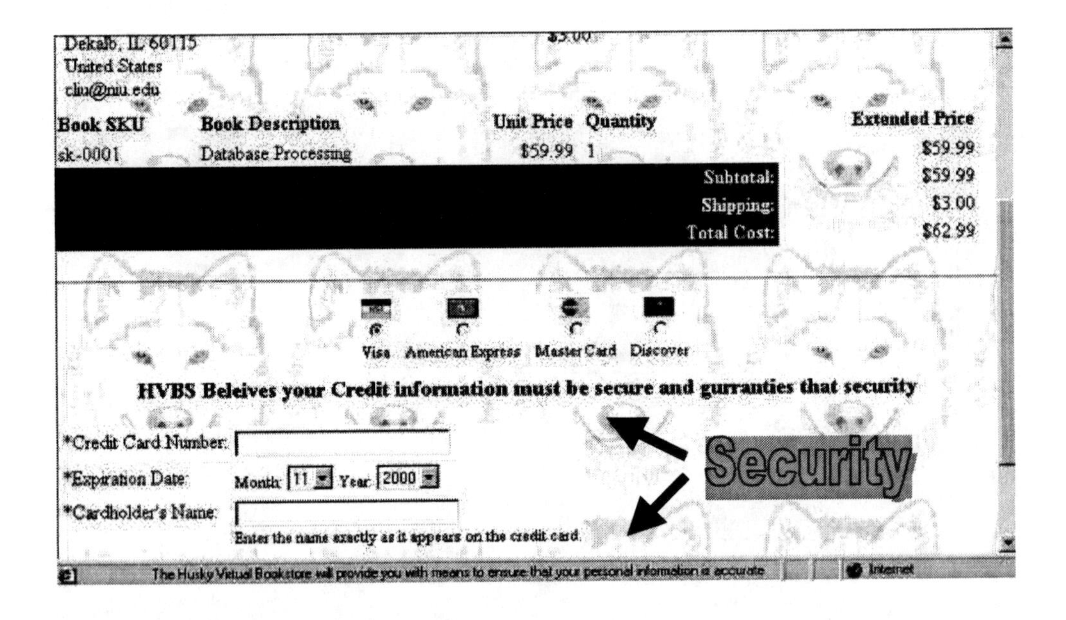

(9) *Choice Page (Figure 10)*: Before subjects left the HVBS Web site, they were given the opportunity to allow the HVBS to share the personal information they provided with external secondary sources. By allowing the subject to "opt-out," the *choice* dimension was supported.

(10) *Order Confirmation Page (Figure 11)*: This page confirms the customer order and re-states the privacy policy for all privacy dimensions of *notice*, *choice*, *access* and *data security*.

The development of the HVBS Web site required a number of revisions to adjust and achieve the best possible effects of an online shopping environment. In addition, many active server page (ASP) techniques were used to combine HTML, client-side and server-side scripts to produce dynamic Web pages for the HVBS site that included privacy elements.

At the final stage of the experiment, an online survey questionnaire was posted at the HVBS Web site. The survey examined the relationship among the privacy, trust and customer behavioral intentions. The research subjects for the experiment were selected on a convenience basis from enrollment in five information system courses.

The survey results strongly suggest that a complete privacy statement leads to improved customers' trust in engaging in a "relationship exchange" with the businesses; also, the trust obtained from customers will help businesses build better

Figure 10: Choice page at HVBS

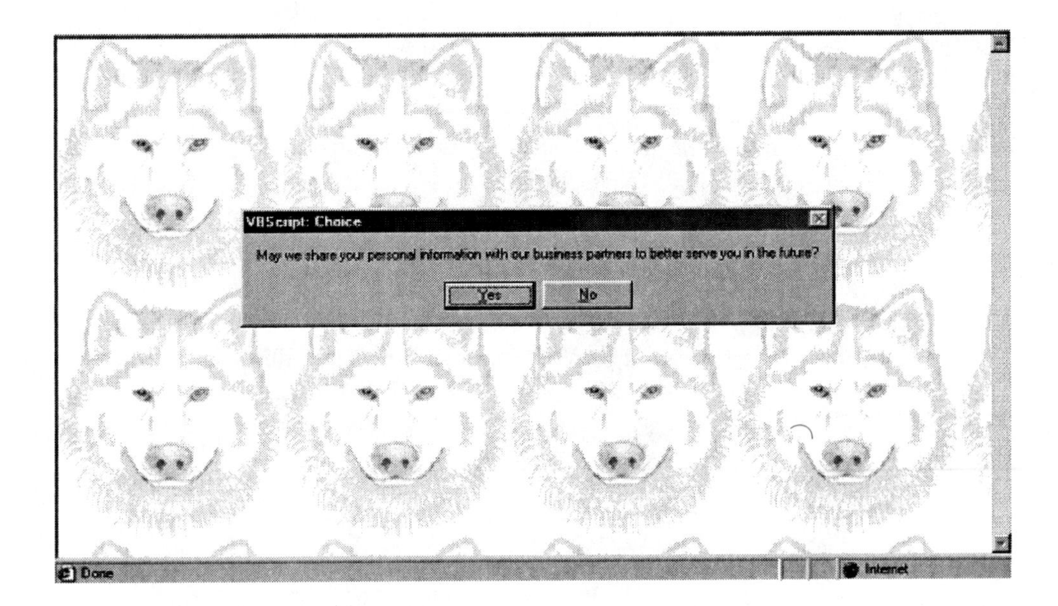

Figure 11: Order confirmation page at HVBS

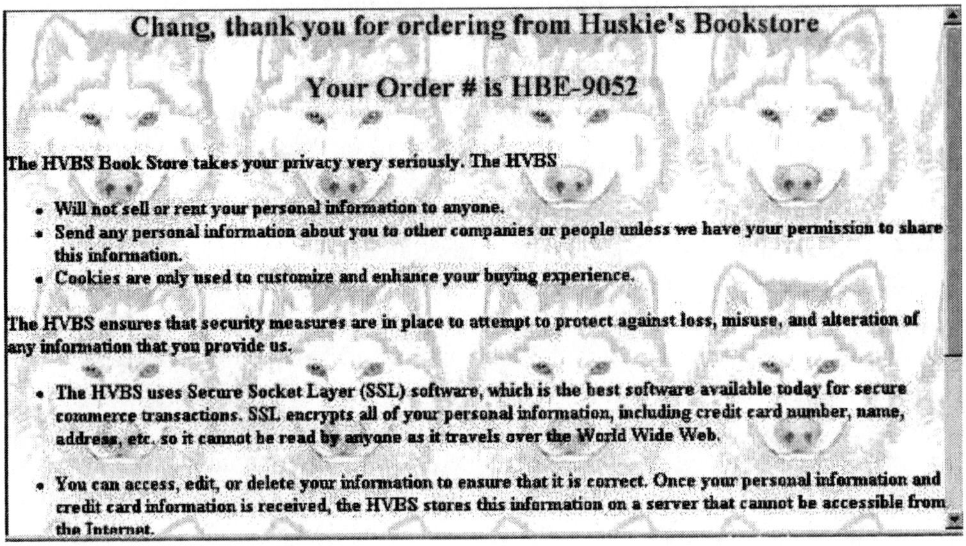

Chang, thank you for ordering from Huskie's Bookstore

Your Order # is HBE-9052

The HVBS Book Store takes your privacy very seriously. The HVBS

- Will not sell or rent your personal information to anyone.
- Send any personal information about you to other companies or people unless we have your permission to share this information.
- Cookies are only used to customize and enhance your buying experience.

The HVBS ensures that security measures are in place to attempt to protect against loss, misuse, and alteration of any information that you provide us.

- The HVBS uses Secure Socket Layer (SSL) software, which is the best software available today for secure commerce transactions. SSL encrypts all of your personal information, including credit card number, name, address, etc. so it cannot be read by anyone as it travels over the World Wide Web.
- You can access, edit, or delete your information to ensure that it is correct. Once your personal information and credit card information is received, the HVBS stores this information on a server that cannot be accessible from the Internet.

results, because customers will have favorable behavioral intentions, such as willingness to recommend the site, re-visit the site, post positive comments and purchase from the site again.

PROJECT DEVELOPMENT ENVIRONMENT

The design effort in this project is focused mainly on the server part of the equation, because the focus here is to implement privacy ingredients for a commercial Web site.

(1) *Hardware Environment*: The server machine that hosts the HVBS Web site has an Intel Pentium III 500 MHZ Processor, 96MB RAM, 2-ten GB hard disks and 3Com EtherLink 10/100 PCI network card. This minimal configuration suggests that any commercial Web server should have no hardware resource constraints for integrating privacy dimensions into its home page.

(2) *Networking Environment*: The Web server chosen for this project was the Internet Information Server 4.0 (IIS 4.0). The IIS 4.0 was available within the Windows NT server package. Windows NT server 4.0 was chosen for the networking environment and its IIS 4.0 is chosen for building Web server for this project.

(3) *Software Tools*: Microsoft Site Server 3.0, Commerce Edition, and Microsoft Visual InterDev 6.0 were the two major software tools for the project development. Site Server Commerce Edition is a comprehensive Internet commerce server, optimized for Microsoft Windows NT Server 4.0 and IIS 4.0. Although Microsoft Site Server 3.0, Commerce Edition, is powerful for site creation, management and maintenance, customization is better supported with Microsoft Visual InterDev 6.0. Visual InterDev provides a design environment to build dynamic Web-based applications through the use of ASP. Therefore, Microsoft Visual InterDev 6.0 was employed for site customization of the Husky Virtual Bookstore that was established using Microsoft Site Server 3.0, Commerce Edition.

(4) *Database Integration*: Because Microsoft SQL Server 7.0 was the premier database engine for the Windows NT platform, it was chosen with Microsoft Access as the database systems to serve and maintain user information at the HVBS site. ODBC (Open database Connectivity) was the chosen method for database connectivity through a system data source name (DSN).

PROJECT DEVELOPMENT PROCESS

The experimental storefronts were custom developed. Therefore, it was important to design these experimental electronic storefronts with a proven systematic approach. So, for example, the selection of a logical development sequence would be chosen first. Then the page design and the look, behavior and attributes of the objects within each page would be designed. Finally, privacy ingredients would be added. Based on this premise, the following stages were set up for the development:

1. Decide on the number of pages, page names, and page content;
2. Design screens and add objects for each page;
3. Define the relationships and hyperlinks among pages and objects;
4. Establish an outline for the project and a time schedule for its completion;
5. Acquire all necessary resources for development;
6. Create the development environment;
7. Use Microsoft Site Server 3.0, Commerce Edition, to build a prototype of electronic book store;
8. Use Microsoft Visual InterDev 6.0 for customization;
9. Test and modify the project; and
10. Present the project for feedback.

CONCLUSIONS

In today's e-commerce environment, many Web sites gather personal information about the company's customers in order to enhance the online buying experience. However, it appears to be good business that the same Web site ensures that privacy concerns are adequately addressed. Posting well-developed privacy policies should lead to more repeat visits and more purchases. Privacy intrusions have been widely publicized. This has made the general public more aware of what may happen with their personal information. Such publicity has placed additional pressure on the Web site provider to address privacy concerns.

The emphasis here has been to provide a research-based method to develop an electronic storefront that includes privacy dimensions to ease customers' concerns. It is hoped that the lessons learned and empirical evidence gathered provide a foundation for other e-business sites. However, more research is needed, and should address issues beyond privacy concerns. More specifically, it is important to establish a connection between privacy policies and business outcomes.

REFERENCES

Benassi, P. (1999). TRUSTe: an online privacy seal program. *Communications of the ACM, 42*(2), pp. 56-57.

Bradner, S. (1999). A perfect example. *Network World, 16*(46), p. 52.

Clark, R.A. (1988). Information technology and dataveillance. *Communications of the ACM, 31*(5), pp. 498-512.

Culnan, M.J. (1993). How did you get my name? An exploratory investigation of consumer attitudes toward secondary information use. *MIS Quarterly, 17*(3), pp. 341-363.

D'Amico, E. (2001). Global e-commerce. *Chemical Week* (September 26, 2001).

Engler, N. (1999). Global e-commerce: How products and services help sites expand worldwide. Retrieved from the World Wide Web: http://www.informationweek.com/755/global.htm.

FTC Congress Report. (2000). Privacy online: fair information practices in the electronic marketplace: a report to congress. Washington, D.C.: Federal Trade Commission.

Garfinkel, S. L. (2000). Private matters. *CIO Magazine*. Retrieved from the World Wide Web: http://www2.cio.com/archive/060100_diff_content.html.

Gilbert, J. (1999). Research points to the growth of online ads. *Advertising Age, 70*(35), p. 6, 25.

Goldstein, H. R., Roth, A. E. & Young, T. L. (2001). International personal data safe harbor program launched. *Intellectual Property & Technology Law Journal, 13*(4), pp. 24-25

Goodman, S. (2000). Protecting privacy in a b2b world. *Mortgage Banking*, (April), pp. 83-87.

Green, H., Alster, N., Borrus, A. & Yang, C. (2000). Privacy: Outrage on the Web. *Business Week,* 3668, (February 14), pp. 38-40.

Harrison, A. (1999). RealNetworks slapped with privacy lawsuits. *ComputerWorld, 33*(46), p. 20.

Hoffman, D.L. & Novak, T. (1999). Building consumer trust online . *Communications of the ACM, 42*(4), pp. 80-85.

Hopper, D.I. (2000). Vendors send personal info to marketing firm. *Marketing News*, (August), p. 19.

IDC - International Data Corporation. (2001). eBusiness is key to sustaining $5.3 trillion in eCommerce by 2005. Retrieved April 24, 2001 from the World Wide Web, Internet keyword *International Data Corporation.*

Kleinbard, D. (2000). Web has its eye on you. Retrieved from the World Wide Web: http://cnnfn.com/2000/03/06/technology/privacy_main.

Liu, C. & Arnett, K.P. (2002). An examination of privacy policies in Fortune 500 Web sites. *Mid-American Journal of Business, 17* (1), 13-21.

Liu, C., Marchewka, J.T., & Ku, C. (2002). A global perspective of the privacy-trust-behavioral intention model of electronic commerce: An experimental study in U.S. and Taiwan. *Proceedings of the International Conference on e-Business*, May 23-26, Beijing, China.

Mendel, B. (1999). Online identity crisis. *InfoWorld, 21*(42), pp. 36-37.

Milberg, S.J., Burke, S.J., Smith, H.J.& Kallman, E.A. (1995). Values, personal information, privacy and regulatory approaches. *Communications of the ACM, 38*(12), pp. 65-84.

Porter, A.M. (2000). Buyers want Web privacy. *Purchasing, 129*(5), pp. 22-25.

Robinson, S. (1999). CD software said to gather data on users. *New York Times.* November 1, 1999.

Smith, H.J, Milberg, S.J. & Burke, S. (1996). Information privacy: Measuring individuals' concerns about organizational practices *MIS Quarterly, 20*(2), pp. 167-190.

Speier, C., Harvey, M. & Palmer, J. (1998). Virtual management of global marketing relationships. *Journal of World Business, 33*(3), pp. 263-276.

Stepanek, M. (1999). Protecting e-privacy: Washington must step in. *Business Week,* July 26, p. EB30.

VanScoy, K. (2000). Get inside your customer's heads (and their wallets too). *SmartBusinessmag.com*, June 2000, pp. 100-103.

Warren, S.D. & Brandeis, L.D. (1890), The right to privacy. *Harvard Law Review, 4*(5), pp. 193-220.

Webster, J. (1998). Desktop videoconferencing: Experiences of complete users, wary users, and non-users. *MIS Quarterly, 22*(3), pp. 257-286.

Chapter XI

An E-Channel Development Framework for Hybrid E-Retailers

In Lee
Western Illinois University, USA

ABSTRACT

Due to the profound impact of e-commerce on organizations, e-channel development emerged as one of the most important challenges that managers face. Unfortunately, studies indicate that managers in most large companies are still unclear about an e-commerce strategy, and tend to lack adequate e-commerce development expertise. Poorly planned and developed e-commerce channels add little value to organizations. Furthermore, these poorly developed e-channels may even have negative impact on their organizations by confusing and disappointing customers who value a seamless cross-channel experience. To develop an e-channel that delivers higher utility to customers and generates sustainable long-term profits, managers need to analyze how an e-commerce channel affects the performance of existing channels and develop a company-wide e-channel development program.

Based on a number of e-commerce case studies, we developed an e-channel development framework that consists of five step-by-step phases: (1) strategic analysis; (2) e-channel planning; (3) e-channel system design; (4) e-channel system development; and (5) performance evaluation and refinement. This framework helps managers evaluate the impact of e-commerce channels on organizational performance and determine the most appropriate channel design and integration mechanisms for the achievement of business strategies. This paper also discusses impact of e-channel structures on organizational performance.

INTRODUCTION

One of the most profound developments of the past decade was the emergence of e-commerce that revolutionized the process of buying, selling, and exchanging products and services on the Internet. According to Forrester research (www.forrester.com), worldwide e-commerce — both business-to-business (B2B) and business-to-consumer (B2C) — will reach $6.8 trillion in 2004, capturing 8.6% of the world's sales of goods and services. While the B2B market has a greater portion of e-commerce business transactions, the growing population of Internet users has provided a large B2C consumer base for e-retailers. Forrester research also predicts that on-line retail sales will reach $269 billion in 2005.

To capture this ever-increasing B2C population, retailers have experimented with a variety of B2C models. Some of the widely used e-commerce models include auction models (e.g., eBay.com); reverse auction models (e.g., Priceline.com); portal models (e.g., Yahoo.com); stand-alone e-retailer models (e.g., Amazon.com); and hybrid e-retailer models (e.g., Walmart.com). While an e-commerce customer base has been growing rapidly, many of the e-commerce models have failed to generate sustainable long-term profits. In times of economic downturn, many stand-alone e-retailers of commodity products suffered the hardest hits due to rising customer acquisition cost, low product differentiation and the lack of financial support of investors. Some e-retailers, such as Garden.com, consolidated with traditional companies to achieve synergy effects. According to Ernst & Young, e-retailers are experiencing growing pains, but the experience will make them stronger in the future. The cost of entering the e-retailing business will rise significantly and become prohibitive for some e-retailers (Shern, 2000).

The success of an e-channel lies largely in developing and implementing a sound e-channel strategy that delivers higher utility to customers and charges lower prices than its competitors. Evidence shows that a misdirected e-commerce channel (e-channel) development leads to costly and frequent revisions of e-commerce strategies. For example, both K-Mart and Wal-Mart experienced a costly revision of their e-commerce strategies. K-Mart initially created a spin-off e-commerce entity, BlueLight.com, in December, 1999, as a joint venture between K-Mart and Softbank Venture Capital. After K-Mart withdrew from a planned initial public offering (IPO) for BlueLight.com in 2000, it acquired all of the interests in BlueLight.com in 2001. By bringing the e-commerce entity home, K-Mart expects to fully coordinate its e-commerce operations with other channels' operations. Walmart.com offers another good example of the costly revision of an e-commerce strategy. Walmart.com was established in January 2000 as an independent company operating as a joint venture between Wal-Mart and Accel Partners. In 2001, Wal-Mart also acquired all the minority interest in Walmart.com, in order to establish the tight integration between its e-commerce entity and physical stores. These costly organizational changes could have been avoided if these retailers had accurately analyzed the impact of the e-channel structures on their business strategy.

This paper presents an e-channel development framework and analyzes the impact of e-channel structures on organizational performance. The framework provides managers with a powerful tool used to fully utilize the opportunities of e-commerce and prevent unexpected threats that would be introduced by carelessly developed e-commerce strategies. The rest of the paper is organized as follows. This chapter discusses the e-commerce channel development framework is presented. It then discusses a computer simulation of e-channel is introduced, and the impact of e-channel structures on organizational performance is discussed. Finally, the implications of the e-channel development framework for managers are discussed.

AN E-CHANNEL DEVELOPMENT FRAMEWORK

Because each organization is uniquely positioned in the market, with a different set of competitive forces and resources, no single e-channel development strategy would be suitable for all organizations. The development of an e-channel requires a comprehensive review of all major business aspects, including business strategy, marketing, IT strategies, business processes, organizational structure, financing and accounting, and vendor/supplier relationship. Poorly planned and developed e-commerce channels, without solid coordination and integration mechanisms, add little value to organizations. Furthermore, these poorly developed channels may even have a negative impact on other channels by confusing and frustrating customers who expect a seamless cross-channel experience. Given the complexity and importance of an e-channel development, managers need to be assisted with a systematic and comprehensive framework that helps them facilitate decision-making processes and minimize strategic and operational mistakes.

Based on a number of e-commerce case studies, we developed an e-channel development framework that consists of five step-by-step phases: (1) strategic analysis; (2) e-channel planning; (3) e-channel system design; (4) e-channel system development; and (5) performance evaluation and refinement. This framework will help managers estimate the impact of e-commerce channels on organizational performance, determine the most appropriate channel and system design and develop coordination mechanisms for the achievement of business strategies.

Because an e-channel affects every aspect of business operations, the e-channel development should be a cross-functional team effort. As shown in Figure 1, the e-channel development process may iterate until the best design of e-channel emerges. Each phase then will be described in detail.

Strategic Analysis

The purposes of strategic analysis is to identify new business opportunities and threats, to solicit senior management support for e-commerce projects, and to develop multi-faceted strategies to achieve business goals. The strategic analysis begins with an analysis of the organization's competitiveness. Understanding the organization's competitiveness requires identifying major internal/external opportunities and threats created by customers, suppliers, competitors, regulatory agencies, employees and other stakeholders involved. Opportunities to take into account include: the e-channel's market expansion in regions in which the company does not have physical stores; on-line market testing for new products and services; and integrated inventory management and product support between physical stores and the e-channel. Threats include: a cannibalization of their own market due to the introduction of an e-channel; dissatisfied customers due to an inconsistent channel image and discriminate prices; and the high switching cost and customer base established by first-mover competitors. Cannibalization among the organization's multiple channels can lead to intense internal channel conflicts.

The bargaining power of customers and suppliers is a critical factor in shaping competitiveness (Porter, 1985). The presence of these powers will remain vital in determining a firm's ability to exploit an e-commerce opportunity (Gallaugher, 2002). The introduction of e-commerce has shaken the landscape of the competitive forces between manufacturers and retailers. Porter's competitive forces model (1985) has been most widely used in analyzing these forces. Some of the changes in the balance of powers come from the disintermediation of the middleman. E-channel allows suppliers to bypass retailers and sell directly to the consumer. This disintermediation raised the bargaining powers of both manufacturers and customers against retailers. Consumers can go directly to a manufacturer's Web site, such as blackanddecker.com, instead of to a retailer, like Lowe's, to buy home improvement equipment. In this case, blackanddecker.com can reduce Lowe's bargaining power with customers. To avoid losing major retailers, Black&Decker also provides hyperlinks to major retailers for customers who want to visit physical stores. The e-channel provides customers with a greater ability to comparison-shop and, thereby, raises the bargaining power of customers.

Another part of the strategic analysis is Critical Success Factors (CSFs) analysis. CSFs facilitate the organization's goal achievement by focusing on the most effective determinants of success. For example, superior customer services and short delivery times are critical success factors that improve customer satisfaction and, ultimately, increase revenue. Once CSFs are identified, channels' roles in

Figure 1: An e-channel development process

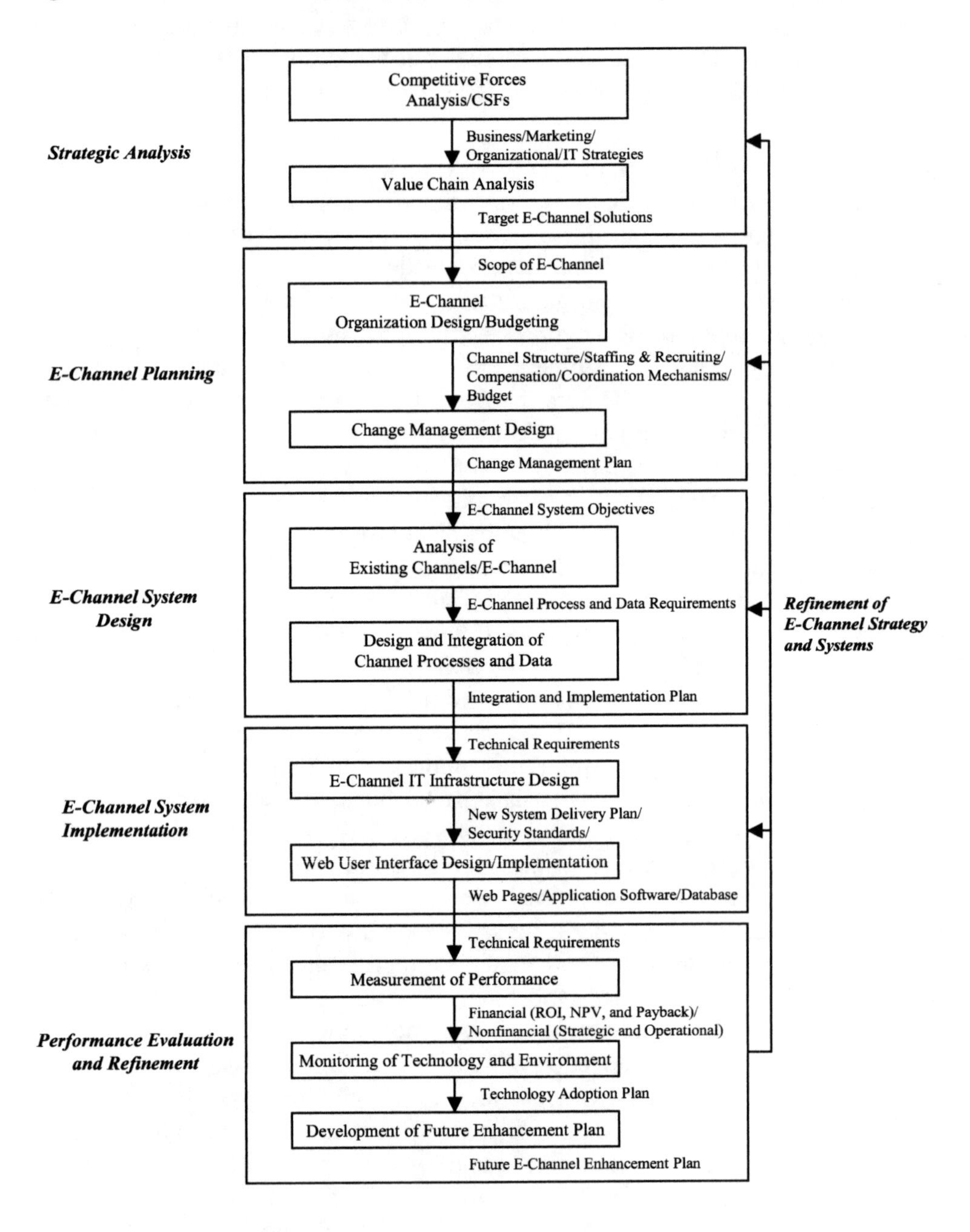

supporting CSFs are defined. At the end of the above-mentioned activities, business strategies are formulated, and the development of organization, marketing, and information technology strategies follows.

Managers should develop business strategies to capitalize on opportunities and reduce threats. If manufacturers develop a direct marketing e-channel, the company needs to develop a business strategy to counter the disintermediating effect of the supplier's e-channel. Raising a switching cost to corporate clients with customized e-commerce software can mitigate the bargaining power of suppliers and customers. Office Depot is a case study in how to raise a switching cost and lock in many corporate clients. Office Depot has set up customized Web pages for 37,000 corporate clients with parameters that provide different employees with various degrees of freedom to buy supplies. The Office Depot's e-channel, in turn, helps corporate clients reduce their purchase costs, because order and billing are handled electronically.

Organization strategy focuses on the creation of an e-retailer organization and the design of a multi-channel coordination mechanism. Strategic options for an e-channel development include on-line division, spin-offs, alliances, partnerships, acquisition of existing dot-com companies and joint ventures. Experimenting with different types of e-channel organizations is very costly. Some retailers, such as K-Mart, Wal-Mart, CVS and Staples, experimented with both on-line division and spin-off e-channel structures. Other retailers, such as Target, Office Depot and OfficeMax, initiated an on-line division and have sustained the same structure. There are two types of multi-channel coordination approaches: strong integration of various channels and loose coordination of multi-channels. The eGM division initially started with a strong integration capability and later let divisions take a part of the e-commerce responsibility. Because Internet technology and the business environment change so rapidly, it is difficult to accurately predict an e-channel structure and cross-channel coordination mechanism that is best for a specific organization. In addition to the creation of e-channel structure, e-retailers need to think about how a partnership can contribute. In a 1999 study of deals among dot-com and traditional companies, McKinsey & Co. found announcements of 13,000 e-commerce alliances (e.g., Rite Aid's partnership with Drugstore.com to extend its retail channel).

The introduction of e-channels requires the development of a new marketing strategy for each channel. While one business strategy applies to physical stores, catalogs, and e-channels, each channel may take a different marketing strategy to achieve the business strategy. A marketing strategy includes the definition of target customers, pricing, distribution channels, positioning of products and services against competitors, etc. Providing a convenient and seamless experience to multi-channel customers is crucial to the success of an e-channel and physical stores. A cross-channel marketing team needs to identify the alternative marketing strategies for a seamless customer experience and evaluate the pros and cons of each marketing strategy for both existing and new customers.

Technology strategy is directed toward setting up an information technology architecture to utilize the potential benefits of an e-channel. It takes into account the capabilities of existing IT architecture to ensure interoperability between new and old

IT systems and prevent a duplicate IT investment. A value chain analysis is widely used to identify the critical technology necessary in achieving strategic goals. Competitors' use of e-commerce solutions that support their value chains are identified and benchmarked. Strategic advantage e-commerce solutions are recommended to achieve a first-mover advantage when their competitors have not yet employed them. Strategic necessity e-commerce solutions are recommended to stay in the market when the majority of competitors have already used the technologies. Because e-commerce technology is developing fast, the timing of the information technology investment is important, in order to gain a competitive advantage and develop a long-term technological competency. A new e-channel technology should maintain interoperability with existing inventory management systems, order-processing systems and customer relationship management systems.

E-Channel Planning

As in the strategic analysis, e-channel planning requires a multidisciplinary coordination among many functions. An e-channel planning team consists of individuals from information technology, marketing, organizational design, channel management, purchasing and distribution functions. Outcomes from the strategic analysis guide e-channel planning. Four major decisions made in this stage are: (1) the scope of an e-channel project; (2) an e-channel organization design and budgeting; (3) multi-channel coordination mechanisms; and (4) a change management design.

The scope of a retailer's e-channel is divided into three major categories: demand-side corporate client e-channel, consumer e-channel and supply-side vendor/manufacturer e-channel. Once the scope of the e-channel is determined, the e-channel planning team designs the e-channel organization. Components of e-channel organization design include e-channel structure, distribution of power and responsibility, staffing and recruiting, compensation and promotion plans, and physical location of e-channel. Once the hierarchical structure of the e-channel is designed, the staffing and recruiting plan provides details on assigning qualified in-house staff to an e-channel project and hiring people who are experienced in e-commerce development. Technical staff assigned to the e-channel project will design e-channel systems and develop technology infrastructure. If employees are afraid of losing their customer base or jobs because customers switch to a new channel, they will show little enthusiasm for or even resist the introduction of a new channel. Therefore, a compensation and promotion plan should be designed to retain core personnel and encourage them to cooperate with individuals working at other channels.

Because an e-channel introduces new business practices and changes almost every aspect of existing channel operations, management needs to establish a change management program to ensure the smooth deployment of an e-channel and to foster an organizational culture of cooperation, innovation and creativity. Change management will be even more important for a bureaucratic organization, where the culture discourages changes. Detailed cross-channel coordination mechanisms should be established for pricing, inventory management, product/service mix, cross selling and distribution network decisions. The following questions guide the development of a coordination mechanism: (1) What are the customer profiles in each channel?; (2) What percentage of the consumers visit e-channel Web sites, but purchase products at physical stores?; and (3) What percentages of consumers purchase products at both the e-channel and physical stores?

Once the e-channel plan is developed, and financial requirements are estimated, the e-channel project is justified, based on quantitative and qualitative evaluation criteria, and a budget and financing plan is approved. Because some of the benefits generated by the e-channel are non-financial, managers need to examine how these non-financial benefits contribute to overall values. These non-financial benefits may need to be transformed into equivalent financial values, in order to apply cost-benefit analyses such as NPV, ROI, and payback period methods.

E-Channel System Design

E-channel system design consists of four major activities: establishment of e-channel information systems' objectives; analysis of existing e-channel and other channels' information systems; e-channel process design and cross-channel process integration; and e-channel database design and cross-channel data integration. The e-channel system's objectives that support specified strategic goals should be established first. E-channel information systems' objectives, such as data collection of customer web activities, information delivery, transaction or customer decision support, guide all the e-channel system development activities that follow.

Analysis and redesign of an e-channel solution focus on the business process and data management. E-channel process (ECP) can be classified into two interrelated categories: ECP in physical flow and ECP in information flow. ECP in physical flow involves designing the physical transfer of products or services to e-channel customers. When the e-channel shares distribution and transportation networks with other channels, a redesign of their physical flow may be needed in order to improve product delivery. Redesigned physical flow may allow e-channel customers to pick up products from nearby physical stores and return products to the physical stores. B2B ECP in physical flow enables just-in-time materials flow between manufacturers and retailers. On the other hand, ECP in information flow changes the ways in which we create, transform, disseminate, and store information. ECP in information flow includes web-personalization processes, order fulfillment processes, payment processes, return processes, etc. The critical role of ECP in information flow is to remove space and time barriers, and lower transaction and coordination costs. For example, ECP in inter-organizational information exchange reduces order and payment costs for customers. Office Depot's corporate clients reported that Office Depot's e-channel reduced an ordering and payment cost from more than $100 to between $15 and $25. Because ECP in physical flow is triggered by ECP in information flow, they should be redesigned concurrently to achieve an optimal performance.

Whether the e-channel is a spin-off or an on-line division, a new e-channel process should be integrated into other channels' existing back-end legacy systems, ERP and CRM systems. Enterprise integration of the new e-channel process with other channel processes is a daunting task. Horizontal process integration integrates e-channel processes with other channel processes (e.g., integration of an e-commerce Web site and physical store processes). Vertical process integration integrates e-channel processes with manufacturer/dealer processes. Integrating e-commerce systems with existing systems is essential to an organization's survival (Capozzoli & True, 2001). Unfortunately, many organizations are not capitalizing on the synergistic advantages of integrated systems.

In preparing a database design, two main questions should be posed: (1) What data or information are our e-channel customers interested in collecting from the Web site? and (2) What data are used for managerial decision-making (e.g., Web-personalization and cross-channel merchandising)? An e-channel's unique data that are difficult to collect from other channels include traffic patterns, customer profile data, repeat purchases and abandoned shopping carts. To minimize the duplication of data, data required for new e-channel systems need to be compared with data maintained in the existing database.

E-Channel System Implementation

E-channel system implementation involves establishing technical requirements, analyzing the existing technical infrastructure, designing a new technical infrastructure, designing a Web-user interface and implementing the new systems. In establishing technical requirements, minimum requirements of CPU, software, storage, telecommunications capacities and security standards are determined based on the e-channel's process and data analysis. After analyzing an existing technical infrastructure, a new infrastructure is designed to achieve scalability for future growth and seamless integration with existing infrastructure.

New system delivery options include in-house development, off-the-shelf software, application service provider, e-retailer acquisition, partnerships and outsourcing. Costs of new systems' options are broadly classified into hardware, software, telecommunications, maintenance and upgrade, and personnel expenses incurred throughout the duration of the systems' life cycles. Managers choose the best system delivery option that meets selection criteria. Selection process usually is iterative and constraints should be incrementally relaxed until a minimum set of delivery options is identified.

There is a growing concern for the security and privacy of the Web-based system (Garfinkel & Spafford, 1997). The system that supports an e-channel should be reliable, accessible when needed and secure. Many firms are installing firewall software. However, the answer to security problems is not in the installation of a single technology. Good security solutions consist of a well-designed set of technological devices and sound security policies and practices. To enhance security, a Web-based system can be developed as a three-tiered architecture: a Web client, network servers and a back-end information system supported by a suite of databases (Joshi et al., 2001). For the e-commerce channel, it is important to have a range of security solutions, because different types of risks are inherent in different types of applications.

Internet technologies are advancing rapidly from a software standpoint. Currently, dominant e-commerce platforms include HTML, Java, Active Server Pages (ASP), relational database and a three-tiered architecture that consists of HTML on the client tier, Active Server Page on the middle tier, and a database on the third tier. While ASP is currently widely used, Extensible Markup Language (XML) is changing the way in which Web content is created, manipulated, stored and disseminated. Recent developments in networking technologies, such as 3G and Bluetooth, further expanded the e-channel into a new dimension — wireless mobile e-channel.

Web user interface design should involve e-channel customers to accommodate their preferences on Web development. Bellman et al. (1999) suggested several guidelines for the design of an on-line shopping Web site: (1) sites should make it more convenient to buy standard or repeat-purchase items (such as the one-click-purchase approach by Amazon.com) and (2) customization should provide the information needed to make a purchase decision. Lohse and Spiller (1998) suggested that the checkout process should be easy for the consumer. Five dimensions of Web user interface design are: information, transaction, communication, community and decision support dimensions. Through a transformation of physical store functions into equivalent e-channel functions, Web user interface delivers a consistent company image to customers. Prototyping and Rapid Application Development (RAD) techniques are widely used to facilitate the Web user interface design process.

Performance Evaluation and Refinement

Performance evaluation has three purposes: (1) to measure financial and non-financial performances; (2) to monitor technology trends and business environments; and (3) to develop a future enhancement plan. According to Ernst & Young's study, retailers rely almost exclusively on financial and operational metrics, such as revenue, gross margin, average sales per transaction and the number of site hits. However, to build a valuable business, a retailer must develop holistic, customer-centric metrics and align every aspect of its organization to meet and exceed measures of success (Shern, 2000).

The objectives of an e-channel should guide the selection of evaluation criteria. While an internal e-channel evaluation is the primary focus in this stage, the interactions between the e-channel and the other channels are also analyzed here. Two major components in financial evaluation are revenue and cost. A challenging aspect of the evaluation is to identify and aggregate dispersed revenue and cost factors, such as cannibalization and hidden costs. For example, due to a cannibalization effect, net revenue increase in an e-channel should be derived by deducting the revenue decrease in other channels. Financial performances, such as net present value (NPV), payback period, and return on investment (ROI), are measured with an appropriate hurdle rate applied. Non-financial evaluation consists of two major dimensions: operational and strategic. Operational dimensions are typically quantifiable and include look-to-buy ratio, data usage rate, Web site access time, accuracy of on-line order process, and security and privacy. Strategic dimensions are mostly nonquantifiable and include customer satisfaction, service quality, market share, new technology development, multi-channel system integration and retention of employees with critical expertise. Appropriate weights may be assigned to nonfinancial performances.

Measuring the influence of the Web site on in-store purchases remains a central issue for retailers (Prior, 2002). The ability to evaluate and refine the e-channel requires that all stakeholders define meaningful metrics in the e-channel planning stage. Although many e-retailers collect revenue/cost and usage data from their e-channel Web sites, few of them understand how their performance compares to that of competitors, and how such information can be used to improve their e-channel performance. Customers' feedback on the Web site experience is an important contribution to the future Web user interface enhancement. The technology and business environment in which an e-commerce channel operates are changing rapidly. The changes in consumers' purchase patterns and demographics create significant opportunities and threats for e-retailers, forcing them to review their current business practices, technology and strategies, and develop a refinement plan.

IMPACT OF E-CHANNEL STRUCTURES ON ORGANIZATIONAL PERFORMANCE

The e-channel development in a multi-channel environment is much more complicated than that in a single-channel environment. If channels fully interact with each other, the complexity of a decision space increases at the rate of K^{mn} where K represents the number of choices in each decision variable, n, the number of decision variables of one channel and m, the number of distinct channels. Due to this exponential nature of the decision-making space, and interactions between decision variables, the addition of an e-channel would make the decision-making processes very complicated and time-consuming. Simply replicating design parameters of one channel to other channels would make the overall performance sub-optimal.

Due to the complex decision-making processes involved in e-channel planning, a prototype e-channel planning support system was developed to facilitate decision-making. One of the features of the planning support system is a simulation capability for organizational performance evaluation. This planning support system is used to investigate the impact of e-channel structures on organizational performance. We attempt to answer two questions: (1) what is the impact of an e-channel structure on profit? and (2) what is the relationship between the growth of a multi-channel customer base and pricing strategy?

Traditional retailers have two options with an e-channel structure: an on-line division and a spin-off. Some arguments for creating an on-line division include: use of complementary assets, sharing of technological resources and ease of channel coordination. Office Depot, an office product supplier, pursued the on-line division strategy and achieved the largest on-line sales among three major office product suppliers. On the other hand, advantages of spin-off strategies include new culture, independent organizational structure, and flexible reward systems that are suitable to the e-commerce environment. To counter the e-commerce initiative of Office Depot, Staples launched Staples.com as a spin-off company. However, Staples' spin-off company was not successful, and eventually merged into Staples in 2001.

To answer the previously mentioned questions, simulation data were generated randomly. Base data ranges were based on price data collected from office supply physical stores and e-channels. To test hypotheses, three decision variables were considered: an e-channel organizational structure (*2*: an on-line division and a spin-off); the proportion of multi-channel customers (*3*: 40%, 60% and 80% of the customers are multi-channel customers who purchase products through e-channels and physical stores); and the number of products sold in the channel (*3*: one, two and three different products are offered to customers). Data generation was based on a full-factorial design. For each parameter value combination, 30 replications were run. All together, 540 simulation runs (i.e., 2*3*3*30) were performed (Note: details of the computational simulation system, research design and statistical analysis are available upon readers' request). From the simulation results, we made the following conclusions at a significance level of 0.05.

1. *The overall profit of an on-line division structure is greater than a spin-off structure.*

Our results show that, across all proportions of multi-channel customers, an on-line division structure performs significantly better than a spin-off structure in terms of profit. On average, the profit of an on-line division structure is 48% higher than a spin-off structure. The result is attributable to the fact that the on-line division accurately reflects customers' preferences and charges high prices to customers who are willing to pay higher prices. On the other hand, the spin-off structure charged lower average prices to customers, due to the lack of price coordination among channels. This result may explain the recent retail industry's trend of merging an e-channel under a parent company. Since the e-commerce shakeout in 2000, many spin-off e-retailers, such as Walmart.com, Staples.com, and BlueLight.com, failed to offer IPOs, and later merged their e-commerce organizations into the parent retail organizations. On the other hand, Target and Office Depot created an on-line-division under the existing organization and sustained the on-line division by successfully leveraging the existing internal resources and price coordination.

2. *As the proportion of multi-channel customers increases, the product price decreases for both an on-line division structure and a spin-off structure.*

While the prediction on the future growth rate of e-commerce varies by researcher, it is clear that e-commerce will be one of the fastest growing market areas. One of the main reasons that people shop on-line is the low price. E-commerce encourages a competitive environment by reducing entry barriers to the Internet startups and by lowering product search costs. E-commerce gives many customers a considerable advantage over retailers by allowing them to search for the lowest prices. Most e-commerce customers also check prices at nearby physical stores. Due to the competitive pressures created by e-commerce customers, as the proportion of multi-channel customers increases, the product price decreases for both an on-line division structure and a spin-off structure. Recent failures of many e-retailers suggest that the zero/low margin e-commerce model will be not sustainable in the long run, due to price competition and lack of financial support from investors. To mitigate the price competition and expand customer bases, e-retailers need to seriously examine what types of e-channel structures and coordination mechanisms are the best for them.

CONCLUSION

While stand-alone e-retailers face increasingly high entry barriers, many traditional retailers successfully launch hybrid e-retailer businesses, leveraging extensive distribution networks, established brand images and strong financial resources. Traditional retailers also leverage existing channels, such as physical stores, call centers and catalogues, for cross-channel collaboration. Let's suppose customers place an order through an e-channel. Instead of receiving products through regular mail, customers may pick up products at the nearest physical store. Customers may also be allowed to return products and get a refund at the physical store. This kind of cross-channel collaboration increases customer satisfaction and helps multiple channels keep their own loyal customers as well as expand customer bases.

The introduction of an e-channel requires the review of a retailer's major business interests, involving a number of stakeholders with different goals, customer bases, cultures and operational constraints. According to a recent survey (Melymuka, 2000), managing an e-channel is one of the most important tasks for marketing managers. Unfortunately, many managers are still unclear about e-channel strategies and lack core e-channel knowledge needed to analyze business environments and benchmark performances, develop strategies, and evaluate alternative e-channel solutions.

Based on a number of e-commerce case studies, this paper presented a comprehensive framework that can be used to develop an e-channel for hybrid e-retailers who are attempting to capitalize on the existing channel infrastructure. The proposed e-channel development framework consists of five step-by-step phases: (1) strategic analysis; (2) e-channel planning; (3) e-channel system design; (4) e-channel system development; and (5) performance evaluation and refinement. The proposed framework is designed to help managers to effectively manage a complex development process.

A cross-channel collaboration provides a unified storefront to customers and helps multiple channels keep their own loyal customers as well as expand their customer bases through cross-channel marketing. The development of a cross-channel coordination mechanism is complicated, due to inherent channel conflicts. The main challenge to hybrid e-retailers is coordinating the product offerings and prices across multiple channels. Channel conflicts are heightened when internal cannibalization of sales occurs among channels. The success of a multi-channel coordination, in large part, depends on the choice of a multi-channel organization design. Retailers have two basic options for an e-channel structure: an on-line division and a spin-off. Some retailers, such as K-Mart, Wal-Mart, CVS, and Staples, experimented with both e-channel structures. Other retailers, such as Target, Office Depot, and OfficeMax, started with one structure and have sustained the same structure. The advantages of an on-line division include the use of complementary assets, sharing of technological resources, and ease of channel coordination. On the other hand, the advantages of spin-off strategies include a new culture, independent organizational structure and flexible reward systems that are suitable for an e-commerce environment.

When traditional retailers establish e-channels, they need to carefully analyze the impact of the e-channel structure on organizational performance and develop a company-wide channel integration program. We performed a simulation experiment to answer the following questions: (1) what is the impact of an e-channel structure on profit? and (2) what is the relationship between the growth of multi-channel customers and price? Our results show that (1) the overall profit of an on-line division structure is greater than a spin-off structure and (2) as the proportion of multi-channel customers increases, the product price decreases for both an on-line division structure and a spin-off structure. These results may explain why many large retailers, such Wal-Mart, Staples and K-Mart, initially created spin-off e-commerce companies and later merged them under parent companies. Recent failures of many e-retailers highlight the importance of profitable e-commerce models. Zero/low margin e-commerce models may not be sustainable in the long run, due to price competition and lack of financial support from investors. To counter price competition and attract customers, e-retailers need to pursue a variety of strategies, including product differentiation and switching cost strategies.

REFERENCES

Bellman, S., Lohse, G.L. & Johnson, E.J. (1999). Predictors of on-line buying behavior. *Communications of the ACM, 42*(2), pp. 32-38.

Capozzoli, E.A. & True, S.L. (2001). An e-commerce systems integration framework. *Southern Business Review, 26* (2), pp. 27-32.

Garfinkel, S. & Spafford, E.H. (1997). *Web Security and Commerce*. Sebastopol, CA: O'Reilly and Associates.

Gallaugher, J.M. (2002). E-commerce and undulating distribution channel. *Communications of the ACM, 45* (7), pp. 89-95.

Joshi, J.B.D., Aref, W.G., Ghafoor, A. & Spafford, E.H. (2001). Security models for Web-based applications. *Communications of the ACM, 44* (1), pp. 38-44.

Lohse, G.L. & Spiller, P. (1998). Electronic shopping: How do customer interfaces produce sales on the Internet? *Communications of the ACM, 41* (7), pp. 81-87.

Melymuka, K. (2000). Survey finds companies lack e-commerce blueprint. *ComputerWorld, 34* (16), p. 38.

Porter, M.E. (1985). *Competitiveness Advantage*. New York: Free Press.

Prior, M. (2002). 12 hot issues facing mass retailing. *DSN Retailing Today, 41* (10), p. 47.

Shern, S. (2000). Retailing in the multi-channel age. *Chain Store Age, 76* (5), pp. 1-3.

Chapter XII

Organization, Strategy and Business Value of Electronic Commerce: The Importance of Complementarities

Ada Scupola
Roskilde University, Denmark

ABSTRACT

Many corporations are reluctant to adopt electronic commerce due to uncertainty in its profitability and business value. This chapter introduces a business value complementarity model of electronic commerce. The model relates high level performance measures, such as business value, first to intermediate performance measures, such as value chain and company strategy, and then to the e-business performance drivers as business processes and complementary technologies. The model argues that complementarities among the different activities of the value chain, corresponding business processes and supporting technologies should be explored to reach a better fit among strategy, business model and technology investments when entering the electronic commerce field. The exploration of such complementarities should lead to investments in electronic commerce systems that best support the company strategy, thus minimizing failures. From a practical point of view, managers could use this framework as a methodology to increase the business value of electronic commerce to a corporation.

INTRODUCTION

The Internet economy is becoming an integral part of many countries' economies, creating new jobs, giving rise to new companies like the dot coms and transforming traditional jobs and traditional companies. The Internet economy is made up of a large collection of global- (IP) based networks, applications, electronic markets, producers, consumers and intermediaries (Barua et al., 1999a). A recent study conducted by the University of Texas' Center for Research in Electronic Commerce (Barua et al., 1999a) estimates the U.S. Internet economy to have exceeded $500 billion in revenue in 1999. In the first half of 2000, the U.S. Internet economy has supported approximately 3.7 million workers, of which 28% are in information technology, while 33% in sales and marketing, which is also the job function generating most Internet-related employment. The study finds that the Internet is increasingly becoming part of the basic business model for many companies, laying the groundwork for even more growth.

There are many definitions of electronic commerce (e.g., Wigand, 1997; Zwass, 1996). In this chapter, a definition by Kalakota and Whinston (1996) is adopted, where e-commerce is "the buying and selling of information, products and services via computer networks today and in the future via any one of the myriad of networks that make up the "Information Superhighway (I-way)" (p.1). In the analysis, I distinguish between physical and digital products. A digital product is defined as a product whose complete value chain can be implemented with the use of electronic networks, for example, it can be produced and distributed electronically, and be paid for over digital networks. Examples of digital products are software, news, journal articles, etc., and companies selling these products are usually Internet-based "digital dot coms" such as Yahoo and America Online. On the contrary, a physical product cannot be distributed over electronic networks (e.g., a book, CDs, toys, etc.). These products can be sold on the Internet by "physical dot coms," but they are shipped to the consumers. The corporations using electronic commerce are distinguished into "bricks and mortar" companies, hybrid "clicks and mortar" companies (such as Amazon.com) and pure dot coms. For further discussion of different types of dot coms, please refer to Barua and Mukhopadhyay (2000).

Electronic commerce can be used to reengineer a corporation's business processes for the electronic marketplace (Kalakota & Whinston, 1996). The broad goals of re-engineering and e-commerce are remarkably similar: reduced costs, lower product cycle times, faster customer response, and improved service quality. Reengineering for electronic commerce is defined here as the redesign (or design) of a corporation's business processes (or part of them) in order to take place on the Internet. We are witnessing the virtualization of value-chain segments as business processes can be moved into the virtual, informational value chains (Zwass, 1996; Rayport & Sviokla, 1995).

Many studies from the early days of deployment of information technology (IT) in organizations have struggled to measure the business value and profitability of information technology (IT) (Barua & Mukhopadhyay, 2000). Many of these studies have showed that productivity gains are small or not existent, and that the effects of information technology and electronic commerce often have to be looked upon from a competitive advantage point of view (Cronin, 1995; Porter & Miller, 1985). Therefore, many companies have difficulty seeing the return on investment (ROI) necessary to justify expenditures in information technology (IT) and electronic commerce technologies, and are skeptical of the business value of e-commerce.

This chapter argues that to increase the business value of electronic commerce to a corporation, it is important to take a radical approach by reconsidering the business processes that have to go on-line. It is also important to shift the focus from whether electronic commerce creates value to a company to *how* to create value. The research question that this chapter attempts to answer is: How can the business value of electronic commerce to a corporation be optimized? To answer this question, this chapter develops a theoretical framework based on existing literature on electronic commerce, strategies and complementarity. Corporation's managers could use this framework to maximize the business value of electronic commerce. Companies, especially those dealing with digital products or services that can be transmitted fully over the Internet, need to investigate which business processes to reengineer for electronic commerce, and how to do it in order to be successful and gain a competitive advantage.

The chapter is structured as follows: the next section presents a short review of the literature on profitability and business value of IT, while the following section gives some examples of sources of business value of electronic commerce. The section "Theoretical Framework" briefly touches upon the theories that are the basis for the main model of the chapter, presented in the section titled "A Business Value Complementarity Model of Electronic Commerce." Finally, the section "What To Expect In The Future?" presents some future trends, while the last section summarizes the findings of the chapter and gives some concluding remarks.

BACKGROUND

Since the early days of IT use in commercial organizations, researchers and professionals have struggled with the problem of measuring the bottom line contribution of IT investments. Barua and Mukhopadhyay (2000) present a thorough literature review of the main research studies in IT value assessment. They distinguish six main areas of research: information economics-based studies, early IT impact studies, production economics studies that did not find positive impacts, microeconomics studies that found positive impacts of IT, business value studies, and studies involving complementarity between IT and non-IT factors. The information economics-based studies date back to the 1960s and, though relevant to the economic contribution of IT investments, they mainly focus on the changes in information due to IT use and their impact on the single decision-maker. Therefore, while the information economics approach is theoretically sound and rigorous, its unit of analysis, which is either the individual or team decisions, makes it difficult to obtain meaningful and insightful results in broader organizational contexts (Barua & Mukhopadhyay, 2000).

In the early 1980s, a stream of research emerged that focused on assessing the contribution of IT investments to performance measures such as return on investment and market share. According to Barua and Mukhopadhyay (2000), one of the most influential studies on IT productivity is the one conducted by Loveman in 1988. The results obtained by Loveman led him to conclude that non-IT factors of production should be preferred to IT for additional investments. The majority of these studies did not find much positive correlation between IT investments and firm performance metrics, up to the early 1990s (e.g., Roach, 1988 & 1989). The lack of correlation between IT investments and productivity made Roach (1988, 1989) coin the term, "IT productivity paradox."

In the 1990s, the research on measuring the economic and performance contributions can be divided into two main streams: one based on production economics and one focused on "process-oriented" models of IT value creation. The IT production studies based on production economics hypothesize that IT investments are inputs to a firm's production function. These studies (e.g., Brynjolfsson & Hitt, 1993, 1996) finally started finding signs of productivity gains from IT. For example, a study by Hitt and Brynjolfsson (1996) identifies three sources of IT value to a corporation: productivity, consumer value and business profitability. The study shows that information technology contributes to increases in the productivity and consumer value, but not business profitability. Simultaneously, process-oriented studies started hypothesizing relationships between IT and other input factors to performance measures at various levels of aggregation. These studies (e.g., Kauffman & Kriebel, 1988) have laid the foundation of the business value approach to the impact of IT on firm performance. This approach, contrary to the production function-based approach, might have the explanatory power to point out where and how IT impacts are created, and where management should act to increase the payoff from IT investments. These explanations are more difficult to find with production function-based approaches, because they operate at a very high level of aggregation, thus making it difficult to distinguish between different types of IT investments and their impacts on specific areas of business. As Barua and Mukhopadhyay (2000) argue, after having dispelled the productivity paradox, new refinements to existing approaches need to be made to measure the contribution of IT to business performance. A number of studies (e.g., Barua et al., 1996; Barua & Mukhopadhyay, 2000; Barua et al., 2002) are emerging that point to the use of complementarity theory to provide a deeper understanding of the interactions between IT and other organizational factors. These studies also show how the way in which these interactions take place can determine the contribution of IT to business performance.

Production economics and business value approaches mostly have ignored the synergy between IT and other related factors, such as the level of fit with business strategies, employee empowerment and team orientation of business processes. Barua et al. (1996), in a formal theory of reengineering, note that isolated changes that do not take into consideration complementary factors, such as business processes and incentive systems, are unlikely to be productive. Barua and Mukhopadhyay (2000) present a generalized business value complementarity model. The basic idea of their business value complementarity model (BVC) suggests that investments in IT should be first related to intermediate performance measures, such as time to market, customer service, response time and extent of product mass customization, to be able to see any positive results from such investments. In a second moment, the intermediate performance measures can be related to high-level performance metrics, such as profitability, return on investment (ROI), market share, etc. According to Barua and Mukhopadhyay (2000), the focal point of a business value complementarity model is the complementarity that potentially exists at each level of the model. However, until recently, the IT business value research was assuming positive results and often was targeting IT applications that were vendor- or technology-specific.

The advent of the Internet, based on open standards and a universal Web browser, raises the question of whether investing more in Internet technology leads to a better financial performance in electronic commerce. Barua et al. (1999b) tests the hypothesis that IT capital contributes more to the performance of digital dot coms than to that of physical dot coms. The results of the study, obtained by using data on 199 publicly traded Internet-based companies, emphasized the need for rapid digitization of both internal and external business processes in physical product companies; the results showed that IT capital contributed significantly to each of the metrics of the digital dot coms. These results imply that the higher the digitization level of a company, the higher the benefits that the company can receive by investing in Internet-based technologies. This calls for more attention to the specific business processes that have to be reengineered for online commerce, and the way they should support the company strategy.

SOURCES OF BUSINESS VALUE OF ELECTRONIC COMMERCE

As already said, many companies do not see the return on investment (ROI) necessary to justify the costs of engaging into e-commerce. Many are skeptical of the value of e-commerce for a company and, often, the Internet is described as the gold rush of the 1800's and e-commerce as a fad (Bloch et al., 1996).

Bloch et al. (1996) argue that the business drivers are an important component of the spreading of electronic commerce. They develop a business value framework identifying 10 sources of value of electronic commerce for commercial organizations, summarized in Table 1. Bloch et al. (1996) basically say that electronic commerce can contribute to the business value of a corporation in three main ways: by improving the business processes, by transforming the organization and by redefining the business models. Electronic commerce can contribute to the business value of a corporation through differentiation strategies that can be implemented with new ways of promoting the product (on-line promotion); new sales channels; by enhancing customer service; by consolidating the brand image with electronic commerce systems; and by lowering costs through on-line delivery.

Electronic commerce can contribute to the organizational transformation due to new customer relations, new technologies and organizational learning. Finally, electronic commerce can redefine a corporation structure by offering the possibility of exploring new business models and new product capabilities, such as digital products. Bloch et al. (1996) identify some sources of business value of electronic commerce to a corporation, but they do not identify or mention how changes should take place in the corporation when organizing for electronic commerce. In order to give some ideas about how to maximize the business value of electronic commerce, this chapter develops a business value complementarity framework of electronic commerce.

THEORETICAL FRAMEWORK

The business value complementarity (BVC) model developed in this chapter is based on the value chain (Porter, 1980), the theory of business value complementarity

Table 1: The sources of the business value of electronic commerce (adapted from Bloch et al., 1996)

The Organization Source of Business Value from Electronic Commerce	
Improve it	Product Promotion New Sales Channel Direct Savings Time to Market Customer Service Brand Image
Transform it	Technological and Organizational Learning Customer Relations
Redefine it	New Product Capabilities New Business Models

(Barua, 1996; Barua & Mukhopadhyay, 2000; Barua et al., 2002) and the concept of strategy (Porter, 1982). The value chain separates the activities of a firm into two main categories: primary and secondary. Primary activities are those involved in the physical creation of the product, its marketing and delivery to buyers, and its support and servicing after sale; secondary activities provide the input and infrastructure that allow the primary activities to take place. Information technology affects all the activities (Porter & Miller, 1985). Here, the value chain framework is used to analyze how electronic commerce technologies can transform each of the primary activities of the value chain from the marketplace to the marketspace, add value to such activities and give the firm a competitive advantage.

The business value complementarity theory is based on the notion of complementarity in economics. Milgrom and Roberts (1990) say that "several activities are mutually complementary if doing more of any one activity increases (or at least does not decrease) the marginal profitability of each other activity in the group." Complementarities among activities imply mutual relationships and dependence among various activities whose exploration can lead to higher profitability. Milgrom and Roberts (1990) have applied the complementarity theory to the fields of manufacturing, management and strategy. In the Management Information Systems (MIS) field, Barua et al. (1996) develops a multi-layered business value complementarity model of reengineering. In this model it is argued that to maximize organizational payoff, complementary factors such as technology, decision authority, business processes and incentives must all be changed in a coordinated fashion — in the right direction by the right magnitude — to move toward an optimal ideal design configuration. According to this theory, it is important to explore complementarities among organizational and technology variables in implementing new business processes or in designing new business models and to avoid considering information technology variables only. The failure to explore such complementarities and to consider all the variables at once is the reason why, Barua et al. (1996) argues, many re-engineering projects fail.

This chapter focuses on the complementarity among the corporation strategy, the primary activities of the value chain, corresponding business processes and supporting technologies in order to maximize the business value that electronic commerce can bring to a corporation.

A BUSINESS VALUE COMPLEMENTARITY MODEL OF ELECTRONIC COMMERCE

In this section a business value complementarity model of electronic commerce is presented (Figure 1), based on previous business value complementarity theory and the value chain. In this chapter, the issues of complementarity between the different variables of the model are addressed in a qualitative way.

Understanding the Variables of the Model

As suggested in the model developed by Barua et al. (2002), also in this chapter, e-commerce performance (the business value) is measured by the same financial measures as the "non- Internet economy," for example, profitability. Furthermore to monitor and influence the business value of e-commerce to a corporation, a manager has a set of e-drivers that he/she can use to influence the intermediate variables of the model, that is, organizational strategy and primary activities of the value chain. The business processes and corresponding complementary technologies are the e-business drivers of the model. According to Barua et al. (2002), "drivers are the actionable part of the model. Managers invest resources to change the levels of these drivers to achieve primary impacts on operational excellence and higher order impacts on financial performance." Furthermore, "these drivers are like knobs that have to be turned in unison by management in the right directions by the right amount" (Barua et al., 2002).

Business value of electronic commerce

The high level (performance) variable of the model of Figure 1 is business value of electronic commerce. Examples of performance measures of the business value of electronic commerce can be: 1) profitability, that is whether electronic commerce contributes to an increase in the profitability of the corporation; and 2) competitive advantage that could be measured as an increase in market share, shareholder value or customer satisfaction. The main objective of the model is to make the business value of electronic commerce as close to optimal as possible in terms of one of the performance measures. This can be done by exploring complementarities among the dependent variables of the model: the company strategy, the activities of the value chain, the corresponding business processes, and the technologies available to transform these activities and processes for the marketspace.

Strategy

Strategic management theories can be classified according to their focus, which can be either on the external environment or on the internal strengths and weaknesses of the corporation. There are many definitions of strategy. The best known is the design view, mainly attributed to Porter (1980), and the process view, mainly attributed to Mintzberg (1994). In the design view, strategy is seen as a planning, rational process through which the company chooses a certain mode of development, among all the possible ones, according to its market position, and maintains that direction through a well-defined period. The process view sees strategy as a process and, even though the strategy goals might have been well defined at the beginning of the strategy implementation period, they might change on the way, giving rise to an emergent strategy. As a result, the realized strategy might be different then the original intended strategy (Mintzberg, 1994). In this chapter, Porter's (1982) classification of strategy as cost leadership, differentiation or focus is adopted.

Primary activities of the virtual value chain

The value chain of the market space or virtual value chain can be redefined as the use of computer and Internet-based technologies to re-organize or reengineer the physical value chain completely on-line. On-line production, on-line distribution, on-line marketing, on-line sales and on-line customer support are the primary activities of the value chain re-engineered or redefined for the electronic market place (Scupola, 1999).

Business processes corresponding to each activity of the value chain

The business processes corresponding to each activity of the value chain are the specific processes into which each primary activity of the virtual value chain can be decomposed. The value chain activities and business processes of on-line marketing, on-line sales and on-line customer service are the same for digital and non-digital products. On-line distribution and on-line production can be applied to digital products only. Furthermore, the business processes of on-line production are specific to the product in question (e.g., software, journals, music, etc.). For example, in electronic publishing, the business processes of on-line distribution are electronic search, electronic selection, electronic retrieval and electronic transmission of the product over digital networks. The business processes corresponding to on-line marketing are on-line advertising, on-line market research, on-line promotions and public relations, on-line pricing models or pricing models for on-line business. Those corresponding to on-line sales are information gathering/recognizing a need, negotiation/search for solutions and settlement/making a purchase. On-line customer

Figure 1: Business value complementarity model of electronic commerce

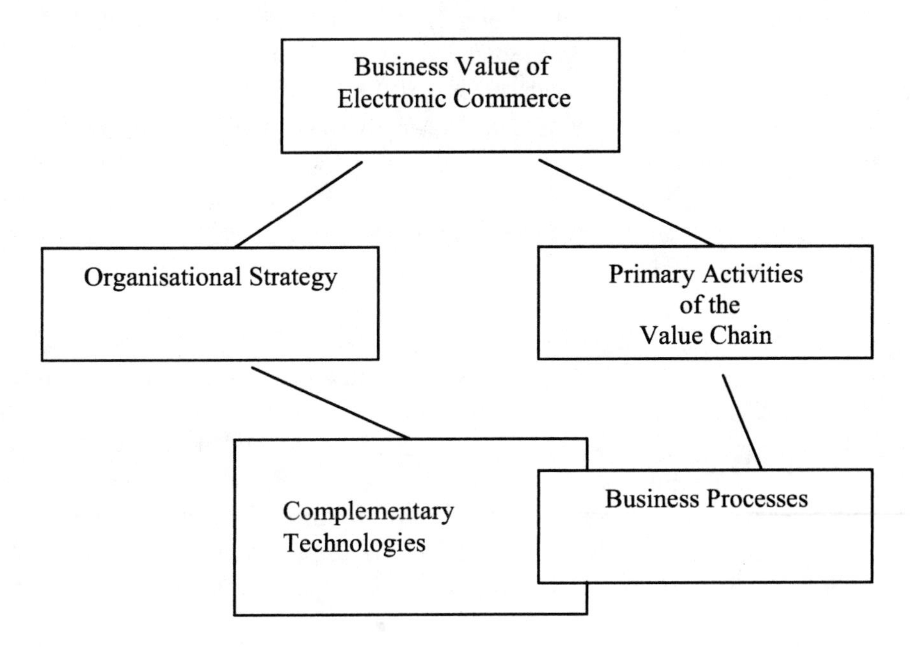

support processes could be defined as customer inquiries and answers to customers (Scupola, 1999).

Complementary technologies

Complementary technologies are those technologies that can be used for the transformation of the business processes from the marketplace to the marketspace. Scupola (1999) divides them into three groups: networking and communication technology, database technology and Data Base Management Systems (DBMS), and application software. Each group includes different technology classes, even though some technology classes might belong to more than one group. For example, networking and communication technologies might include Internet, the World Wide Web, client/server computing and Web-database integration, while database technology and DBMS might include repositories, object-oriented databases, inverted file and relational databases, etc. Each activity of the value chain has, as complementary technologies, a subset of all the technology classes.

Explanation of the Model

In this paper, it is hypothesized that complementarity exists among the variables of the same level and different levels of the model of Figure 1. It is furthermore hypothesized that the exploration of complementarities and possible synergies among the company strategy, the primary activities of the value chain, corresponding business processes and supporting technologies in the business value complementarity model of Figure.1, should: 1) maximize the business value of electronic commerce to a corporation; and 2) lead to a better fit between the overall organizational strategy, the business processes that have to be transformed for the on-line market place, and the ICT's system that should be designed and implemented to support these strategies. The exploration of complementarities, it is hypothesized, also can help businesses avoid the following: investing in an information system that could not be used at a later point, if new e-business processes should be added to the system, and implementing a business model that does not correspond to the corporation's strategy. It is argued that, to succeed in electronic commerce, it is important to reengineer the parts of the value chain and the corresponding business processes relevant to the product in question and the company strategy.

Complementarity at strategy level

It is hypothesized that the strategy or combination of strategies a company wants to pursue is relevant for the primary activities of the value chain, and the corresponding business processes that have to be implemented on-line. The strategy is also relevant to the classes of technologies that have to be chosen to enter the electronic market place. A company should therefore first decide which of the three fundamental strategies (cost leadership, differentiation and focus), or any combination of them (Porter, 1980, 1982), to pursue, and then explore how electronic commerce can support such strategy. For example, a company can use electronic commerce to implement a cost leadership strategy, or to become the low-cost producer in the industry (Bloch et al., 1996). Electronic commerce can contribute, in fact, to lower costs by promoting the products directly to the customer (thus saving promotion costs), sending the product over the Internet (thus saving distribution costs), and by lowering marketing costs through on-line marketing and one-to-one marketing. Electronic commerce can be used to create new substitute products, enhance some product attributes, or give different customized versions of the same product, thus supporting a product differentiation strategy. Companies can use e-commerce also to implement a focus strategy. For instance electronic commerce provides the possibility of offering the customer highly tailored, one-to-one marketing campaigns, or products highly customized to the taste, needs and preferences of the single user (Bloch et al., 1996). Once the strategy has been decided, it is important to explore complementarities between the strategy and the value chain activities, in order to implement on-line all those activities that would support an optimal implementation of the strategy chosen.

Complementarity at value chain activity level

The number of primary activities and corresponding business processes that should be transformed for the marketspace depends on the company's type of product and strategy. It is argued, however, that it is important to take into consideration complementarities among the different activities of the value chain when reengineering for electronic commerce. It is here hypothesized that the more activities of the value chain are simultaneously conducted on-line, the more likely it is that the business value of electronic commerce will be optimized. This chapter suggests that the adoption of a holistic approach in redesigning the primary activities of the value chain for electronic commerce would be a more successful strategy than reengineering only one or some at a time. This is due to potential complementarities among the different activities, which leads to a better performance in one if the others are reengineered also for on-line commerce. This hypothesis is supported partially by a study conducted by Barua et al. (1999b). Their study hypothesizes that IT investments contribute more to various output measures (e.g., revenue, revenue per employee, gross margin, etc.) for digital dot coms than for physical dot coms. The analysis showed that IT capital (computer hardware, software and networking equipment) contributed substantially to each of the metrics for the digital dot coms, while the contribution for the physical dot coms was uniformly insignificant. Teece (1988) provides a similar argument by showing the importance of exploring complementary relationships among different assets when a company wants to start marketing an innovation in the marketplace.

Complementarity at business process level

Theoretically, each business process corresponding to any activity of the value chain could be reorganized for e-commerce independent of the others. It is here hypothesized that the exploration of complementary relationships among these processes, and the simultaneous reengineering of all the complementary processes of a particular activity for e-commerce, would lead to a higher business value than if only some of the processes were reorganized on-line. In distribution, for example, a company that provides for search, retrieval, selection and physical transmission of a product over a network will be better off than a company that delivers through conventional distribution channels only, even though it allows customers to search and order the product on the Internet. In marketing, the complete implementation of an on-line marketing program, from advertising to market research, to promotions, public relations and ad-hoc on-line pricing models, will increase the net benefit to the company, compared to a program that only has reengineered some of these processes for electronic commerce. For example, market research data collected at the company Web site, combined with data from more conventional market research channels, would contribute to a more effective on-line advertising program than data gathered only in the marketplace. This is because the data collected on-line can give more accurate customer profiles than data gathered only in the marketplace with conventional market research tools. In on-line sales, providing for electronic payment/settlement, in addition to information searching and gathering of the company's product selection, would make the shopping process easier and decrease the chance that the customer closes the Internet connection without having downloaded or bought the product on-line (Scupola, 1999).

Complementarity between business processes and supporting technologies

In the design phase, it is important to consider potential complementarities between the business processes that have to be redesigned for on-line commerce and the supporting technologies. It is here hypothesized that the exploration of this complementarity should lead to an optimal system design that also offers possibilities for further expansion, if other on-line business processes should be added in the future. For example, the faster and more advanced the search engine is – and the better built the user interface and repository systems, the more accurate and efficient the electronic search of the company's information.

Consequently, it is important to decide on the database system by taking into consideration the structure and nature of the product to be stored and the level of granularity desired (in turn, depending on the company strategy).

Complementarity at technology level

At this level, it is hypothesized that the exploration of complementarities between the different technologies used to implement the system for electronic commerce could lead to a more robust and flexible computer system than a system built without the exploration of complementary relationships between the different component technologies. For example, end user interfaces and repositories are complementary technologies in the sense that the better designed the repository system, the simpler the user interface can be. In on-line distribution, a system with both advanced search engines and user-friendly search forms would be much more effective than a system where the user interfaces are not so friendly or the search engine is not so powerful. Finally, the total value of a system using both a repository supporting a very high level of granularity and a sophisticated micro-payment system would be much higher than that of a system not providing for micro-payments, where the user has to use the credit card, even for small transactions. Generally, advancements in security, networking technologies and software developments are required today in order to offer effective Internet shopping, as well as good and informative home pages from the side of the company (Jarvanpaa & Todd, 1996, 1997).

WHAT TO EXPECT IN THE FUTURE?

The studies on IT productivity and business value conducted throughout the last decade have showed positive impacts of IT investments on firms' productivity, both with respect to labor and other non-IT capital used by organizations (Barua & Mukhopadhyay, 2000). However, as already said at the beginning of the chapter, Internet-based technologies, with their open standards and wide applicability, raise again the issue of profitability and business value of investing in such technologies. Furthermore, as Barua et al. (1999b) states, the fact that the Internet is giving rise to a "new economy" raises a number of questions, among which: how productive are the players in this new economy? Does e-commerce increase the profitability and business value of brick and mortars and hybrid click and mortars companies? For dot coms, do more investments in Internet commerce technologies necessarily lead to a better performance of the company? If all the companies have equal access to Internet-based technologies, what are the factors that differentiate their performance in e-commerce?

As discussed in the previous section, recent literature investigating the business value and profitability of electronic commerce is focusing on the exploration of complementary relationships between electronic commerce technologies and other factors in order to see positive returns from investments in these technologies. For example Barua et al. (2002) develops a framework of electronic commerce business value that identifies linkages between performance drivers such as Internet applications, processes and electronic business readiness of customers and suppliers and operational excellence and financial metrics. They argue that "firms engaged in electronic business transformation must make synergistic investments and commit resources not only in information technology, but also must align processes and customer and supplier readiness to maximize the benefits."

Similarly, an empirical investigation of the business value of e-commerce in small, medium and large companies across Europe and the USA, conducted by Barua et al. (2000) at the Center for Research in Electronic Commerce, identifies some key e-drivers and some critical links among them. These key e-drivers are: system integration; customer orientation of IT (informational and transactional);

supplier orientation of IT, informational (quality, supply continuity and relationship management) and transactional; internal orientation of IT; customer related processes; supplier related processes; customer e-business readiness; and supplier e-business readiness. "IT–related drivers include the level of integration between online and back office applications, and informational and transactional capabilities of online systems. Business- process drivers include standards and procedures for interactions with customers and trading partners. And an external environment conducive to e-business success on the readiness of customers and suppliers" (Barua et al., 2000). The study also shows that high performance companies have invested more effort and resources in these e-business drivers than companies who have not benefited from e-business.

Barua and Mukhopadhyay (2000) mention a study conducted in 1998 by Grenci, Barua and Whinston that develops and tests a complementary-based theory and a business value complementarity model of the impact of technology-enabled customization in the financial service sector. This study finds that complementarity indeed exists between the business strategy of customizing financial services, such as mortgage loan processing, and the sales process characterized by the number of alternatives generated and the speed of handling.

To conclude, these studies show that ignoring complementarities in research on business value measurement might lead to misleading results. On the other hand, from a managerial point of view, the non-exploration of complementary relationships between IT and related factors, such as strategy, business processes, business models, incentives, etc. might lead to failure of investments in sophisticated electronic commerce systems and ventures. These considerations point to the need for more empirical, as well as normative, prescriptive research on complementarity and business value of IT in general, and electronic commerce technologies in particular.

CONCLUSIONS

Many companies are very skeptical about investing in electronic commerce technologies due to the lack of profitability, (or at least the difficulty in showing positive return on IT investments) that, until now, often has characterized the investments in IT and electronic commerce. This chapter has presented a framework that can be used as a methodology to analyze organizational strategies and technology choices in reengineering for electronic commerce. The chapter has argued that companies should explore the potential complementarities existing among strategy, value chain activities, and business processes and supporting technologies when entering the field of electronic commerce. This should lead to investments in electronic commerce systems that best support the company strategy, thus minimizing failures. In organizing a business for electronic commerce, a number of choices have to be made at every step of the process. It is important to decide what strategy to adopt; which value chain activities and corresponding business processes have to be organized on-line; and what electronic commerce technologies should be used in order to implement such a strategy successfully. Finally, how to acquire the competencies and technologies necessary to establish an online presence, and therefore formulate a technology strategy, must be determined. Such strategy should include not only what kind of computer hardware, software and networks to use, but also decisions regarding how to acquire such resources and competencies. Should the system be developed in-house or bought on the market? Should the company outsource the resources necessary to the building, operation and maintenance of the system? Should the company form alliances or partnerships with corporations that have complementary technology assets and skills (Teece, 1988)? During this process, it is important to explore potential complementarities between the different variables. The business value complementarity model of electronic commerce presented in this chapter, however, is only theoretical, and limited to the value chain of a single company. Further research is necessary to empirically test the model in specific industries (by conducting surveys) and corporations or departments within a corporation (by qualitative case studies focusing on complementarities or synergies between different variables). Further research is also necessary to investigate the relationship between the digitization of a single company value chain and the digitization of the value chains of all companies involved in an activity or transaction.

REFERENCES

Barua, A. & Mukhopadhyay, T. (2000). Information technology and business performance: Past, present, and future. In R. Zmud (Eds.), *Framing the Domains of IT Management, Projecting the Future Through the Past.*

Barua, A., Konana, P., Whinston, A. B. & Yin, F. (2002). Managing e-business transformation: Opportunities and value assessment. Forthcoming in *Sloan Management Review*. Retrieved June 19, 2002, from the World Wide Web: http://www.utexas.edu.

Barua, A., Konana, P., Whinston, A. B. & Yin, F. (2000). Making e-business pay: Eight key drivers for operational success. *IT Pro*. IEEE Publisher, November-December.

Barua A., Lee, S. C.H. & Whinston, A. B. (1996). The calculus of reengineering. *Information Systems Research, 7* (4), 409-428.

Barua, A., Pinnell, J., Shutter, J., Wilson, B. & Whinston, A.B. (1999a). The Internet economy indicators; Part II. Retrieved June 19, 2002, from the World Wide Web: http://www.internetindicators.com.

Barua, A., Whinston, A.B. & Yin, F. (1999b). *Not All Dot Coms are Created Equal: An Explanatory Investigation of the Productivity of Internet-Based Companies*. (Working Paper, University of Texas at Austin: Center for Research in Electronic Commerce). Retrieved June 19, 2002, from the World Wide Web: http://www.utexas.edu.

Bloch, M., Pigneur, Y. & Segev, A. (1996). *On the Road of Electronic Commerce: A Business Value Framework.* (Working Paper 96-WP-1013). Berkeley: University of California at Berkeley.

Brynjolfsson, E. & Hitt, L.M. (1993). Information technology and the productivity paradox: review and assessment. *Communications of the ACM, 35* (December), pp. 66-77.

Brynjolfsson, E. & Hitt, L.M. (1996). Paradox lost? Firm-level evidence of the returns to information systems spending. *Management Science, 42*, pp. 541-558.

Cronin, M. (1995). *Doing More Business on the Internet: How the Electronic Highway is Transforming American Companies*. New York.

Jarvanpaa, S. & Todd, P.A. (1996/1997). Consumer reactions to electronic shopping on the World Wide Web. *International Journal of Electronic Commerce, 1*(2), 60-88.

Kalakota, R. & Whinston, A. B. (1996). *Frontiers of Electronic Commerce.* Addison-Wesley Publishing Company, Inc.

Kauffman, R.J. & Kriebel, C.H. (1988). Modeling and measuring the business value of information technologies. In P.A. Strassman, P. Berger, E.B. Swanson, C.H. Kriebel & R.J. Kauffman (Eds.), *Measuring the Business Value of Information Technologies*. Washington, D.C.: ICIT Press.

Milgrom, P. & Roberts, J. (1990). The economics of modern manufacturing: Technology, strategy and organization. *American Economic Review*, 511-528.

Mintzberg, H. (1994). *The Rise and Fall of Strategic Planning*, Prentice Hall.

Porter, M. (1980). *Competitive Advantage.* The Free Press.

Porter, M. (1982). *Competitive Strategy.* The Free Press.

Porter, M. & Miller, V. (1985). How information gives you competitive advantage. *Harvard Business Review.*

Rayport, J. F. & Sviokla, J. (1995). Exploiting the virtual value chain. *Harvard Business Review,* 75-85.

Roach, S.S. (1988, June 20). [Interview with G. Harrar, Editor] *ComputerWorld Extra.*

Roach, S.S. (1989). The case of the missing technology payback. Presentation at the *Tenth International Conference on Information Systems.* Boston, MA.

Scupola, A. (1999). The impact of electronic commerce on the publishing industry: Towards a business value complementarity framework of electronic publishing. *Journal of Information Science, 25* (2).

Teece, D. (1988). Profiting from technological innovation: Implications for integration, collaboration, licensing and public policy. *Policy Journal.*

Wigand, R.T. (1997). Electronic commerce, definition, theory and context. *The Information Society,* 13, 1-16.

Zwass, V. (1996). Electronic commerce: Structures and issues. *International Journal of Electronic Commerce, 1*(1), pp.3-23.

Chapter XIII

Continuous Demand Chain Management: A Downstream Business Model for E-Commerce

Merrill Warkentin
Mississippi State University, USA

Akhilesh Bajaj
Carnegie Mellon University, USA

ABSTRACT

The demand side of supply chain management has drawn considerable research attention, with focus on disintermediation and syndication models. In this chapter, we evaluate new business models for establishing a continuous demand chain structure to streamline the logistics between the vendor and its direct consumers. The Continuous Demand Chain Management (CDCM) model of E-Commerce is one in which the physical products for sale are delivered directly to the customer without the use of a third party logistics provider, such as a common carrier, and in which the physical product may be continuously "pulled" from the seller. We present three submodels of CDCM. The CDCM Model A *applies to business-to-consumer (B2C) online sellers of physical goods who own or control their own delivery vehicles, and may provide further services to extend the value proposition for the buyer. The online grocer is a typical example of businesses in this category. The* CDCM Model B *applies to business-to-business (B2B) sellers of physical goods, who also own a significant portion of their delivery fleet and deliver goods on demand to local distributors or business customers. Office supply E-Merchants provide an example of this model. The* CDCM Model C *applies to businesses that typically provide virtually instantaneous delivery of third party goods to consumers or businesses. Businesses in this category own or control their own delivery fleet and add value by delivering items within very short periods of time, usually one-hour delivery. In order to analyze these models, we conducted structured interviews with key senior managers of one representative business each in the* CDCM Model A *and* Model B *categories. We extensively surveyed recent literature on companies in the* CDCM Model C *category. We use the results of our study to analyze different aspects, such as revenue streams, cost structure, and operational peculiarities of businesses following the CDCM model and, finally, discuss the long-term viability of the sub models.*

BACKGROUND: E-COMMERCE BUSINESS MODELS

Traditionally, E-commerce activities may be business-to-consumer ("B2C," such as direct book sales to the general public by Amazon.com); business-to-business ("B2B," such as corporate procurement or supply chain management using a secure extranet); consumer-to-consumer ("C2C" such as a public auction at Ebay.com); or, within a business, (such as an employee intranet or an enterprise resource planning (ERP) system), where an electronic environment enables organizations to reengineer their internal and external functions and activities, increasing both efficiency and effectiveness. This taxonomy has been more recently extended with B2G (business-to-government), A2B (unattended appliance-to-business) (Charny, 2000), B2E (business-to-employee, as in the corporate intranet), and others.

As with older ("brick and mortar") markets, e-buyers must find sellers of products and services; they may need expert advice prior to purchase and for service and support afterwards. Similarly, e-sellers must find buyers, and they may provide expert advice about their product or service. Both buyers and sellers may automate handling of their transaction processing and "electronic financial affairs." Several categories of new types of businesses have evolved to take advantage of the unique opportunities within this new environment. There are a number of ways these new business models can be viewed. The following sections categorize these emerging business models.

Content, Community and Commerce Strategies

Business models that have emerged in this era of e-commerce activity have been categorized in a number of ways. Most models explicitly leverage the ubiquitous and universal availability of the World Wide Web as a platform for communication between and among organizations and individuals. One fundamental taxonomy of Web-based e-commerce models is based on whether they are oriented toward the purpose of *Content, Community* or transactional *Commerce,* as shown in Figure 1. Note, however, that many sites incorporate two or all three of these objectives to increase their attractiveness to visitors.

The *Content* category includes sites offering news, reports, publications ("e-zines"), clipart, music, information or other soft goods. Classic examples include NYtimes.com, Yahoo.com, and activebuyersguide.com. Many such sites "give away" the content for free, and generate revenue by displaying banner ads. Others charge a subscription. Sites with a *Community* purpose include those that provide focused niche discussions, forums, newsgroups, message boards, or chatrooms, plus sites that offer filesharing or image sharing. Community sites may display banner ads, charge a subscription, or generate revenues through affiliate programs or other advertising. An active community sites offers dynamic content which draws virtual visitors back frequently at a low cost, given that the visitors provide the content! Classic examples include tripod.com and myfamily.com.

Sites for the purposes of *Commerce* include all sites that directly sell either "soft goods" (documents, reports, clipart, music, software, etc.) or "hard goods" (requiring common carriers or shippers) by implementing order-entry, order-fulfillment, and payment processing online. Classic examples include Amazon.com, Dell.com and Egghead.com. Commerce-oriented sites usually seek to generate revenues by

Figure 1: Fundamental Web-based e-commerce models

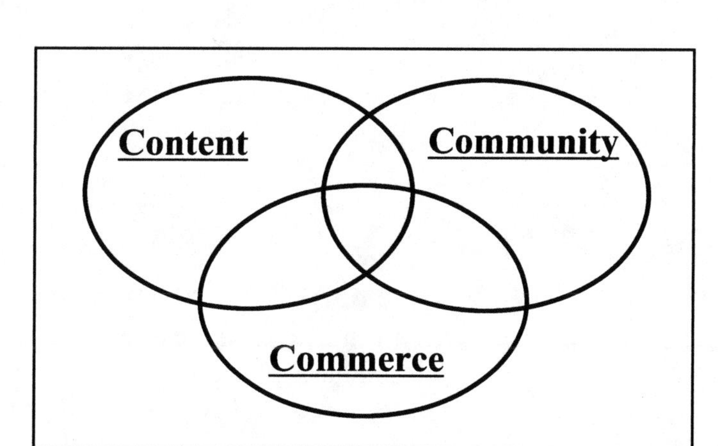

selling items at a price marginally higher than the total associated cost (fixed plus marginal) for each item.

Strategies for Organizational Structure and Linkages

One classification scheme identifies business models by the inter-organizational structure and linkages that are facilitated. Examples include intermediaries, infomediaries, disintermediaries, auctions, buyers' unions, aggregators, consolidators, and others. Timmers (1998) has identified ten e-commerce business models, which are extended here to include several other models:

1. E-shop (individual seller at one Web site)
2. E-procurement (direct or through exchanges organized by seller, buyer, or third party intermediary)
3. E-auction (for B2B or for C2C purposes)
4. E-mall (consolidates many sellers under one umbrella which handles payment processing, billing, shipping, etc.)
5. 3rd party marketplace
6. E-communities
7. Value chain service provider
8. E-integrators
9. Collaboration Platforms
10. Third Party Business Infrastructure sites: information brokers, trust and other services, ratings, standards certifiers, comparison sites, agents, etc.
11. E-exchanges or industry spot markets for commoditized products
12. E-reverse auctions that allow buyers to request competitive pricing offers from multiple sellers
13. "Name your own price" sites
14. E-aggregators that consolidate demand (group purchasing), quantity discounters.

In Figure 2, we place these models into a framework that identifies each as a B2C model, a B2B model, or a C2C model.

Having surveyed the selected e-commerce business model, we next present a business model, which we call the Continuous Demand Chain Management (CDCM) model.

THE CONTINUOUS DEMAND CHAIN MANAGEMENT BUSINESS MODEL

Supply chain management, which evaluates models and methods for coordinating and streamlining the logistics of incoming supplies and inventory, has been the focus of considerable research in the Electronic Commerce field. The demand side

Figure 2. E-commerce business models (source: Warkentin et al., 2001)

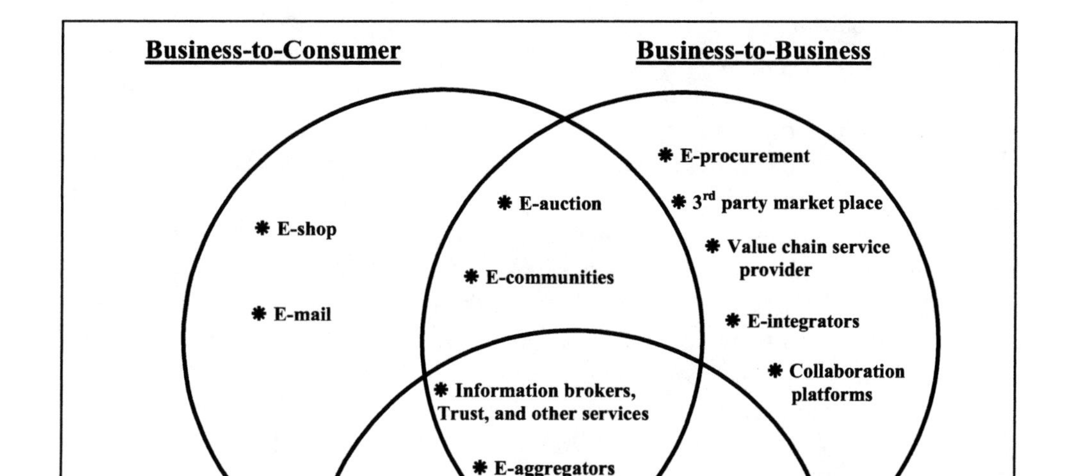

of the supply chain also has drawn considerable attention, with focus on disintermediation and syndication models.

Most enterprises engaged in E-Commerce practice a large degree of outsourcing. While concentrating on core competencies, they develop strategic alliances with partner firms in order to provide activities such as payment processing, order fulfillment, outbound logistics, Web site hosting, customer service and so forth. Many "etailers" are virtual organizations with nothing more than a marketing function under direct control. Partyka and Hall (2000) suggest that the Internet has created three classes of home delivery models, based on route characteristics – substitute, next day and same day. The Continuous Demand Chain Management model of E-Commerce is one in which *the physical products for sale are delivered directly to the customer without the use of a third party logistics provider, such as a common carrier, and in which the physical product may be continuously "pulled" from the seller*. This implies the ownership or control of a fleet of delivery vehicles by the business, a situation that has several very important implications. First, it may offer the ability to exercise greater control over a significant cost of conducting business. Second, the delivery function can be used as a source of distinct competitive advantage; promising next-day delivery, for example, can be a powerful value proposition for the customer. Finally, the direct connection to the customer may enable stronger connections, which, in turn, may enable a greater ability to understand the customer, provide improved services to the customer or create a sense of loyalty. This leads to "stickiness" and switching costs for some customers. Despite the added costs associated with ownership of a delivery fleet, it must not be outsourced if it is central to the value proposition of the firm – if it constitutes one element of the firm's core competence.

The following sections subcategorize the CDCM model into three different forms called the CDCM Model A, CDCM Model B and CDCM Model C. A case study of each is presented and discussed. The CDCM model is diagrammed in Figure 3, which indicates the control the organization extends over the outbound logistics function: a major connection to the customer.

Figure 3: Continuous demand chain management

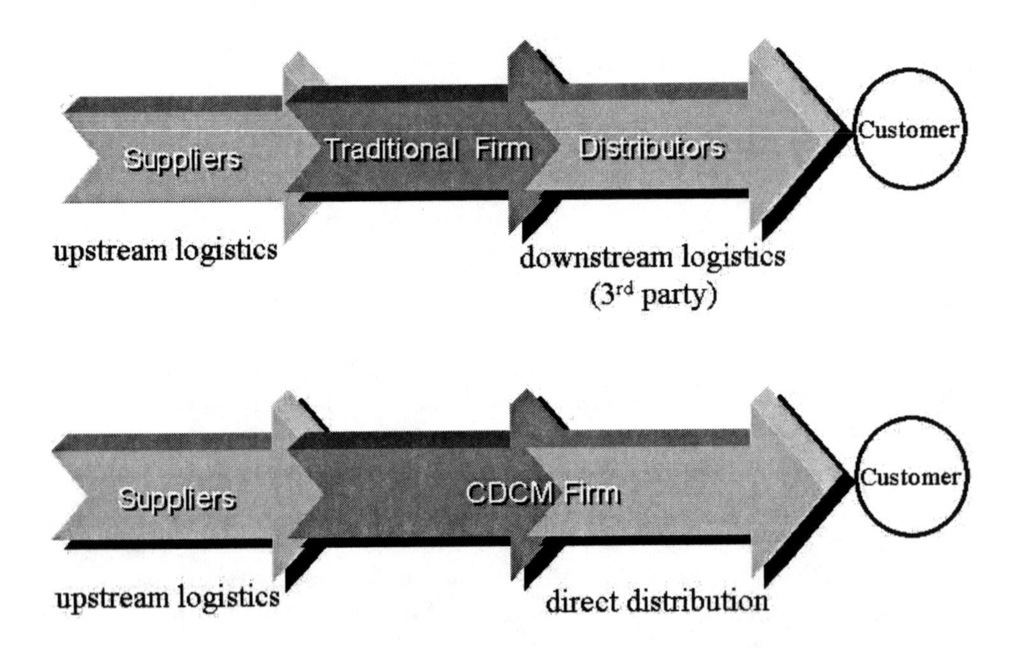

CDCM Model A

The *CDCM Model A* applies to B2C online sellers of physical goods who own or control a fleet of delivery vehicles, and who may provide further services to extend the value proposition for the buyer. Online grocers like ShopLink.com, Streamline.com and PeaPod.com are examples of businesses in this category. In this first form of the CDCM model, the firm integrates repeated or continuous delivery as an added service to channels they may use already (such as a brick-and-mortar storefront or a catalog business).

CDCM Model B

The *CDCM Model B* applies to B2B sellers of physical goods, who also own a significant portion of their delivery fleet and deliver goods on demand to local distributors or business customers. Online office supply E-merchants, such as Staples.com and onVia.com provide examples of this model.

CDCM Model C

The *CDCM Model C* applies to businesses that typically provide virtually instantaneous delivery of third party goods to consumers or businesses. Firms in this category own or control their own delivery fleet and add value by delivering third-party items within very short periods of time, such as one-hour delivery. Some of the firms in this industry failed during the shake-out period for dot-com sellers; others have reorganized. Examples of businesses in this category included Kozmo.com, PinkDot and UrbanFetch.com.

Having introduced the three types of CDCM E-Commerce business models, we now present an analysis of these models, based on in-depth field research.

RESEARCH STUDY: THREE CDCM FIRMS

To analyze the three CDCM models, this project utilized a comparative case study methodology to evaluate three e-commerce organizations, each employing one of the CDCM business models. Structured interview data from key senior managers was used in two cases, and extensive secondary data was compiled and evaluated for the third case. The key managers (Chief Operating Officer and Director of E-Commerce at Shoplink.com and VP of Operations for Staples.com) were interviewed at length to identify key integrating factors related to their separate business models. The list of questions asked of each manager is presented in Table 1. These questions represent dimensions of the CDCM model that, in our judgment, differentiate the CDCM model from other business models. This list is based on our earlier interactions with employees of companies that followed this model (these interactions are not reported in this study).

Each interview used in this study lasted approximately 45 minutes. As can be seen from Table 1, the questions were relatively open-ended in nature, so that the managers were free to voice their own insights. The interviews were transcribed and analyzed by the authors for insights on the part of the subjects, who had many years of experience at senior management levels in the area. These insights are used in this work to analyze the CDCM models.

The first business, ShopLink.com, provided unattended weekly delivery of groceries and other items to tens of thousands of consumer households in New England and New York. Customers paid a monthly subscription, and ordered once (or occasionally twice) per week on the World Wide Web. ShopLink.com's fleet of trucks delivered the orders the next day in chillpacks to the customers' garages, using a numeric keypad to open the garage door, taking away dry cleaning, refundable bottles and cans, UPS packages and the empty chillpacks from the week before. Thus, ShopLink.com was a *CDCM Model A* type of business. The authors worked closely with ShopLink.com, and developed extensive material on the business model of this firm and its competitors, including Peapod and Streamline.com.

The second organization also offers next-day delivery, but for business customers. This firm, Staples.com, is a B2B office supplies delivery provider which owns and operates its own fleet of trucks, but also uses third-party logistics providers. Their clearly stated value proposition is "next day delivery guaranteed." Staples, Inc., the second largest office supply superstore company in the U.S., sells office products, computers, supplies, furniture, and printing and photocopying services at

Table 1: Specific structured interview questions

Could you briefly describe your Customer Relationship Management (CRM) operations? What kind of software do you use to enhance your CRM? How do you see your organization's business model as being different from other business models, as far as the use of CRM is concerned?
Could you briefly describe your outbound logistics management? What kind of software do you use for outbound logistics management? How do you see the your organization's business model as being different from other business models, as far as the use of outbound logistics are concerned?
What are your sources of revenue? Where do you see the company making revenue in the long term? What are your sources of cost? Which are fixed? Which are variable? How do you see the cost/revenue model in your organization as being different from other business models?
Suppose your customer base were to triple. How would that affect • Your operations, • Your revenue/cost model, and • Your ability to provide service?
Where do you see yourself 5 years from now? What are the top 3 factors (in descending order of importance) that make your organization unique, in your opinion?

more than 110 stores in the US, through telephone catalog sales, and with their Staples.com website. They offer more than 8,000 office products primarily to small- and medium-sized businesses (Hoovers, 2000b). The Staples.com website creates greater flexibility for these business customers, and values its brand and repeat business from these customers. Staples.com is a *CDCM Model B* type of business.

The final organization, Kozmo.com, provided one-hour delivery of CDs, DVDs, ice cream, juice, videotapes, candy, snacks and other items to individuals in many large cities. They also operated "drop boxes" in convenience stores and Starbucks coffee retailers throughout these cities for individuals to return rented movie videotapes and DVDs, or borrowed chill-bags. While this firm's business model was based on repeat business, in which many customers "pulled" products regularly, Kozmo.com did not implement true continuous demand chain management. In fact, it often satisfied demand for impulse buying ("want" not "need") for consumers, and employed less customer-relationship management activity. Many purchases were one-time events, rather than part of a pattern of ongoing purchases. Kozmo.com was a *CDCM Model C* type of business.

Analysis of CDCM Model A

The domestic grocery market is valued at more than $300 billion annually. The online grocery market has been reduced, with the departure of WebVan and other high-profile E-Grocers, but with new entrants, there is some prospect for renewed growth. The recent forecast for online grocery sales is $7 billion annually in the U.S. by 2005. However, it is unclear what this future will be, given current economic conditions. There is clearly a significant demand for this service, given the tens of thousands of customers that signed up. There were more than a dozen competitors in the grocery e-tail business, which now has consolidated and is undergoing structural changes. The most notable E-Grocery providers were WebVan (which bought HomeGrocer), Peapod.com, ShopLink.com, Streamline.com, and NetGrocer. All offered the ability to order items online and have them delivered to the consumer's house. Some offered regular "unattended" weekly delivery (to your garage, for example) based on a monthly subscription model. Others offered on-demand deliveries (if you were home) with a surcharge on the grocery bill (and sometimes an additional delivery charge). Many also offered additional services, such as dry cleaning pickup and delivery. One sold only nonperishable items shipped via common carrier. Other unique features included "don't run out" automatic continual ordering of weekly staples, fresh flower delivery, movie rental and pickup, shoe shines, meal planning, recipe tips, multimedia information, and nutritional information.

An extensive survey conducted by the Consumer Direct Cooperative (Orler, 1998) (a cooperative research group which included PeaPod, Coca-Cola, Streamline.com, Harvard Business School, and others) pointed to six major groups of *potential* online grocery shoppers, some of which were more likely to use online grocers than others. These included:

1. shopping avoiders, who dislike grocery shopping,
2. necessity users, who are limited in their ability to shop,
3. new technologists, young, comfortable with technology,
4. time-starved, will pay to free time in their schedules,
5. responsibles, gain sense of self-worth from shopping, and
6. traditionals, older individuals, enjoy shopping in stores.

ShopLink.com, incorporated in 1996, was an online provider of groceries, household consumables, and services. They utilized an "unattended" direct home delivery logistics plan, which means the consumer does not have to be home (as with other online grocers). Its two facilities in Massachusetts and Connecticut served more than 175 suburban communities. In 2000, Gomez Advisor, Inc. (www.gomez.com) ranked the company first of 11 online grocers in the "Selective Shopper" and "Online Resources" categories, and number three overall. Their value proposition was providing a total solution for today's busy consumer by creating a pleasant experience for time-starved, stress-laden, or supermarket-averse shoppers.

ShopLink customers generally were repeat customers who ordered week after week in a tight, ongoing relationship with the grocer. The interaction with the Web site was much more substantial than with other B2C websites, and user feedback also was more prevalent than for most Web sites. "It's a very sophisticated purchase versus if you're going to Amazon you might be buying one to three to four items — our average order has 54 different items in it cutting across five temperature zones"(Kruger, 2000).

After a series of information conversations with ShopLink.com's Director of Marketing and Director of E-Commerce, we conducted a structured interview with the Chief Operating Officer, with additional customer relationship management information provided by the Director of E-Commerce. We next describe the insights provided as a result of the interviews, following the sequence in Table 1.

It is clear that CRM was very important to ShopLink.com. Their "Delivery Specialists" were a critical link in the communication chain that started at different parts of the organization and extended to the customer. Delivery Specialists often were used to learn more about customers' needs and behavior, and this information was used by departments, such as customer service and operations, to modify their activities to provide better CRM. A number of different software systems were used to capture different information about their customers, and information from customer records was given to Delivery Specialists en route. The buying process of groceries is different from buying other retail items, such as books or music, for example, because the customer typically purchases many items together (often 50 or 60 items). Furthermore, the customer had an ongoing relationship with ShopLink.com, characterized by routine weekly purchases automatically charged to the credit card account on file, weekly communication to the customer, and personalization of the interface with shopping lists and suggestions. The product quality and selection is of vital importance to customers, because food is a basic need.

Most customers ordered nearly all their groceries online, so ShopLink.com mined the data to create valuable information, which was aggregated and sold to food manufacturers. (Individual data was never disclosed to third parties.)

Further, because all online activity was captured, the level of detail was finer than that developed by traditional grocers, which enabled ShopLink.com to create even more value to customers in the form of personalization of customer interaction and information. Ongoing efforts in the area of CRM included better integration of systems and providing customers the ability to track their drivers so as to be able to estimate their own personal delivery time more accurately for that particular day. ShopLink.com was in the process of enhancing their ability to evaluate customer needs by expanding their use of data mining. (A brick-and-mortar store doesn't know if you pick up a can, read the nutritional label, then put it back on the shelf without purchasing it, but the online grocer does!)

The outbound logistics model for ShopLink.com started with the customer placing an order online for groceries by 2 p.m. on a business day, in order to get delivery the next business day. ShopLink.com delivered on Tuesdays through Fridays in eastern Massachusetts and parts of Connecticut and New York. ShopLink's commitment was to deliver groceries the next business day before a pre-determined time (usually 6 p.m.). Delivery Specialists were routed using intelligent routing software. Important parameters that determine the cost of delivery included: a) the latest time promised for delivery (extending the promised delivery time to 8 or 10 pm. can significantly reduce delivery costs); b) the geographical distribution of customers; and c) the mean time taken by a delivery specialist to physically transfer the groceries from his or her truck to the customer's door. Most customers gave ShopLink.com the code to the garage door keypad, and chillpacks were placed within the garage. Employment of their own Delivery Specialists allowed for greater reliability in the delivery process — something that is important to consumers of online grocers. Tight control of delivery costs was critical to the success of their business model. Controlling their own outbound delivery system (consisting of a fleet they owned and Delivery Specialists they employed) was critical, because it allowed flexibility in routing intelligently, in making changes en route, and for the provision of ancillary services like dry cleaning and package pickup. These factors combine to allow greater control of cost, while also allowing the offering of extra services, reliability and timeliness. For example, the dynamic routing algorithm planned the delivery schedules each day using current statistics to ensure maximum route efficiency. The routes formed concentric circles, with distant routes dispatched first, followed by closer routes. In some cases, Delivery Specialists could deliver nearby routes and return to pick up and deliver a second or third truckload of groceries in one day. Capacity utilization was a critical cost component contributing to profitability.

ShopLink.com offered a guarantee of "next day delivery" as one value proposition. Other online grocers would deliver the same day, but only if the customer was home to receive the groceries. The interview data provided evidence for a clear cost/benefit tradeoff between promised delivery time and total cost for delivery. Thus, while delivering groceries in an hour or the same day may have been be marginally more valuable to the customer, the costs associated with such a model made it prohibitively expensive, because the costs had to be born by the customer, or absorbed by the grocer, into the overall operating costs. The optimal time for delivery seemed to be in the range of 18-36 hours after ordering, which necessitated an unattended delivery model. While some customers were home for deliveries and spoke with the Delivery Specialists, others frequently communicated with ShopLink's Customer Service specialists and left notes in the return bins (empty Chillpacks). The Web site also facilitated communications by providing a categorized message area with each order and a menu item to communicate with ShopLink about any issue.

The revenue sources for ShopLink.com included an industry-standard markup on the actual sale of grocery goods, a subscription fee of $25-$35 per month, and the sale of aggregated data on consumer buying practices (without revealing information on any particular customer) to well-known manufacturers of consumer goods. Its primary cost components included the cost of goods sold, labor, the cost of delivery, warehousing costs, and Web site and database maintenance. Delivery costs were very sensitive to fuel costs, so that increasing fuel prices may have required some adjustments to the other cost factors. Inventory maintenance and warehousing costs were significantly lower than for brick-and-mortar grocers, because demand was better understood. In fact, because orders were submitted by 2 p.m. on the day before delivery, there was sufficient time to acquire many items from third parties, eliminating the need for any inventory of those items! Fixed costs included the purchase of delivery vehicles, office infrastructure, a Web site and a warehouse. Variable costs included the cost of goods sold for each order, delivery costs, and holding costs for their limited inventory. Though the firm grew very rapidly, and was reaching the economies of scale necessary for break-even, ultimately, it ran out of time and money to continue operations until profitability was reached. Its goal of covering the variable delivery costs through the subscription revenues was almost reached. ShopLink's demise caught most industry analysts off guard, given the apparent success of their operation up to that point (Regan, 2000).

From a capacity fulfillment standpoint, it is clear that warehouse capacity and fleet capacity were crucial to meeting increased demand and growth. Possible methods of dealing with increased demand include the construction of new warehouses (a large fixed cost), increasing the size of the fleet, limiting the geographic area that can be serviced, and the hiring of more personnel. The reality is that it is essential always to be aware of market size and market share, so that there are no surprises with regard to suddenly increased demands on capacity. This also allows us to conclude that the CDCM Model A probably is less suitable for the sale of seasonal goods, like holiday items, which are characterized by sharp increases in demand, unless customer expectations are lowered with regard to delivery schedules and price competitiveness.

The sources of competitive advantage among firms in the online grocer industry appear to be in: a) the adoption of methods to maximize the value of services to busy customers, including the unattended delivery model; b) the maintenance of high standards for delivery quality while controlling total costs of outbound logistics; c) the accurate execution of order fulfillment to ensure that customers accurately receive the correct 50 or so items per order; and d) the product quality and selection of the groceries sold. Further, recent entrants into this business (Safeway, Royal Ahold and FreshDirect) have sought to leverage a brick-and-mortar presence for various integrated strategies (O'Connell, 2002).

The managers thought that wider adoption of "broadband" Internet access by consumers and the increasing number of Web users meant that the overall market for on-line grocers would increase, so that it would consist of late technology adopters, as well as early adopters. They felt it was imperative to adopt financially responsible models by charging for delivery or requiring minimum order size, so that the number of customers required to break even was in the tens of thousands and not the hundreds of thousands. The future of the online grocery business is unknown – economic recovery in the U.S. may be accompanied by new attempts to capture this known demand, but only if the overall business model, including cost control, can be perfected. An industry analyst recently suggested that "the future of the online grocer market belongs to the grocery stores ... They know the business, they can mix (sales) channels, and they can take their time" (Farmer & Sandoval, 2001).

Analysis of CDCM Model B

Staples.com, a wholly owned subsidiary of Staples, Inc., is a B2B office supplies delivery provider that operates its own trucks, but also uses third-party logistics providers. Staples.com provides an alternate business channel to the firm's businesses customers, in addition to their extensive nationwide chain of brick-and-mortar stores and their mature catalog sales channel. Their clearly stated promise to their customers is "next day delivery" of office supplies. We conducted a structured interview of the Vice President for Operations of Staples.com, asking the open-ended questions listed in Table 1. This section presents the findings and reports the insights obtained from this research.

Customer Relationship Management (CRM) forms a key component of Staples.com's strategy. Data relating each customer's shopping patterns are gathered and stored in a central database that links information from all channels of Staples, Inc., including their telephone catalog services, Staples.com, and their brick-and-mortar stores. In all, information on about 6 million customers is stored. This aggregation of data across channels allows key managers to see customers as they change channels, and gives them a complete picture of each customer. The overall profitability of each customer is also interpolated from statistical analysis of purchase patterns. It also allows them to see the comparative strength of each channel, and to set rational incentives for managers of different channels, so that they are not penalized if the customers in their area shift to another channel (for example, from brick-and-mortar to Staples.com). Staples, Inc. also incorporates CRM with their outbound logistics by providing information on every unfulfilled order to customer service, so that customers can call in and get complete information. Problem orders are logged permanently, so that customers can be offered better service in the future. High levels of customer service are key to Staples.com's strategy. Their system also enables their small business customers to market their own products and services to each other in a virtual marketplace (Internet.com, 2000).

In the area of outbound logistics, Staples.com owns and operates 29 warehouses, called "order fulfillment centers," through which all orders flow. Merchandise sold by Staples.com is either inventoried at these centers or ordered from a third party. Because Staples.com promises next day delivery, third party vendors often will ship their part of an order directly to the customer, so the customer receives all the parts of an order the next day. This means that customers may receive their orders in multiple packages. The entire outbound logistics are controlled by a sophisticated system built in house by Staples. Staples, Inc. maintains its own delivery fleet, and, in some cases, outsources delivery to third party common carriers, such as United Parcel Service (UPS). They view the growth of their own fleet as important, and also try to control delivery costs by restricting Staples.com to densely populated areas. Staples.com also views as advantageous their long-term relationships with third party carriers, though they see their own delivery fleet taking a larger share of deliveries in the densely populated areas in the future. However, the fact that their customer can be a small business anywhere, (including a rural area out of reach of their fleet) necessitates their continued use of third party carriers for the future. Senior management at Staples.com has carefully evaluated the costs and benefits of same-day or one-hour delivery mechanisms and has determined that such delivery options are not economically feasible. The optimum delivery period appears to be next day, in terms of value to customers and current estimates of costs for such delivery. It should be noted that one of their smaller competitors, W.B. Mason, recently has begun same-day delivery of certain office supplies – orders must be placed by 9 a.m. and they are delivered before 5 p.m. This competitive development and potential reactions will be observed in the next year.

The revenue sources for Staples.com include the actual sale of office supplies and the delivery fee that is added to each order. The cost sources include the cost of goods sold and the cost of delivery. The fixed costs are the creation and maintenance of the order fulfillment centers and the delivery fleet. Variable costs include the cost of goods for each order, as well as the cost of delivery. Their business is a low-margin, high-volume business and is therefore highly dependent on customer retention, necessitating a strong customer-orientation throughout the organization.

From a capacity standpoint, it is clear that Staples.com believes in steady, planned growth and careful monitoring of past trends and sales, so as to manage demands on capacity. Because their value proposition is next-day delivery and accurate fulfillment of orders, they avoid marketing schemes that cause sudden short-term spikes in demand. This approach has been successful to date, and is likely to ensure continued controlled growth and profitability.

The sources of competitive advantage among firms in this industry appear to be: a) the successful offering of complementary multiple channels; b) the ability to control total costs and purchase inventory at a discount due to the low-margin, high-volume nature of the business; and c) the offering of value-added services and a brand that their businesses' customers can trust. The latter may be instrumental in enabling somewhat higher margins than competitors, thereby achieving greater profitability.

Analysis of CDCM Model C

Kozmo.com and similar firms competed in the convenience food and entertainment retail industry. Kozmo was a consumer-oriented (B2C) Web site that sold and provided one-hour delivery of CDs, DVDs, ice cream, juice, videotapes, magazines, snacks and other items to individuals in many large cities. They also operated "drop boxes" in convenience stores and StarBucks coffee retailers throughout these cities for individuals to return rented movie videotapes and DVDs, or borrowed chill-bags. Traditional in-store shopping offers consumers immediate access to desired items — but often with problems associated with driving, parking, and loss of valuable time. Founded in 1997, "with a handful of bike couriers delivering goods ordered online by New Yorkers, Kozmo.com grew into an army of "Kozmonauts" that served about a dozen major cities" (Hoovers, 2000a). It attracted significant funding from several high-profile Venture Capital firms and from Amazon.com, but could not achieve the necessary economies of scale before venture funding ran out.

Kozmo's business model, which also was based on the demand delivery services value proposition, was primarily built on the satisfaction of consumer demand for impulse purchases or cravings, and not planned, repeating purchases, like ShopLink. The firm had fewer overall repeat customers and employed less customer-relationship management activity.

Kozmo.com employed carriers, generally known as "Kozmonauts," who owned their own vehicles. Kozmo.com's roots as a bicycle courier service in New York City offers some insights into their corporate culture. While there were some repeat customers, a large percentage of purchasing was "single time" or non-routine purchasing to satisfy cravings or impulses. While impulse purchasing is a powerful and proven driver for marketing, it does not provide sustained and reliable customer purchase patterns and income streams necessary for capacity planning and long-term growth.

Our analysis, based on numerous independent articles that describe the failure of Kozmo.com, indicates some factors that contrast with CDCM Models A and B. Key lessons include finding an optimal delivery schedule that is sustainable by revenues (mainly, delivery charges or added cost of product), and being the delivery agent for one or two businesses. Thus, Kozmo evolved from being a delivery agent for several businesses to being one only for Starbucks.com. Cost control will be critical to make the model viable.

Because order sizes were significantly smaller than for the two previous CDCM models, the cost of delivery as a percent of sales was much larger. Yet, for a long time, Kozmo.com did not charge its customers any delivery charge, nor have any minimum order size, resulting in orders for a single candy bar! The entire outbound logistics system of personnel represented a major cost component, but their lack of ability to recover costs from this major source was a primary source of their failure.

Other firms that operated with this business model were UrbanFetch (in New York City and London) and PDQuick (formerly called Pink Dot), a Los Angeles-based company that has operated a fleet of Volkswagen Beetles to deliver goods since 1987 (Helft, 2000). Another example is iToke (www.itoke.co.uk), a British delivery service that promises to make deliveries in Amsterdam within 30 minutes. The twist is that their customers will be able to order marijuana using computers, fax, phone, or WAP-enabled wireless phones (Evangelista, 2000). The Web site, designed after Starbucks Coffee's Web site, accepts smart credit cards that can be refilled at Amsterdam kiosks. The "iTokkerista" couriers ride green-and-white bicycles. Amsterdam brick-and-mortar coffee shop owners are concerned with the "Amazon-ization" of their business.

It seems evident that cost control is a key factor for the success of CDCM-based businesses in the future. Same-day delivery is suspect itself as a long-term strategy for success. As Laseter et al. (2000) suggested, WebVan, Urbanfetch and other same-day transporters are based on economic models that "won't deliver for long." Future entrants into this segment might seek to expand their product selection, in order to appeal to a broader audience, enter new markets and increase the average order size to achieve profitability.

Having presented the insights from the research study, we next discuss the findings and provide recommendations for managers of managers of e-commerce organizations.

DISCUSSION AND RECOMMENDATIONS

The growth of the volume of electronic commerce activity will rely not only on the continued growth in the size of the online community and the increase in Internet access, but also in the confidence of consumers and business customers in the ability of online firms to provide clear value propositions vis-à-vis brick-and-mortar sellers. Such value propositions will come from technological innovations in some cases and, in other cases, from traditional business practices modified for the online experience. For the purchase of ongoing or routine products, purchasers will demand consistency and reliability in the delivery mechanism, as well as personalized services. The development of highly unique customer interaction experiences will be a key component to retaining loyal customers that build a solid base for growth and profitability.

Our research indicates that the CDCM model, when implemented correctly, can constitute a strong basis for customer retention, growth and profitability. From our findings above, it is clear that, while the three CDCM models have clear differences, they also have common features that differentiate them from other E-Commerce models. The key common features of all the CDCM models include:

a) The control of delivery costs is necessary to make the business model viable. Factors that play a major role in controlling the delivery costs include the minimum order size, a delivery fee, a subscription fee, the density of customer population, fuel costs, and the "time to delivery" deadline. One key manager said they looked at one hour and same day delivery options and "it doesn't make sense to us ... You'd have to charge a significant delivery fee to make it work";
b) The importance of CRM in general, and its linkage to outbound logistics. The control of one's own delivery personnel and/or fleet presents a unique opportunity for face-to-face customer interaction and information gathering which can be utilized to enhance CRM if a feedback loop from delivery to CRM is incorporated in the company. Further, strong incentives for loyalty will ensure a strong basis for a reliable income stream. This is extended with a variety of value-added services that enhance the interaction experience, contribute to the convenience of online purchasing and create "switching costs" as barriers to entry by potential competitors.

Friedman (2000) suggested that another key to success for businesses in this sector is the ability to develop "networks of cross-industry partners that provide products and services that relate to the customer's basic life objectives." He emphasizes the personalized point of interaction and expertise at assimilating and leveraging customer information. He said the use of the Web enables network partners (such as an online grocer delivery service and the grocery product manufacturers) to collect and refine information about each customer to create customer intelligence for all partners. The Web also enables interactive, continuous and real-time customer dialog. Several key differences between the CDCM models are summarized in Table 2.

Based on our findings, we propose the following set of recommendations for organizations that wish to adopt a CDCM model for e-commerce.

a) The control of delivery costs is crucial to the long-term viability of the model. Some methods of controlling these costs include: targeting only densely populated areas, requiring a minimum order size, extending the promised delivery deadline, and using intelligent routing software. Further, geographic growth into areas with demographic characteristics consistent with larger order sizes or lower delivery charges will contribute to profitability.

Table 2: Strategy dimension for three continuous demand chain management models

Models Dimension	CDCM – Urban 1	CDCM - National	CDCM – Urban 2
Customer base	General Consumers. Geographically confined (usually metropolitan and densely populated)	Businesses. Geographically dispersed across the country.	Usually consumers but sometimes businesses. Usually metropolitan and densely populated
Outbound delivery mechanism	Usually ownership of delivery personnel and/or fleet	A mixture of own delivery personnel and/or fleet, and third party carriers for geographically scattered customers in rural areas.	Usually ownership of delivery personnel and/or fleet.
Delivery deadline	Same or Next Day	Usually Next Day	Usually same day
Type of purchase	Frequently recurring purchase, like groceries, or other regularly consumed items	Business purchase, less frequently recurring than in Model A, like office supplies.	Impulse purchase like pizza or ice cream or DVD rental, usually not frequently recurring.
Cost of Delivery	Small as a percentage of overall purchase	Small as a percentage of overall purchase	Large as a percentage of overall purchase.

b) An important benefit is the information that delivery personnel can get on face-to-face interaction with the customers. This information can be fed into the CRM system to provide better service. It can be combined with purchasing behavior information, aggregated and then sold to third parties (while preserving individual anonymity of customers) to yield significant revenue.

c) It is useful to link the CDCM model to existing channels in the organization (such as retail outlets, catalogs, and field sales forces). This gives the customer flexibility and convenience. If this is done, it is important to change the organization's incentive structure, so that all the channels work together to provide a positive experience for the customer, rather than compete with each other. One way to avoid channel conflict is to reward managers in all channels based on the zip code of customers through a sharing of credited revenues.

In the future, there may be even greater opportunities for tight linkages between the CDCM provider and the customer. One technological development that may enable this customer intimacy is the wireless or mobile Internet access device. In the future, customers will be able to "access the routing program and their data on a remote server" (Partyka & Hall, 2000). Accessing each order anytime, from anywhere, will enhance the loyalty and value proposition for CDCM customers.

The business models developed for success on the Internet are appearing as fast as the technologies to support them. Many will not be successful models in the long run, while others will create entirely new marketspaces and industries. The

Continuous Demand Chain Management model for customer satisfaction, when implemented according to strict cost-benefit guidelines, and with tight cost controls, will likely become a profitable long-term model for conducting certain business online. Its strength is the provision of reliable delivery of essential products and valuable related services to consumers or business customers. When combined with a personalized interaction experience, and expertise at assembling and analyzing customer information for customized offerings, it should contribute to the retention of a loyal, predictable customer base, which will lead to overall profitability.

REFERENCES

Charny, B. (2000). A2B: Eliminating the human element. Retrieved from the World Wide Web: http://netscape.zdnet.com/framer/hud0022420/www.zdnet.com/zdnn/stories/news/0,4586,2646086,00.html): ZDNET News.

Evangelista, B. (2000, August 31). Founders dream of high demand on iToke net size. *San Francisco Chronicle*. Retrieved November 2, 2000 from the World Wide Web: http://itoke.co.uk/pressdown.html.

Farmer, M.A. & Sandoval, G. (2001). WebVan delivers its last word: Bankruptcy. Retrieved July 9, 2001 from the World Wide Web: http://news.com.com/2100-1017-269594.html?tag=rn.

Friedman, J. (2000). Best intentions: A business model for the e-economy. Retrieved November 2, 2000 from the World Wide Web: http://friedman.crmproject.com: Anderson Consulting.

Helft, M. (2000). Expand and deliver. *The Industry Standard.* Retrieved from the World Wide Web November 2, 2000: http://www.thestandard.com/article/display/0,1151,18925,00.html.

Hoovers. (2000a). Kozmo, Inc. *Hoover's Company Capsule.* Retrieved November 2, 2000 from the World Wide Web: http://www.thestandard.com/companies/display/0,2063,61706,00.html.

Hoovers. (2000b). Staples, Inc. *Hoover's Company Capsule.* Retrieved November 2, 2000 from the World Wide Web: http://www.thestandard.com/companies/display/0,2063,14790,00.html.

Internet.com. (2000). Staples.com enhances its marketplace for small businesses. Retrieved November 2, 2000 from the World Wide Web: http://ecommerce.internet.com/ec-news/article/0,,5061_325921.00.html.

Kane, M. (2001). Web grocer seeks stateside foothold. Retrieved June 29, 2001 from the World Wide Web: http://news.com.com/2100-1017-269231.html?tag=rn.

Kruger, J. (2000). *Reponses Given During Structured Interview.*

Laseter, T., Houston, P., Chung, A., Byrne, S., Turner, M., & Devendran, A. (2000). *The Last Mile to Nowhere*. Booz-Allen and Hamilton. Retrieved November

2, 2000 from the World Wide Web: http://www.strategy-business.com/bestpractice/00304/.

Muehlbauer, J. (2000). Delivery merger couldn't go the last mile. *The Industry Standard.* Retrieved November 2, 2000 from the World Wide Web: http://www.thestandard.com/article/display/0,1151,19270,00.html.

O'Connell, P. (2002, September 24). Can FreshDirect bring home the bacon? *BusinessWeekOnline.* Retrieved from the World Wide Web: http://www.businessweek.com/smallbiz/content/sep2002/sb20020924_1399.htm.

Orler, V. (1998). *Early Learnings From the Consumer Direct Cooperative.* Anderson Consulting.

Partyka, J. G., & Hall, R. W. (2000, August). On the road to service. *OR/MS Today,* 27, 26-30.

Regan, K. (2000, November 17). Dot-com death by degrees. *E-Commerce Times.*

Timmers, P. (1998). Business models for electronic markets. *Electronic Markets,* 8, 3-8.

Warkentin, M. & Bajaj, A.. (2000) An investigation into online grocer selection criteria. *Proceedings of the 2000 Annual Northeast Conference of the Decision Sciences Institute*, 155-157.

Warkentin, M., V. Sugumaran, V., & Bapna, R.. (2001) Intelligent agents for electronic commerce: Trends and future impact on business models and markets. In Syed Mahbubur Rahman and Robert J. Bignall (Eds.), *Internet Commerce and Software Agents: Cases, Technologies and Opportunities.* Hershey, PA: Idea Group Publishing.

Chapter XIV

Web-Based Supply Chain Integration Model

Latif Al-Hakim
University of Southern Queensland, Australia

ABSTRACT

This chapter discusses various business process supply-chain models and emphasizes the need for organizations to apply CRM concepts and to integrate the Internet within the functions of the supply chain in order to be able to gain good customer expectations in the era of e-commerce. This chapter outlines a framework for developing a Web-based supply chain integrating model based on SCOR and key features of CPFR, and attempts to link this model with Business Process Reengineering, and with traditional productivity improvement programs. The development of a Web site at two levels is suggested. The first level is within the public domain and the other is limited to supply chain partners. The chapter incorporates fuzzy set theory into the dynamic of production scheduling to allow the integrating model to deal with vague constraints, and to enable conflicting multi-criteria objectives to be managed effectively in the production environment.

INTRODUCTION

Traditionally, firms did not consider the potential for their suppliers or customers to become partners. Instead, they may have competed with their suppliers and customers, fearing that they would be taken advantage of by them (Frendendall & Hill, 2001). As a result, firms were constrained by their customers' or suppliers' lack of collaboration and unresponsiveness. These attributes prevented firms from responding quickly to changes in the market or to customers' requirements.

Changing conditions of competition have forced organizations now to adopt a different strategy. Lambert and Copper (2000) point out that one of the most significant paradigm shifts of modern business management has been that individual businesses no longer compete as autonomous entities, but rather as supply chains. Managing the supply chain has become a means of improving competitiveness (Chantra & Kumar, 2000; Lee, 2000). Proactive supply chain managers begin to view the supply chain as a whole, and promote customer-focus, supplier partnership, co-operation and information sharing (Jayaram et al., 2000). Three major developments in global markets and technologies have brought the emerging supply chain management (SCM) to the forefront of management's attention (Handfield & Nichols, 1999):

1. The information revolution;
2. Customer demands in areas of product and service cost, quality, delivery, technology, and cycle time brought about by increased global competition; and
3. The emergence of new forms of interorganizational relationships.

Today, the effectiveness of an organization's response to rapidly changing market conditions will be determined by the capability of trading partners (Power & Sohal, 2001). Members within the supply chain should "seamlessly" work together to serve the end consumer (Towill, 1997). The notion of supply chain management (SCM) is therefore holistic, rather than functional, and of strategic, rather than tactical, importance.

The association of supply chain management with e-business offers new challenges for marketing. In addition, the explosion of the Internet and other telecommunication technology also has made real-time, on-line communication throughout the entire supply chain a reality. The Internet allows companies to interact with customers and collect enormous volumes of data, and manipulate it in many different ways to bring out otherwise unforeseen areas of knowledge (Abbott, 2001). The concept of Customer Relationship Management (CRM) is one of the new ways of interacting with customers (Galbreath & Rogers, 1999). The aim of this chapter is to explain the possibility of developing a Web-based supply chain integration model that enhances CRM and collaboration among supply chain entities by incorporating the advantages of existing supply chain reference models. The chapter defines SCM and introduces CRM and its dimensions. CRM, e-commerce and information sharing are considered as three major constituents of SCM (Figure

1). The significant role of e-commerce in the competitive advantage of organizations is referred to, and traditional computational commerce tools are compared with the current status of e-commerce. The chapter also highlights information as a strategic asset, and argues that the real time flow of information among supply chain entities will enhance the competitive advantage of the whole supply chain.

Because of the complexity and overwhelming number of processes in the supply chain, process mapping and modeling are essential for supporting the supply chain integration. Modeling is the first step of implementing the integration. This chapter provides a section relating to the most well-known business process supply chain reference models: SCOR, CPFR and DAMA. The three models are compared and contrasted. The chapter highlights the importance of modeling for the improvement of supply chain processes. It develops a model which is based on SCOR and key features of CPFR. The integrating model deals with the continual dynamics of scheduling, and considers the potential to synchronize process improvement throughout the supply chain.

Figure 1: Outline structure of the chapter

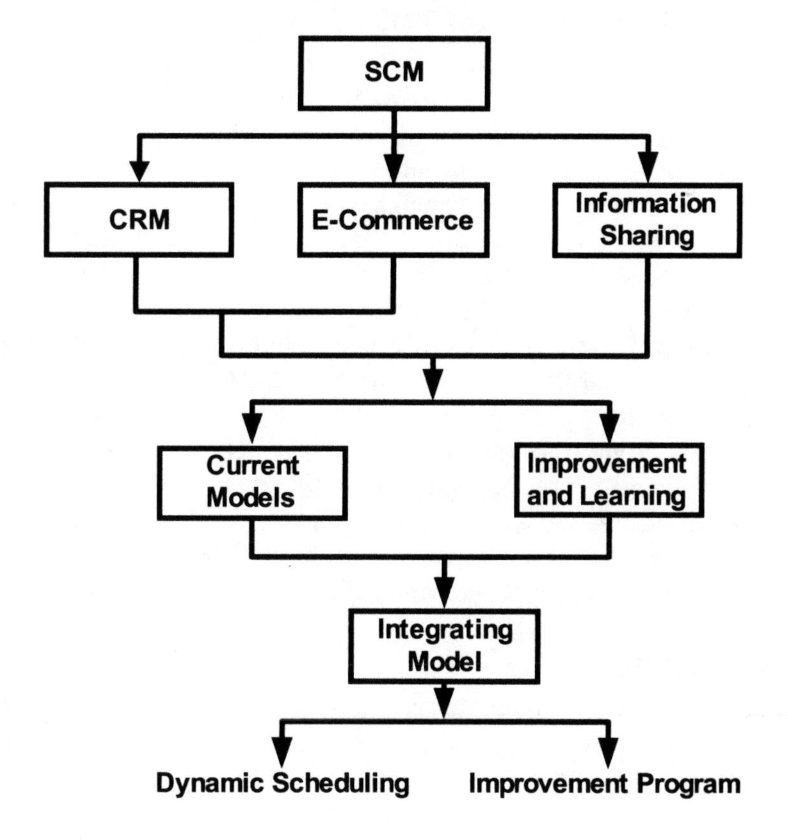

WHAT IS SCM?

Ellram and Cooper (1993) describe supply chain management as an integrating philosophy to manage the total flow of a distribution channel from supplier to ultimate customer. The APICS Dictionary explains the supply chain as (Cox et al., 1995):

1. the processes from initial raw materials to ultimate consumption of the finished product linking across supplier-user companies; and
2. the functions within and outside a company that enable the value chain to make products and provide services to the customer.

Lummus and Alber (1997) illustrate the supply chain as the network of entities through which material flows. The Supply Chain Council (2001) describes the supply chain as encompassing every effort involved in producing and delivering a final product, from the supplier's supplier to the customer's customer.

According to Poirier and Bauer (2000), the use of Internet connections within the supply chain is an inevitable action, as electronic commerce (e-commerce) is destined to be the medium of communication. They refer to the term "e-supply chain" as a reference for "the natural combining of supply chain and e-commerce."

CUSTOMER RELATIONSHIP MANAGEMENT

CRM concerns the management of technology, processes, information resources, and people needed to create an environment that allows a business to take a 360-degree view of its customers (Galbreath & Roger, 1999). CRM environments, by nature, are complex, and require organizational change and a new way of thinking about customers and about business in general. Creating such an environment also requires a new form of leadership. Within the framework of supply chain, such an environment is grounded in three core principles: (1) customer satisfaction, (2) collaboration and partnership, and (3) continual improvement and learning. While CRM environments improve business performance, initiatives undertaken in this new management field require sound leadership, as well. The current focus of CRM tends to be almost entirely on the front office. Extending CRM into multiple media means interacting the front office (and aspects of back office, where appropriate) with different communication channels (Bradshaw & Brash, 2001).

Creating the optimal CRM environment within the supply chain is, and will increasingly become, a rapidly growing critical business challenge. CRM is a major strategic challenge for the twenty-first century (Fredendall & Hill, 2001).

Dimensions of CRM

Poirier and Bauer (2000) highlight three constituents in the preparation and execution of e-supply chains:

1. E-network: Business networks should satisfy customer demands through a seamless (fully connected end-to-end) supply chain to serve the end consumer (see also Towill, 1997);
2. Responses: Customer responses form the central theme of the supply chain strategy. The market value of the supply chain can be enhanced dramatically by jointly creating profitable revenue growth through integrated inter-enterprise solutions and responses; and
3. Technology: Each of the above constituents can achieve the purposes and goals of the supply chain by being supported with leading-edge technology, particularly e-commerce.

The three constituents: e-network, customer responses and technology could be seen as the "input" into the CRM environment, working together to achieve the ultimate aim (output) of the supply chain, i.e., customer satisfaction. In synergy with the model developed by Meade and Sarkis (1999) for agile manufacturing, Figure 2 models the dimensions of the CRM environment for SCM.

In order to achieve the CRM output in the SCM environment, members of the supply chain need to engage in comprehensive partnerships. A supply chain partnership, according to Yu et al. (2001), is "a relationship formed between two independent members in supply channels through increased levels of information sharing to achieve specific objectives and benefits in terms of reductions in total costs and inventories." Mason-Jones and Towill (1997) rejected the assertion that information should be considered as a source of market power and argue that the "real power of information only becomes evident when it is utilized throughout the supply chain." Ashankas et al. (1995) argue that effective partnership outsourcing requires information and resources movement across the boundaries of the firms in the value chain.

Process management involves planning and administering the activities necessary to achieve a high level of performance in a process, and identifying opportunities for improving quality and operational performance. Ultimately, it includes translating customer requirements into product and service design requirements (Evan & Lindsay, 2002).

Goldman et al. (1995) recognize the significance of employees as a company asset, and emphasize the importance of leveraging the impact of people and information for an agile enterprise. Evans and Lindsay (2002) show direct correlation between employees' (people) satisfaction and customer satisfaction, and argue that 'people' are the only organization asset that "competitors cannot copy; and the only one that can synergize, that is, produce output whose value is greater than the sum of its parts." Evans and Lindsay (2002) also emphasize the two key components of service system quality: employees and information technology. Meade and Sarkis (1999) state that people and information are the most valued resources. It follows that the mechanism that converts the input of CRM in the SCM environment to its output, that is, customer satisfaction, includes three constituents:

process, people and information sharing. In an analogy with agile enterprise dimensions (Goldman et al., 1995) the leading mechanism for CRM in SCM environment is leveraging the impact of process, people and information for the entire supply chain.

In an agile environment, employees' skills, knowledge and information are no longer enough for achieving or enhancing competitiveness without "the ability to convert the knowledge, skills, and information embodied in its personnel into solution products for the individual customers" (Meade & Sarkis, 1999). "Ability to convert" is what CRM is really relying on to achieve customer satisfaction. Such ability should be maintained via continual process improvement and learning. Accordingly, improvement and learning form the control dimension of the CRM environment for SCM, as shown in Figure 2.

E-COMMERCE, INTERNET AND WEB SITES

Electronic commerce (e-commerce) plays a significant role in the development of supply chains. The Internet and e-commerce technologies have become a strategic necessity for many business organizations. The World Wide Web is a distributed information system that is based on a hyper-text paradigm, and is now regarded as the principal navigational tool for accessing the resources of the Internet (Dholakia & Rego, 2002). Web browsers become the de facto standard of interface to data (Turban & Aronson, 2001).

Figure 2: Dimensions of CRM for supply chain management environment

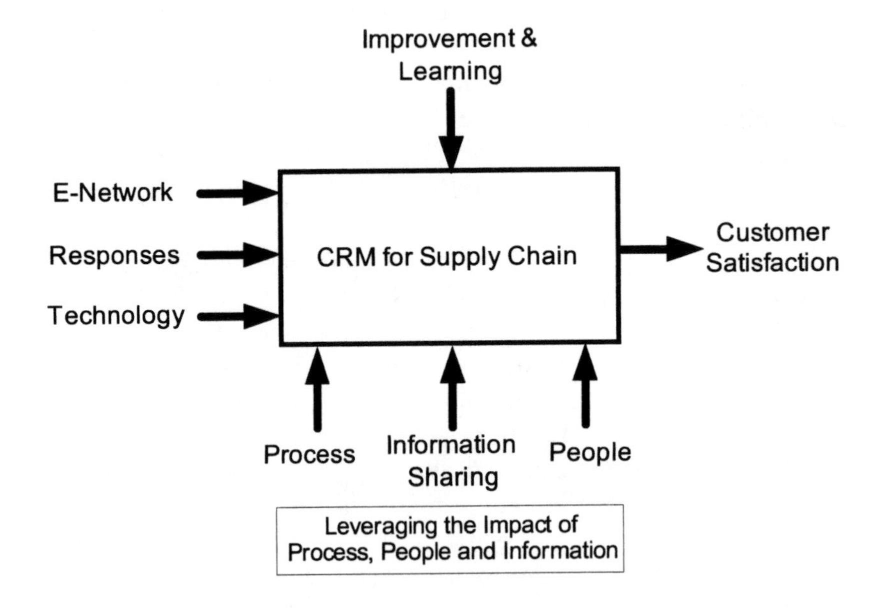

Through the combination of interactivity, networking, multimedia and data processing, the Internet offers a wide variety of new generation e-commerce opportunities. It brings with it ubiquitous connectivity, real-time access, and a simple universal interface provided by Web browsers (El Sawy, 2001). The current evolution in the Internet changes the way businesses communicate internally and externally with their customers, suppliers and partners, and creates a new environment for competitiveness. Enterprises are transforming themselves into e-commerce by reinventing the way they carry out their business processes to take full advantage of the Internet's capabilities, mainly because of several characteristics of this medium (El Sawy, 2001). Some of these characteristics are:

1. The exponential increase in the utilization of the Internet;
2. The capacity of the Internet to accommodate product or service customization inexpensively (Kling, 1994);
3. The Internet facilitates an interactive multimedia "many-to-many" communication network, (instead of traditional "one-to-many" communication model), thus radically altering the way in which firms can do business with customers and suppliers (Hoey, 1998);
4. With effective Web sites, the Internet not only supports application-to-application e-commerce similar to that of traditional electronic data interchange (EDI), but also person-to-person and person-to-application forms of e-commerce;
5. The Internet allows manufacturers, distributors and retailers to set up Web sites selling essentially the same products and services (Dholakia & Rego, 2002). This aspect raises the importance of information sharing with all the supply chain entities;
6. The customer is in greater control of choosing and processing information about the firm. It is entirely within the customer's power to decide which Web pages to browse, for how long, and how much information to obtain (Dholakia and Rego, 2002);
7. In general, the Internet provides businesses with a global presence, and customers with global choices and easy access to information (Simchi-Levi et al., 2000);
8. The Internet allows wider availability of hard-to-find products and wider selection of them (Hoffman et al., 1995);
9. The Internet allows organizations to reduce costs by helping divisional managers of an organization select suppliers having long-term relationships with other divisions of the organizations (Turban & Aronson, 2001); and
10. Finally, the Internet is a convenient and cost-effective media for lodging complaints. This allows organizations to create more effective complaint management and has a dramatic impact on customer retention rates.

The above characteristics create a revolutionary improvement in the communication and the exchange of digitized information between business-to-business partners. Electronic Data Interchange (EDI) is the traditional form of electronic communication between organizations. It allows one-on-one linkages between the computers of organizations' partners. The popularity of EDI increased dramatically in the 1980s; however, it has not been unilaterally, by business and industry (Lankford & Johnson, 2000). Many companies have adopted EDI, often with results inconsistent with the expectations of the company (Walton & Gupta, 1999). It is expected that the EDI growth in the next few years could be nurtured significantly with the prolific expansion of Internet access and use (Lankford & Johnson, 2000). Lankford and Johnson also show that the Internet is attractive to small businesses because of its very low costs, its handling of EDI and the opening of a new marketplace for linking with suppliers. Table 1 summarizes the main differences between traditional EDI and the current status of e-commerce.

Applications

The significant growth of the e-commerce applications is notable (Al-Mashari, 2002). The following are but some remarkably successful examples. Courier or distribution services, such as Federal Express (FedEx), allow customers to track their packages over the Internet (www.fedex.com, 2002). Now a subsidiary of Federal Express, Road Package System (RPS) has become the second largest business-to-business small package carrier. It effectively uses the Internet for controlling and monitoring its operations. The company even provides free Decision Support System (DSS) software to customers for report generation and work monitoring (Turban & Aronson, 2001). In the early 1990s, the Atlantic Electric Company of New Jersey was losing its monopoly and, to survive, the company had to become the least expensive provider in its territory. By using the Internet (see www.atlanticelectric.com) and Intranet facilities, the company provides employees with the information they need to make up-to-date decisions, and the information the customers need to attract them to join the company. By 2000, the company had developed the DSS applications on its Intranet (Turban & Aronson, 2001). Dell's suppliers share demand and inventory information online with the company. Netscape and Shockwave fill and deliver orders over the Internet through downloads of electronic products (www.netscape.com, 2002; www.shockwave.com, 2002). Transtec uses the Internet to communicate with its partners and is currently testing a "virtual warehouse" which accesses warehouse systems and databases belonging to all European warehouses operated by Transtec partners (Al-Mashari, 2002). Companies can use the Internet for negotiations and auctions to set prices of products and services. Companies like eBay allow people to auction products over the Internet (www.ebay.com, 2002).

Amazon is one of the largest online booksellers. Its website, (www.amazon.com), allows customers to receive constructive evaluation about the contents and selling rate of the books. It allows customers to make online payments by credit, provides

Table 1: Traditional EDI and current e-commerce status

Issue	Traditional EDI	e-Commerce
Software	Very expensive in price and in implementation	Dramatically reduces costs of software/hardware and the implementation of the technology
Volume of Transactions	Must be very high transaction volume	Large or small transaction volume
Company size	Because of costs, EDI has been used by large organizations	Small businesses can afford to use Web-based technology
Format	Multi-database architectures. Businesses must agree on certain format for transferring the data between them	Deals with diverse data formats and structures
Maintenance	Expensive	Relatively low-cost
Networking	Computer-to-computer data exchange between business partners only	Business-to-business and business-to-customer data exchange with the potential of information sharing along the entire supply chain
Paper work	There is a need to re-enter data from paper documents, high possibility of clerical error	Dramatically reduces the re-entering of data by the business and thus reduces paper work and time elements (Attaran, 2001)
Clerical error	High possibility of clerical error during data entering (Attaran, 2001)	The need for re-entering data from paper documents could be eliminated (Attaran, 2001) and thus clerical error could significantly be reduced
Delivery time	Delivery time is 12 to 24 hours (Lankford & Johnson, 2000)	Minutes in the Internet
Security	High data security	Relatively low data security. The Internet is perceived as an unsafe medium for sensitive information (Lankford & Johnson, 2000), however, this area is under intensive research and developments
Reliability	High -- there is responsibility over EDI reliability by the EDI developers	Relatively low, a serious vulnerability of the Internet is its lack of organization; no one owns it, and no one responsible for its reliability (Lankford & Johnson, 2001)
Utilization	Requires high and specific skill	User friendly
Simplicity	Requires large number of basic concepts and commands Users need specific skills for understanding and analyzing information	Minimal number of non-overlapping basic concepts and commands. Users can understand and analyze information in a more convenient and broad way.
Translation	Requires specific programming for translating certain entities and relationships	Ability to translate entities and relationships in the required source data with minimal effort
Accessibility	Could be accessed by certain staff and via certain divisions of the business	Could be accessed directly by any manager or employee
Requirement technology	Businesses have to be tied to client/server technology (Attran, 2001)	Just need a Web browser; product/services information can be called up from online catalogue and ordering can be done by sending e-mail
Empowerment	Usually requires management approval	Usually employees are empowered to make transactions and complete the transaction from their desktops (Attran, 2001)

options for shipping books and also allows customers to check the status of their orders on-line. Statistics show that Amazon's growth has been extraordinary. Its number of customers grew from 0.34 million in early 1997 to 4.5 million by the end of 1998, and passed the 10 million mark by 1999 (Al-Mashari, 2002).

However, it is not enough for a company to develop Web sites and provide its customers with easy access to its products world wide but not be able to deliver on time, due to poor customer service or delivery delays caused by a lack of integration among supply chain elements.

INFORMATION SHARING

Stalk and Hout (1990) expressly point to the problems associated with the slow flow of information between the organizations within the chain, which "causes amplifications, delay, and overhead." They emphasize "the only way out of this disjointed supply chain system between companies is to compress information time so that the information circulating through the system is fresh and meaningful." Along the same thinking as that of Stalk and Hout, Towill (1997) emphasizes that, to maximize competitive advantage, all members of the supply chain should "seamlessly" work together to serve the end consumer. One critical factor in achieving such a work environment is the "seamless" flow and sharing of information between the supply chain members. Mason-Jones and Towill (1997) conclude that tremendous benefits can result from adopting an holistic approach to the seamless supply chain, if the attitudinal problems associated with information sharing can be overcome. From the comparison of results of inventory reduction and cost savings of supply chain, Yu, et al. (2001) deduce that the Pareto improvement is achieved, in respect to the entire supply chain performance, with an increased level of information. A comparison between traditional information systems and those required for SCM allows identification of the attributes and characteristics of the information-sharing factor (Table 2).

BUSINESS PROCESS MODELS

Lee (2000) surmises that supply chain complexity grows exponentially because companies use the services of multiple suppliers, and often, each of them has its own suppliers. Gunasekaran et al. (2001) emphasize that, because of the increase in the complexity of the supply chain, companies rarely manage the supply chain without a framework. As a result, such a framework (model) provides the following in addition to the CRM core principles mentioned earlier:

(1) continual improvement of the entire supply chain business processes;
(2) valid line of customers' feedback to orient the improvement;
(3) seamless flow of information along a company's supply chain; and

Table 2. Comparison between traditional information systems and contemporary supply chain information systems

Attribute	Traditional information systems	Supply chain information systems
Transactions between organizations	Visualizes information as a set of repetitive transactions between different organizations, e.g., buyers, suppliers, distributors and retailers	Visualizes information as transactions of one virtual entity organization, by centralized coordination of information flow for the SCM
Information flow	Information flows between functional areas within an organization	System based on information that crosses organizational boundaries (Bakos, 1991)
Internet Access (Al-Hakim, 2002)	Web sites provide customers with access to the organization's products and to a certain extent allow transactions via Internet	Two level Web sites -- first level allows consumers to track their orders in real time, second level allows partners and suppliers to access internal information
Uncertainty (Yu et al., 2001)	Uncertainties arise due to lack of information, which may create "bullwhip" effect: excessive inventory management, poor customer service, lost revenues, misguided capacity plan, etc. (Lee et al., 1997)	With information sharing (partnership), the negative impact of "bullwhip" and uncertainties can be reduced or eliminated
Timely information	Slow flow of information between organizations	All member links in the supply chain are continuously supplied information in real time (Balsmeier & Voisin, 1996)
Communication between managers	Communication between partners and supply chain managers is individual	Communication between partners and supply chain managers is on-line and joint
Power of information (Mason-Jones & Towill, 1997)	Many organizations protect their information regarding it as a source of market power	The real source of information only becomes evident when it is utilized throughout the chain

(4) a channel that makes it possible for information to be located at a central source, such as the seller's Web server.

Modeling the processes for the supply chain is also necessary, because companies have an overwhelming number of processes that require integration. Modeling will help expose patterns among business units and supply partners (Alvarado & Kotzab, 2001). There are several process reference models available in the market. The most well known ones are the Supply-Chain Operations Reference model (SCOR), the Collaborative Planning, Forecasting and Replenishment model (CPFR), and the Demand Activated Manufacturing Architecture model (DAMA).

SCOR was developed by Supply-Chain Council (SCC) in 1996. SCC member companies include diverse industry leaders, such as Dow Chemical, Merck, Texas Instruments, and Compaq. SCOR is an extension of business process reengineering (Stewart, 1997). It classifies business processes into four basic processes - plan, source, make and deliver. It classifies these processes into detailed sub-processes, and offers benchmark metrics to compare process performance to objectives and to external points of reference. SCOR also describes the best-in-class management practices. This allows companies to prioritize their activities, quantify the potential benefits of specific process improvement, and determine financial justifications. Each process is a customer of the previous process and a supplier of the next (Figure 3). The "Plan" process manages these customer-supplier links and balances the supply chain.

The supply chain of an individual firm includes both its upstream supplier network and its downstream distribution channel. Supply chains are, within this context, essentially, a series of linked suppliers and customers. Every customer is, in turn, a supplier to the next downstream organization, until a finished product reaches the ultimate user (Handfield & Nichols, 1999). SCOR provides a list of best-in-class practices for each process element, and comprises four levels of process detail (SCC, 2002):

Level 1: At this level, the organizations define the scope content and the strategy for their operations.

Level 2: At this level, supply chain partners select the configuration of the supply chain and implement their operation strategy through the chosen configurations.

Figure 3: Configuration of supply chain. Adapted from SCC (2001)

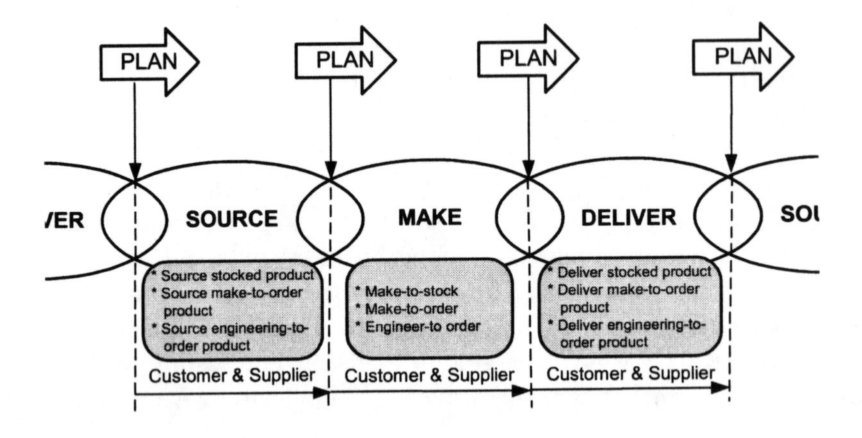

Level 3: This level identifies and defines the organization's ability to compete successfully in the market. At this level, organizations 'fine tune' their operations strategy. They define their process elements and the required information input and output. This allows the organizations to identify the process performance metrics, best practices and system capabilities required to support the best practices.

Level 4: At this level, the organizations implement the specific management practices and adapt the changing business conditions.

However, SCOR lacks in the area of external supply chain collaborative planning and long-term forecasting. Although, Level 4 of SCOR focuses on implementation, the specific elements of the level are not defined within the industry-standard model (Steward, 1997; SCC, 2001). SCOR lacks a roadmap for process improvement.

CPFR is a strategic, cross-industry initiative designed to improve supplier and retailer partnerships through co-managed processes and shared information. CPFR involves collaboration and joint planning to make long-term projections that are constantly updated based on actual demand and market changes (Stank et al., 1999). CPFR originated in 1995 as an initiative of five American companies: Wal-Mart, Warner-Lambert, Benchmarking Partners, and two software companies, SAP and Manugistics. The goal was to develop a business model to collaboratively forecast and replenish inventory (Harrison & Hoek, 2002). In 1998, the Voluntary Inter-Industry Commerce Standards Committee (VICS) was involved in developing computer enablers for CPFR. The VICS Web site, www.cpfr.org. shows various elements and modules of CPFR. A nine-step business model has been developed for CPFR as follows (www.cpfr.org, 2002):

Step 1- Develop Front-End Agreement
Step 2- Create Joint Business Plan
Step 3- Create Sales Forecast
Step 4- Identify Exceptions for Sales Forecast
Step 5- Resolve/Collaborate on Exception Items
Step 6- Create Order Forecast
Step 7- Identify Exceptions for Order Forecast
Step 8- Resolve/Collaborate on Exception Items
Step 9- Order Generation

The steps reflect the fact that initiatives in CPFR involve collaboration among the entities of the supply chain on an external basis. However, CPFR has several barriers (Barrat & Oliveira, 2001) which make the system sensitive to the agile environment:

(1) Ineffective replenishment in response to demand fluctuations;
(2) Ineffective planning using visibility of POS customer demand; and

Table 3: Comparisons of the three models

	Feature	Model		
		SCOR	CPFR	DAMA
1	Enter-enterprise collaboration	-	*	*
2	Demand-supply link	-	*	*
3	Intra-enterprise processes	*	-	-
4	Performance measurements	*	*	*
5	Benchmarking	*	-	-
6	Best-in-class practices	*	*	-
7	POS feedback / analysis	-	*	*
8	Generically	*	*	-
9	Real-time demand-production link	-	-	-
10	Improvement program	-	-	-
11	Web site feature	-	-	-

* *Note: The asterisks indicate that the model adequately incorporates the particular feature.*

(3) Difficulty in managing the forecast exception/review process (sales and order forecast).

The DAMA project was created in 1993 as part of the American Textile (AMTEX™) partnership (www.dama.tc2.com). DAMA represents an effort focused on increasing the competitiveness of the fiber, textile, sewn product or apparel, and retail industries (see also Mitchiner, 1999). Similar to CPFR, DAMA focuses on inter-enterprise collaborative business across the entities of the supply chain and does not provide any sub-process details of the industry processes.

The strengths and weaknesses of each model are apparent in Table 3. For instance, SCOR provides detailed intra-enterprise processes and allows for benchmarking, while it lacks enter-enterprise collaboration. By combining the features of both SCOR and CPFR, we can have a generic model which deals with both the inter-enterprise collaboration and intra-enterprise processes. However, all models lack real time demand-production link, improvement programs and Web site features.

IMPROVEMENT AND LEARNING

Collaborative supply chains will not be successful by only sharing the information. Success requires changes in the attributes of products and services throughout the entire supply chain. High quality, competitive prices, and just-in-time (JIT) delivery are but three attributes, among many others, that enhance the ultimate aim of CRM for the collaborative supply chain, that is, the customer satisfaction. These attributes can only be enhanced by continual improvement of processes (Figure 4). In addition, JIT delivery attributes necessitate coordination, communication and collaboration among suppliers. It follows that an improvement of a process in an organization will not have the balanced attributes, unless such improvement is consistent with the level of related processes of the other organizations in the chain. This requires linking the improvement programs of the supply chain in a manner that affects the attributes of the whole supply chain, rather than limited entities of the chain.

Needless to say, sharing information plays a significant role in dealing with improvement of the supply chain processes; but, improvement also requires real-time information sharing, and a means of analyzing and benchmarking the processes. One way is to develop an integrated model for this purpose.

Figure 4: Attributes of the collaborative supply chain

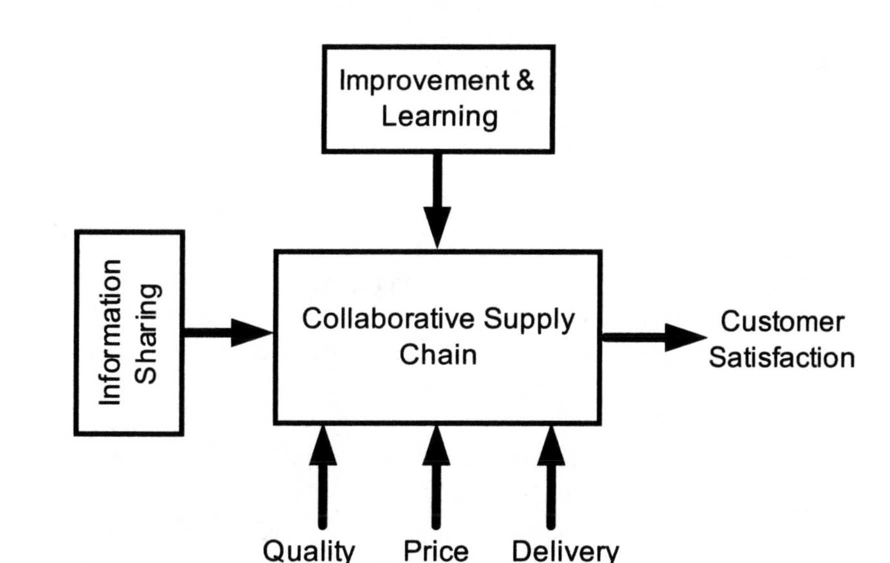

WEB-BASED MODEL

To create a more comprehensive model, we found that it is possible to enhance the SCOR with key features of CPFR, in order to achieve the following purposes:

1. Incorporate fuzzy set theory into dynamics of production scheduling, in an attempt to deal with vague constraints and to enable conflicting multi-criteria objectives to be managed effectively in the production environment; and
2. Identify potential processes in the chain that need improvement, and to synchronize the level of the improvement throughout the entire supply chain.

This can be managed within an integrating Web site comprised of two levels. The first level of the Web site is in the public domain. It allows consumers to track their orders in real time, to lodge complaints and to provide their input on a regular basis. This level can be used to obtain answers to questions such as:

- Which products/services meet your expectations? Which do not?
- What products/services do you need that you are not receiving?

The second level is accessible only to supply chain partners and suppliers. This level uses the collaboration features of CPFR. It allows partners and suppliers to access internal information, POS data and partners' specific activities. It also helps in identifying the core competence of each partner.

The integration between the two levels of the Web site is used to achieve the purposes of the integrated model.

Improvement Program

The implementation of a productivity improvement program requires mapping the "as-is" state of supply chain processes and deriving the desired "to-be" state. This necessitates:

1. defining the processes of the supply chain;
2. identifying the process element information input and output, activities, and "as-is" process capability;
3. identifying the best-in-class practices required to achieve the activities of the process;
4. mapping the logical flow of activities and information for each process of the supply chain, in association with the best-in-class practices;
5. determining the gap between "as-is" and the desired "to-be" state; and
6. defining the system requirements to change.

SCOR is a very useful tool to achieve Steps 1, 2, and 3 above. This will allow the SCM mapping the activities and information of the process in association with the best practices. Figure 5 illustrates such mapping (Al-Hakim, 2002). The mapping also makes possible the identification of the gap between the "as-is" and "to-be"

Figure 5. Mapping SCOR process element S1.3

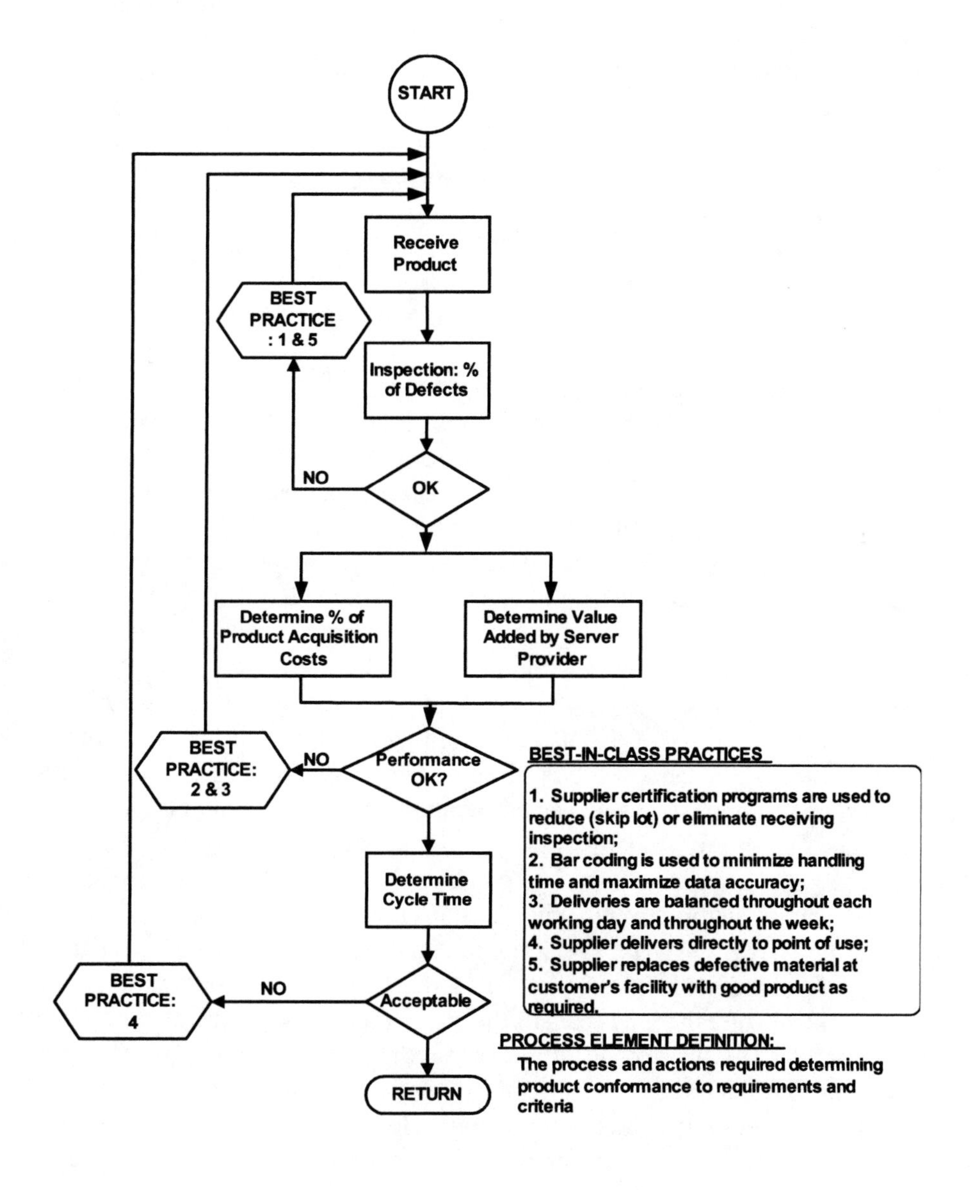

status, and then the definition of requirements for bridging the gap. The magnitude of the gap is an important issue to the SCM. If the gap is small, then an ordinary improvement of the process is needed; otherwise, a radical change of the process may be required, and under such circumstance, Business Process Reengineering (BPR) should be considered.

Dynamic Production Scheduling

The difficulty in today's manufacturing environment is that manufacturers can no longer rely on forecasting and entertaining high demand and market stability. Further, manufacturing firms must now compete by offering variety, and must strive to get their products to the market place in ever decreasing times. All of this means that manufacturing firms must put sequencing and scheduling jobs in place that are sufficiently flexible to accommodate (simultaneously) the following criteria (Al-Hakim & Al-Haj, 2000):

1. Respond quickly to unanticipated orders changes;
2. Meet orders due dates; and
3. Satisfy customers according to predetermined priorities.

The above criteria form a multi-criteria objective for the Master Production Schedule (MPS) of the Material Requirement Planning (MRP) system. Many organizations implementing MRP have been faced with disappointment. According to Al-Hakim and Jenny (1991), this is due to inability to cope with the dynamics of shop floor activities. Al-Hakim and Al-Haj introduce a fuzzy logic, in an attempt to solve the problem for a single organization. In a collaborative supply chain environment, Al-Hakim and Al-Haj's system can be extended to handle production scheduling for the entire supply chain by considering the entities of the chain as a virtual organization.

In conventional modeling and reasoning, the way of representing an object is crisp. By crisp is meant that the attributes of that object are either "yes" or "no" rather than "more" or "less." Much of human knowledge processing in manufacturing is based on concepts, ideas and associations that are neither crisply defined nor described in terms of probability distributions. Fuzzy set theory (Zadeh, 1965, 1968) provides a systematic way of combining and propagating measures of vagueness or imprecision through a production rule system.

According to Kerr (1991), every late job on the shop floor has some degree of 'lateness' associated with it, in contrast to the conventional set theory, which considers only whether the job is past its due date or not. The degree of membership of a job provides a measure for the number of days past-due of the late job. The same concept is applicable to the closeness of the job from the due date. The degree of membership becomes higher as the job become closer to its due date. Here we have two fuzzy sets: 'lateness' and 'closeness' in reference to the due date. All late jobs are members of the set 'lateness' with various degrees of membership. The sequencing of a late job becomes more significant, or higher, if its degree of membership becomes higher. In a similar manner, the sequencing of a job becomes more significant if its degree of membership in the set 'closeness' becomes higher, i.e., as the job becomes closer to its due date.

The membership function of each set is subjective, and depends upon user opinion. Figure 6 represents one possible distribution of the degree of membership for both sets. A job may be a member to one of these sets or member of both. Both sets can form one set referred to as 'completion' that defines the relative completion day from the due date (Al-Hakim, 1999). Figure 7 on the following page illustrates the fuzzy set of the job's 'completion' set. Job priority is another fuzzy set that forms part of the developed procedure. Where more than one job has the same degree of 'completion,' customer priority plays a role in deciding the most significant job. The significance or urgency of scheduling a job is directly proportional to its degree of membership in both completion and priority sets.

The procedure assumes that customer priority is not a subjective matter based on sales personnel. It is part of the manufacturer's competitive strategy. Quantity required, order frequency and the customer market share are but just three factors for assigning customer priority. However, the priority could be changed by top management as a result of changes in the market requirements and relations with the customer.

At any given time, if an order is received and entered into the system, the procedure sequences the received order according to its membership in priority and completion sets. The degree of membership of a job of the fuzzy set of completion improves as the due date gets closer and, accordingly, the order sequence relative to other orders could also be improved, provided that no orders with a higher priority are received and are close to their due dates. The new entered order with a high degree of membership of fuzzy set priority may affect the sequence of orders currently in the system.

The overall degree of satisfaction of the conditional parts of the rules of both completion and priority fuzzy sets could be represented by the degree of membership of the job of the fuzzy set 'sequence.' Fuzzy set 'sequence' could be represented as a matrix. The elements of this matrix determine the urgency, or priority, of scheduling jobs. Elements with value equal to "0" represent jobs with very low sequence degree and those with value "1" represent jobs with very high sequence degree.

Figure 6: Fuzzy set mapping function for closeness and lateness

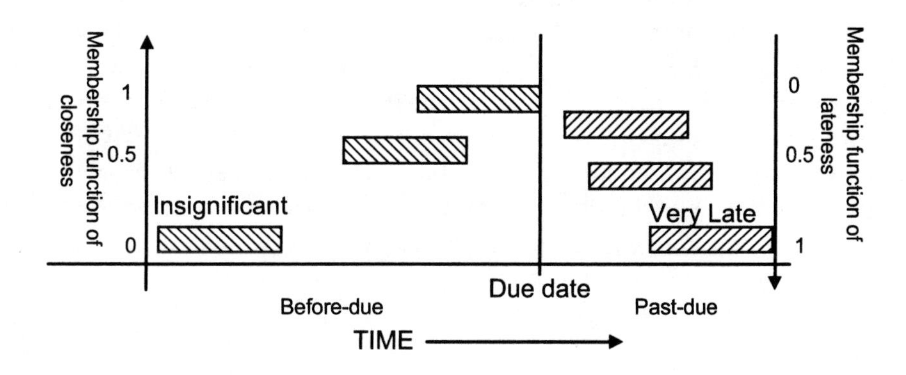

Figure 7: Fuzzy set of job completion

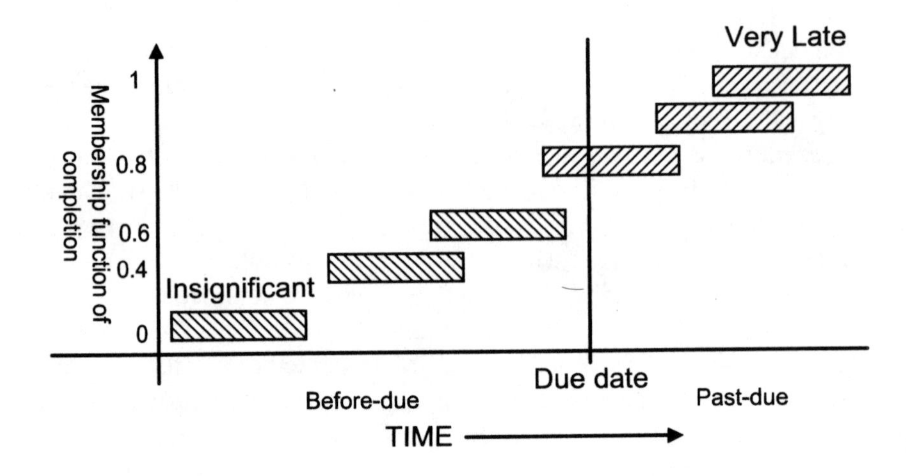

The system monitors the completed jobs, removes them from the job sequence list and re-sequences the current jobs according to their degree of membership of the fuzzy set 'sequence.' The fuzzy set 'sequence' is changed dynamically with every new order entered in the system. Unanticipated order change could be considered by removing the former order and replacing with a new order. The system lists the jobs in descending order according to their degree of membership of the fuzzy set 'sequence.' It governs the information flow from the MRP in relation to the gross requirements, scheduled receipts and net requirement for the parts and operations of the jobs, according to their degree of membership of the fuzzy set 'sequence.'

Integrating Model

The first level of the Web-based model connects the Web with the dynamic production scheduling module. This allows a real-time update of sequence of the received orders, according to their importance, priority and closeness to due dates. With the aid of CPFR, direct information from points-of-sale (POS) and the level of inventory across the supply chain can be obtained. This information determines the actual production requirements of the supply chain. The model allows automatic input of this information into the MRP systems of the various entities of the supply chain.

The flow of material and information are constantly monitored throughout the supply chain, and any bottlenecks are detected. Complaints received internally or from customers help to define the bottlenecks. With the aid of the second level of

Figure 8. A schematic view of the Web-based integrating model

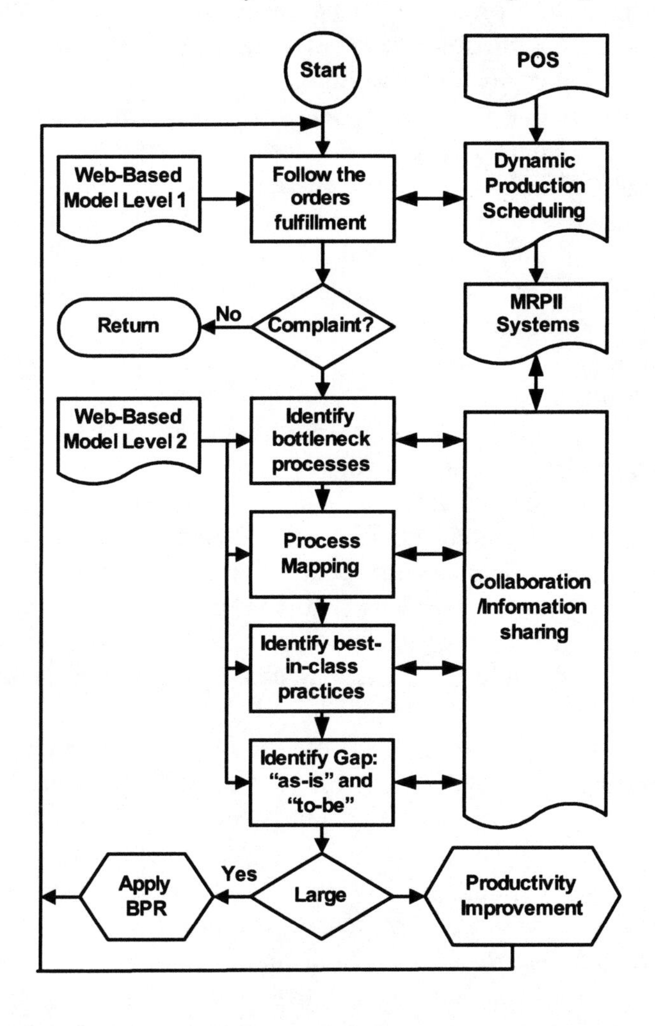

the Web-based model, the bottleneck process is mapped, and the gap between the "as-is" and "to-be" is defined. This allows the SCM to decide whether an ordinary improvement program is needed, or Business Process Reengineering (BPR) should be used to change the process radically. Figure 8 illustrates a schematic view of the integrating model.

CONCLUSION

Today's market is increasingly dynamic, unpredictable and volatile, forcing organizations to continuously focus on change and innovation, in order to maintain competitive advantage. Customer Relationship Management (CRM) and Supply Chain Management (SCM) have emerged as strategic solutions. Proactive managers begin to view customer satisfaction as their ultimate aim and to look to the supply chain as a whole, as well as promoting customer-focus, supplier partnership, co-operation and information sharing. The association of SCM and CRM with e-business offers new challenges for marketing.

The explosion of the Internet and other telecommunication technology also has made real-time, on-line communication with customers and throughout the entire supply chain a reality. This chapter introduces the concepts of SCM and CRM and discusses the dimensions of CRM for the supply chain. The chapter also emphasizes that the leading mechanism for CRM in the SCM environment is leveraging the impact of process, people and information. It also concludes that process improvement and learning form the control dimension of CRM.

The chapter introduces three well-known business process supply chain reference models: CPFR, SCOR and DAMA. The three models are compared and contrasted, in order to find the core features of each model. The chapter finds that SCOR and CPFR can complement each other, and develops a two level Web-based model based on SCOR and key features of CPFR. The first level of the Web site is in the public domain. It allows consumers to track their orders in real time, to lodge complaints and to provide their input on a regular basis. The second level is accessible only to supply chain partners and suppliers. This level uses the collaboration features of CPFR. It allows partners and suppliers to access internal information, POS data and partners' specific activities. It also helps in identifying the core competence of each partner.

To overcome the difficulties arising from the fluctuation in demand and the reliance on forecasting, the chapter introduces fuzzy set theory into the dynamics of production scheduling, in an attempt to deal with vague constraints and to enable conflicting multi-criteria objectives to be managed effectively in the production environment. The model simultaneously accommodates three criteria: (1) quick response to unanticipated changes in orders; (2) meeting due dates for orders; and (3) satisfying customers according to predetermined priorities. The model allows automatic input of this information into the MRP systems of the various entities of the supply chain.

The flow of material and information are constantly monitored throughout the supply chain, in order to detect bottlenecks. By mapping the bottleneck processes, the gap between the "as-is" and "to-be" can be defined. Analysis of the gap allows SCM to answer the question as to whether there is a need for an ordinary improvement program or for Business Process Reengineering to change the process radically.

REFERENCES

Abbott, J. (2001). Data data everywhere-and not a byte of use? *International Journal of Qualitative Market Research. 4*(3).182-192.

Al-Hakim, L. (1999). Intelligent scheduling. Working paper. Melbourne: Monash University.

Al-Hakim, L. (2002). Web-based supply chain integration model. In Khodrowpour, M. (Ed.), *Issues and Trends of IT Management in Contemporary Organizations.*, Hershey: Idea Group Publishing, 907-909.

Al-Hakim, L., & Al-Haj, H. (2000). Fuzzy logic aided dynamic scheduling with conflicting multi-criteria objective. *16th International Conference on Production Research (ICPR16).* 29 July- 3 August, Prague, Czech Republic.

Al-Hakim, L. & Jenney, B. (1991). MRP: An adaptive approach. *International Journal of Production Economics,* 25, 65-72.

Al-Mashari, M. (2002). Electronic commerce: A comparative study of organizational experiences. *Benchmarking: An International Journal, 9*(2), 182-189.

Alvarado, U.Y. & Kotzab, H. (2001). Supply chain management: The integration of logistics in marketing. *Industrial Marketing Management,* 30,183-198.

Ashkenas, R., Ulrich, D., Jick, T., & Kerr, S. (1995). *The Boundaryless Organization.* San Francisco: Jossey-Bass.

Attaran, M. (2001). The coming edge of online procurement. *Industrial Management & Data Systems, 101*(4), 177-180.

Bakos, Y. (1991). Information links and electronic marketplaces: The role on interaorganizational information systems in vertical markets. *Journal of Management Information Systems*, Fall 1991, 15-34.

Balsmeier, P.W. & Voisin, W.J. (1996). Supply chain management: A time-based strategy. *Industrial Management*, September-October 1996, 24-27.

Barrat, M. & Oliveira, A. (2001). Exploring the experiences of collaborative planning initiatives. *International Journal of Physical Distribution & Logistics, 31*(4), 266-289.

Bradshaw, D. & Brash, C. (2001). Managing customer relationship in the e-business world: How to personalize computer relationships for increased profitability. *International Journal of Retail & Distribution Management, 29*(12), 520-529.

Chantra, C. & Kumar, S. (2000). Supply chain management in theory and practices: A passing fad or a fundamental change? *Industrial Management & Data Systems, 100*(3), 100-113.

Cox, J.F., Blackston, J.H., & Spencer, M.S. (Eds.). (1995). *APICS Dictionary.* 8th edition. Falls Church, VA: American Productivity and Inventory Control Society.

Dholakia, U.M. & Rego, L.L. (1998). What makes commercial Web pages popular? An empirical investigation of Web page effectiveness. *European Journal of Marketing, 32*(7/8), 723-736.

Ellram, L. & Cooper, M. (1993). Characteristics of supply chain management and the implications for purchasing and logistics strategy. *International Journal of Logistics Management, 4*(2), 1-10.

El Sawy, O.A. (2001). *Redesigning Enterprise Processes for e-Business.* Singapore: McGraw-Hill.

Evans, J.R. & Lindsay, W.M. (2002). *The Management and Control of Quality. 5th edition.* Ohio: South-Western.

Fredendall, L.D. & Hill, E. (2001). *Basics of Supply Chain Management.* Florida: CRC Press LLC.

Galbreath, J. & Rogers, T. (1999). Customer relationship leadership: A leadership and motivation model for the twenty-first century business. *The TQM Magazines, 11*(3), 161-171.

Goldman, S.L., Nagel, R.N., & Preiss, K. (1995). *Agile Competitors and Virtual Organisations, Strategies for Enriching the Customer.* New York: Von Nostrand Reinhold.

Gunasekaran, A., Patel, C., & Tirtiroglu, E. (2001). Performance measures and metrics in a supply chain environment. *International Journal of Operation & Production Management, 21* (1/2), 71-87.

Handfield, R. & Nichols, Jr., E. (1999). *Introduction to Supply Chain Management.* New Jersey: Prentice Hall.

Harrison, A. & Hoek, R.V. (2002). *Logistics Management and Strategy.* Essex: Pearson Education Limited, Prentice Hall.

Hoey, C. (1998). Maximizing the effectiveness of Web-based marketing communications. *Marketing Intelligence & Planning, 16*(1), 31-37.

Hoffman, D.L., Novak T.P., & Chatterjee, P. (1995). Commercial scenarios for the Web: Opportunities and challenges. *Journal of Computer-Mediated Communication, 1*(3).

Jayaram, J., Shawnee, K., Dorge, V. & Droge, C. (2000). The effect of information system infrastructure and process improvements on supply-chain time performance. *International Journal of Physical Distribution & Logistics Management, 30*(3/4), 314-330.

Kerr, R. (1991). *Knowledge-Based Manufacturing Management.* Sydney: Addison-Wesley Publishing Company.

Kling, A. (1994). The economic consequences of the World Wide Web. *2nd World Wide Web Conference*. Geneva, Switzerland.

Lambert, D.M. & Cooper, M.C. (2000). Issues in supply chain management. *Industrial Marketing Management,* 29, 65-83.

Lankford, W.M. & Johnson, J.E. (2000). EDI via the Internet. *Information Management & Computer Security, 8*(1), 27-30.

Lee, H.L. (2000). Creating value through supply chain integration. *Supply Chain Management Review, l4* (4), 30-37.

Lee, H.L., Padmanabhan, V. & Wang, S. (1997). The bullwhip effect in supply chains. *Sloan Management Review,* 38, Spring 1997, 93-102.

Mason-Jones, R. & Towill, D.R. (1997). Information enrichment: Designing the supply chain for competitive advantage. *Supply Chain Management, 2*(4), 137-148.

Meade, L.M. & Sarkis, J. (1999). Analyzing organizational project alternatives for agile manufacturing processes: An analytical network approach. *International Journal of Production Research, 37*(2), 241-261.

Mitchiner, J. (1999). *DAMA Supply Chain Architecture.* Chicago, Illinois: 1999 Supply Chain World Conference.

Lummus, R.R. & Alber, K.L. (1997). Supply chain management: Balancing the supply chain with customer demand. Falls Church, VA: The Educational and Resource Foundation of APICS.

Poirier, C.C. & Bauer, M.J. (2000). *E-Supply Chain.* San Francisco: Berrett-Koehler Publishing, Inc.

Power, D. & Sohal, A. (2001). Critical success factors in agile supply chain management. *International Journal of Physical Distribution & Logistics Management, 31*(4), 247-265.

Simchi-Levi, D., Kaminsky, P., & Simchi-Levi, E. (2000). *Designing and Managing the Supply Chain.* Singapore: McGraw-Hill.

Stalk, G. & Hout, T. (1990). *Competing Against Time.* Free Press.

Stank, T., Daugherty, P., & Autry, C. (1999). Collaborative planning: Supporting automatic replenishment programs. *Supply Chain Management, 4*(2), 75-85.

Stewart, G. (1997). Supply-chain operations reference model (SCOR): The first cross-industry framework for integrated supply-chain management. *Logistics Information Management, 10*(2), 62-67.

Supply Chain Council. (2001). Supply-Chain Operations Reference-model: SCOR Version 5.0. Pittsburgh: Supply Chain Council, Inc. Retrieved from the World Wide Web: http://www.supply-chain.org.

Towill, D.R. (1997). The seamless supply chain: The predator's strategic advantage. *International Journal of Technology Management, 13*(1), 37-56.

Turban, E. & Aronson, J.E. (2001) *Decision Support Systems and Intelligent Systems.* New Jersey: Prentice Hall.

Walton, S.V. & Gupta, J.N. (1999). Electronic data interchange for process change in an integrated supply chain. *International Journal of Operations and Production Management, 19*(4), 372-388.

Yu, Z., Yan, H., & Edwin Chen, T. (2001). Benefits of information sharing with supply chain partnerships. *Industrial Management & Data Systems, 101*(3), 114-119.

Zadeh, L. (1965). Fuzzy sets. *Information Control,* 8, 338-383.

Zadeh, L. (1968). Probability measures of fuzzy events. *Journal of Mathematical Analysis and Applications,* 23, 421.

Chapter XV

A Cooperative Communicative Intelligent Agent Model for E-Commerce

Ric Jentzsch and Renzo Gobbin
University of Canberra, Australia

ABSTRACT

The complexities of business continue to expand. First technology, then the World Wide Web, ubiquitous commerce, mobile commerce, and who knows. Business information systems need to be able to adjust to these increased complexities, while not creating more problems. Here, we put forth a conceptual model for cooperative communicative intelligent agents that can extend itself to the logical constructs needed by modern business operations today and tomorrow.

INTRODUCTION

For more than ten years, the changes to the way business is being conduced has been accelerating. The acceleration is based on a range of developments in information technology and the World Wide Web infrastructure. The developments in information technology and the infrastructure presented by the World Wide Web (WWW) are well documented and discussed every day. With these changes, a more complex business environment, in particular, a global electronic business environment, now exists and continues to expand. Electronic business (e-business), with its essential partner, electronic commerce (e-commerce), continues to increase in number of users and data transactions. Computerworld estimated that, by the end of 2001, there were to be 450 million users of the Internet (Computerworld, 2001). Forrester Research projected that world-wide B2B (business-to-business) e-commerce alone will be valued at $6.9 trillion U.S. dollars by 2004 (Computerworld, 2001). Similar estimates in B2C (business-to-consumer) transactions (in both value and number of transactions) show similar increases to those of B2B estimates (Gartner, 2001; eMarketer, 2001). In this changing and growing environment, businesses continue to look for, and need, better and newer ways to deal with the increasing transaction volumes and the management of the complex e-commerce business environment. Individuals, in and out of organizations, need better ways to interact with the growing complexities that the internationalized business world is bringing to them, in order to achieve their goals and tasks.

This chapter describes a research project into the development of a cooperative communication intelligent agent conceptual model for e-commerce. We begin with the development of the model by illustrating a conceptual framework and architecture. The model, in its conceptual framework, applies to Internet commerce, mobile commerce and related electronic business areas, as well as today's wired e-commerce. The primary research emphasis is on business efficiencies, in particular labor efficiencies, with lateral research into such areas as, but not limited to: information dissemination, decision-making, and business intelligence.

Research Focus

Franklin and Graesser presented a taxonomy of software agents in 1996 (Franklin & Graesser, 1996). Figure 1 shows their taxonomy. Figure 2 illustrates the extension to Franklin and Graesser's taxonomy by Klusch (Klusch, 2000).

Our research flows along the lines of the cooperative agent branch of the software agent categories, as proposed by Franklin and Graesser (Franklin & Graesser, 1996) and extended by Klusch (Klusch, 2000). However, our conceptual model is directed toward the generic aspects of the e-business environment, including the transaction models of business-to-business (B2B), business-to-consumer (B2C), and business-to-employee (B2E) environments.

We are defining e-business in the broadest sense, which includes government, for-profit and not-for-profit organizations. We envision that our conceptual model will be able to be applied to the various types of electronic business, including mobile commerce and silent commerce, that go beyond the current transactions business model encompassing the entire spectrum of ubiquitous commerce. From the conceptual model, a more precise logical model will be derived. This perspective will be directed at specific-Internet based e-business environments and their use of intelligent agents. A complex example of this is given later in this paper.

Research has focused on (among others things) two areas that the dynamic complex world of e-commerce business comes with: 1) the ability to accurately filter user requests (business and individuals) of the globally available information and 2) finding ways that businesses and individuals can more efficiently utilize the resources available through the Internet, while achieving user efficiencies. The first area, information filtering has, and continues to be, researched by both practitioners and researchers (Jentzsch & Mohammadian, 2002). This area requires a concentration of technological efficiencies. The second area, user efficiencies, is a complex business and social issue. User efficiencies has been looked at from various perspectives for many years, but is only beginning to be addressed in current research in e-business.

The costs to business in doing e-commerce alone continue to increase. One of the major cost factors is the labor intensive nature of many of the current e-business activities. Thus, the need for the research and development of models and architectures to support businesses and individuals in achieving their goals and objectives in the e-commerce, m-commerce, s-commerce and total e-business environments. These models, with their frameworks and architectures, can lead to

Figure 1: Franklin and Graesser agent taxonomy

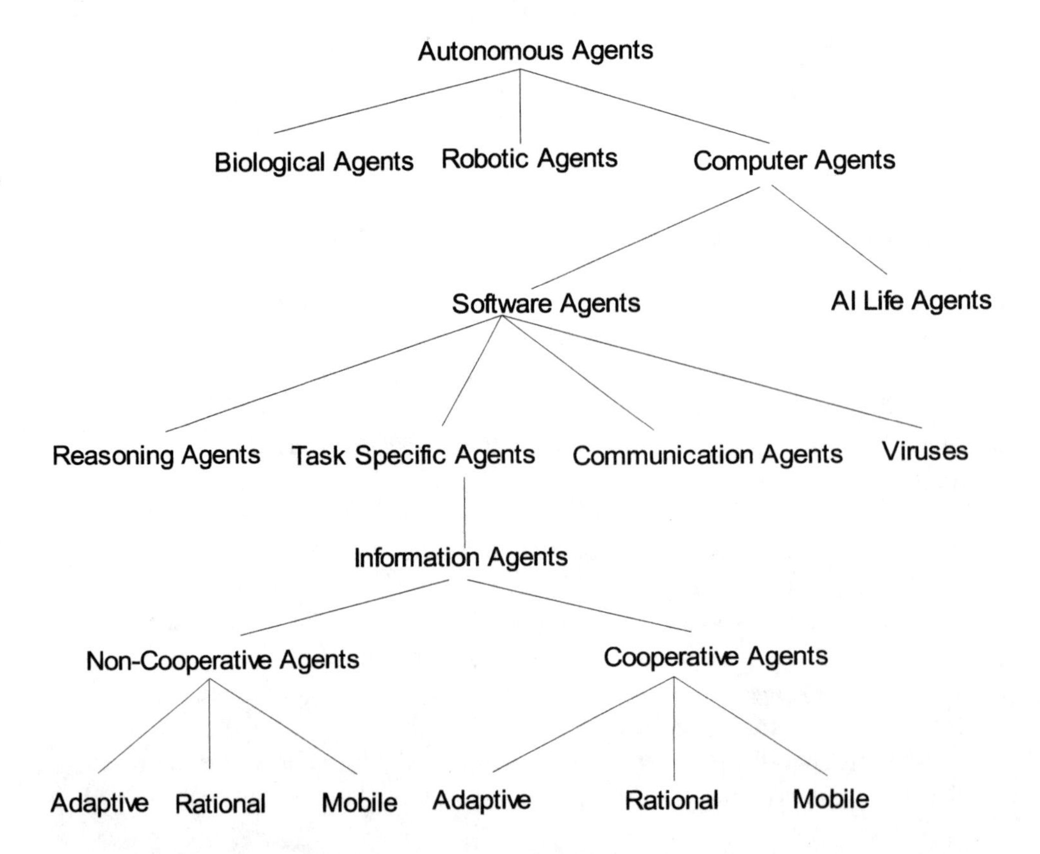

the development of applications to address the labor-intensive operations that the vast resources of the Internet are imposing on businesses and individuals. The first step in the development of these applications is the design of appropriate models and architectures that can work with and for businesses and individuals. To this end, intelligent agents can be considered a viable and essential player in aiding business and individuals in labor efficiencies.

Agent Taxonomies

In 1996, a software agent taxonomy was proposed by Franklin and Graesser (Franklin & Graesser, 1996). This taxonomy is considered to be a classic example of software agent categorization. This taxonomy provides a mapping for researchers in their development of agent models and architectures, as well as broadening into other research areas.

Klusch in 2000 (Klusch, 2000) made an addition to Franklin and Graesser's (1996) software agent taxonomy diagram, as seen in Figure 2. A new category was added to the cooperative agent's branch. This additional subcategory resides at the same level as adaptive, rational and mobile agents. This subcategory, "*communicative agents,*" reflects recent research activities into software agents.

Klusch's addition to Franklin and Graesser's taxonomy brings up an important point. While non-cooperative agents, such as search robots and other software agents, are considered to be software tools, communicative agents need to display a cooperative behavior. This behavior depends on the particular context of the agent's activities. This behavioral component provides us with a means to branch out and further develop agent cooperative concepts for modeling a cooperative communicative intelligent agent framework and architecture. However, this does add complexity. This complexity includes cooperation among agents and intelligent agent communication that requires the use of one or more ontology's.

Figure 2: Klusch's agent taxonomy extension

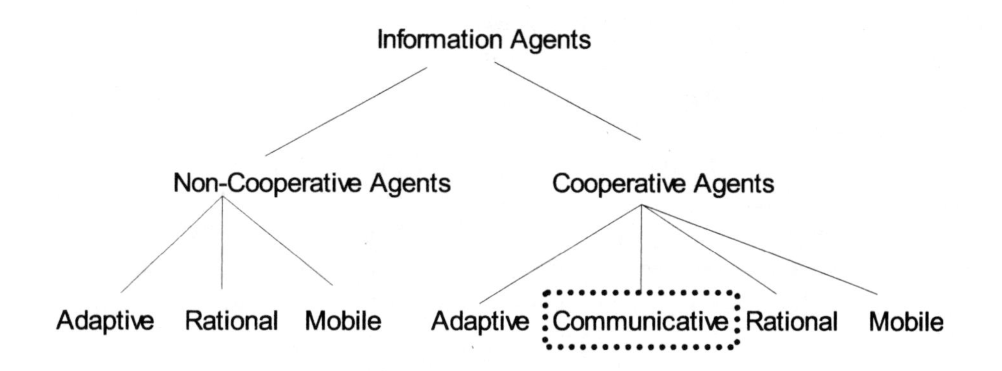

Agent Concepts

Agents operating, on the Internet and other environments, are being used for a range of business and individual applications. These applications include information dissemination, simple decision-making and business intelligence, all of which relate to labor efficiency. These applications complement and enhance the activities conducted by businesses, as well as those conducted by individuals. Our research perspective is first the development of a conceptual model and, second, a logical model. The logical model applies to, and is to be illustrated, based on an application that is applied to individual and complex business related activities. The application is directed at improving labor efficiencies in dealing with the ever-growing complex business environment that has resulted from the Internet being used as an expanded business medium.

The pace of recent technological advancements in the area of Internet data transactions, labor costs and information retrieval has generated a lot of interest in the use of *agent* and *intelligent agent* technology for business and, to a lesser extent, for individuals. The terms *software program*, *software agent* and *software intelligent agent* are often used and, far too often, used interchangeably. The term agent also is used to refer to concepts that have been applied to either or both types of agents, as well as non-software agents.

Thus, the question: what is the difference between a software program and an agent? Franklin and Graesser (1996) put that question in a codified perspective. Two of the most important distinctions between software programs and agents are awareness and control of behavior. A software program does what it is told to do, and goes where it is told to go. An agent can incorporate the software program characteristics, but adds self-awareness and self-control in relationship to its environment. Self-awareness includes knowing where the agent is and who it is in its environment. Before delving too deep into this, let's better define an agent.

In general, an *agent* represents or acts on the behalf of another agent. An agent may be some type of software agent, an entity, or a human agent. The term *software agent* (i.e., *agent*) means that the software application has more "capabilities" than the conventional information technology software application. The term *software intelligent agent* (i.e., *intelligent agent or i-agent*) not only acts on behalf of another agent, but also has the capability to make one or more decisions, within its awareness of its environment, on behalf of another agent. These decisions can range from very simple to extremely complex. Overall, *intelligent agents* can be thought of as being "more clever" than traditional software applications (Klusch, 2000).

Agents that are in production, and those being researched, are designed for a specific task or purpose. For example, software agents are being used for maintenance and reporting of problems with and status of telecommunications, servers, and network management. In the business arena, agents are being used to improve the efficiency of business practices, especially labor-intensive practices, such as constantly checking for available resources before committing the entity to those resources.

We are using the term agent to refer to the greater body of agents, thus incorporating agents and intelligent agents.

Intelligent Agents

Intelligence is not a physical quality, but rather an abstract property that usually is associated with an entity. Although a clear and satisfactory definition for intelligence is still being debated, and will most likely be debated for many years, instances of intelligent activities have been characterized by the various disciplines (Sperber, 1994). The general view is that whatever the term intelligent agent refers to, these intelligent agents will have a basic set of attributes and facilities that distinguish them from other software agents. For example, an intelligent agent's *state* must be *formalized* by its knowledge and its ability to *act* within the limits of that knowledge. Furthermore, if an agent is intelligent, it must be able to *interact* and *cooperate* within its environment. Expanding on the description of an intelligent agent, we add its ability to be communicative. A communicative intelligent agent needs to be able to associate with other agents (human or electronic) using some type of agreed upon communications. An i-agent must be aware of its own existence, its environment and its status at any point in time. Researchers have long been using an understanding of how humans think as a model for designing software intelligent agents (Odell, 2000).

In addition to the attributes mentioned above, i-agents need to incorporate many operations. These operations involve creative reasoning and inferences that need to integrate a range of existing and new data in information identification. This integration is on the basis of inferential processes that are deemed to be *intelligent*. If a process is just a data-driven, automatic reaction to a specific environment, or input from that environment — such as a thermostat — it would be a software program or a tool, rather than an i-agent (Franklin & Graesser 1996).

Cooperative Communicative Agents

Cooperation is defined as a multiple agent activity, whereby two agents work together to achieve their goals. A communicative capability is needed in order for the agents to exchange information about their goals, nature and status. The agents share and evaluate each others' information, so that they can achieve their goals. We are not, at this time, considering belief, desire or intention to the extent that Georgeff, Pell, Pollack, Tambe, and Wooldridge did (1999). Although we believe these are important, these areas are not currently within the boundary of this research.

To accomplish cooperative communicative activities between agents, there are four high level properties that agents need to incorporate. These four properties are:

- mediate activities using one or more symbolic languages;
- subjective and objective qualities;

- internalization and externalization abilities; and
- mediate activities using communicative tools.

To Klusch's taxonomy, we have extended the cooperative communicative agent's taxonomy description, as illustrated in Figure 3.

Mediated activities require communicative tools that utilize one or more symbolic languages. Mediated communication activities and symbolic languages provide a means for agents to establish an agreed upon common medium of exchange. Agents need the ability to use flexible communicative tools and symbolic patterns in order to facilitate mediated cooperative communicative agent activities. Agent research and developed prototypes have used a number of symbolic languages. As of 2001, no one agreed upon standard exists. DARPA, FIPA, and OMG are trying to lead the way by proposing standardized symbolic languages. FIPA-ACL (FIPA, 2002) and KQML (Cohen & Levesque, 1995; Fuchi & Yokoi, 1994) are two symbolic language standards. Even though these standards are a step in the right direction, agent models do not, at this time, sufficiently integrate the dynamics of the existing tools in mediating communicative intelligent agent activities.

Subjective and objective qualities are required by communicative intelligent agents to perform bi-directional and multiple communicative activities. Agents need to have the ability to internalize knowledge representations for communicative patterns and to externalize internal patterns for knowledge representation. In so doing, the intelligent agents can extract from software programs, as well as associate and communicate with other agents of all types. This provides a basis to facilitate inter- and intra-agent communicative activities.

Figure 3: Cooperative communicative agent

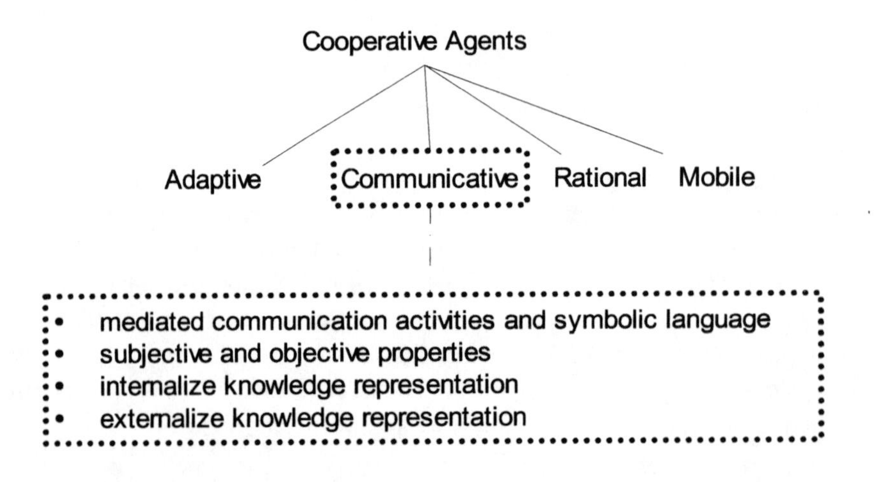

Internalization is about how the agent refers to itself, its self awareness and its accessible knowledge base. Externalization is how the agent, knowing about itself and its status, references itself in relationship to its external environment. The incorporation of these properties, and those shown in Figure 3, provide a theoretical basis for investigating a range of intelligent agent models with their frameworks and architectures in cooperative communicative intelligent agents.

For an agent to be fully cooperative and communicative, self-awareness is an important issue. Activity theories, as proposed by Vygotsky (Vygotsky, 1978; Vygotsky, 1986), provide a basis for the explanation of cognitive development. Activity theories provide a theoretical framework for investigating cooperative communicative agent activities. By linking communicative activities to a developmental learning process, Vygotsky proposed new directions in the area of cognitive sciences involving learning, thinking and language development. These areas are being extended into the development of intelligent agent models and architectures. A number of variations of Vygotsky's theories have been used in the research of human computer interaction to model the flow of communication tools for human-computer mediated activities.

THE ALMA MODEL

ALMA (Agent Language Mediated Activity) is a conceptual cooperative communicative intelligent agent model within a framework and architecture. The model is based on a theoretical mediated activity framework. The model is used to describe a range of cooperative communicative intelligent agent activities that are applied to e-business and individual activities.

The Agent Language Mediated Activity (ALMA) differentiates between processes at various categories and levels, taking into consideration the goals and objectives of these processes. Objectives include such areas as activities and actions. *Activities* are oriented toward *motives*. Each motive is a conceptual knowledge-based system's object that satisfies one or more of an agent's need. *Actions* are the processes functionally and hierarchically subordinated to activities that are driven by the specific intelligent agent's own goals and objectives. Actions are realized through step-by-step operations that are determined by the actual condition of the activity in its environment. The range of agent activities (agent evoking agent) involves agents' *motives* that are mainly of environmental origin, while agent subject *actions* are mainly goal-driven. Agent operations involve formal interaction of the agents' subjectivity and objectivity with the mediation of the communicative tools (Kaptelinin, 1996; Kutti, 1996). In short, motives are driven by environment and actions are driven by goals and objectives.

ALMA follows Vygotsky's work, in its classic form, and takes into account the subjectivity and objectivity dialectics of agent mediated communicative activities. The ALMA model will be used as an initial step in the research-in-progress into the development of a cooperative communicative multi-agent agent prototype. The development is intended to display subjective and objective traits, and to perform internalization and externalization of conceptual representations.

In describing the research that has produced the initial design of the ALMA model and its architecture, an activity model is presented. This model is derived from activity theory, and is presented as if the agent were autonomous. Next, agent concepts are expanded to incorporate communicative properties into the model. Finally, the model incorporates intelligent agent characteristics, thereby showing the current state of the research and development of the ALMA model and its architecture.

Activity Model

Figure 4 is an agent activity model. This model considers an agent as if it were autonomous. Within the oval is the environment of the activity. Within this environment, all mediated activities for a communicative agent occur.

The *activity model activities* take into account an agents' *subjective* (subject), as well as *objective* (object), characteristics. This model can be used to describe rational, inferential and potential intentional agent activities. The agent's *mediated activities* are mediated using a communicative tool, such as FIPA's Agent Communication Language (ACL).

Subject or subjective properties of the agent have a fuzzy boundary. The subject illustrates the agent's capability to act or perform an action. It can be an originator of an act or can act on behalf of another agent. Some electronic agent or human agent evokes this initial action or act, as if the agent were autonomous.

The communicative tool, or just, tool, mediates the subject's act or action. This mediation can be static or dynamic, and adapts to the particular environment that the agent is in. The tool performs its mediation activity in influencing how the goals and

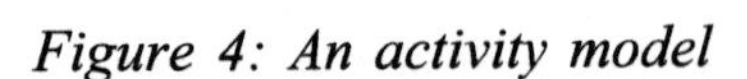
Figure 4: An activity model

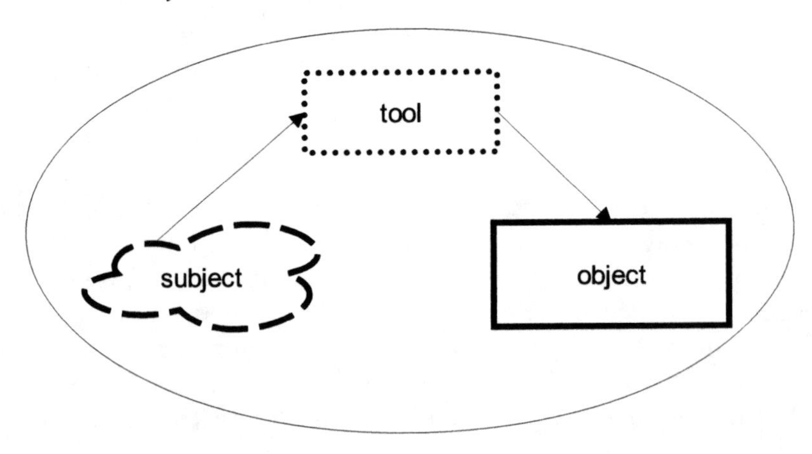

objectives are accomplished. The resulting object represents the final objective of the originating subject's act or action.

By using a tool in the activity model, we can analyze agent subjectivity, its internal and external processes, and its objectivity in an asymmetric fashion. The current object-oriented approach to model agent activities is directed toward the design of the agents' internal modules. However, we note that object oriented (OO) modeling provides an environment where an agent and its communicative tools and goals are symmetric. This creates a situation where the dialectics of an intelligent agent's *subjective* and *objective* communicative activities, in an environmental context, are not adequately taken into account.

When a subjective entity is dialectically related by tool mediation to a goal, as in an objective cognitive concept, then their subjectivity and objectivity can be dialectically identified as a single agent entity. We can derive, from Figure 4, that whenever agents are internalizing communicative concepts, a modification of their knowledge base using some type of ontology is needed.

As an agent is a substitute for a range of human activities in a particular instance, commercial or industrial context, mediated activity tools are ideal for modeling agent activities where agents' individual actions can be analyzed in a contextual environment. Agents' activities are also *dynamic,* and under continuous development in an historical time-related environment (Kutti, 1996). The more important aspect of mediated activity is the use of tools as *mediators* in performing an activity, hence the term *mediated* (Sperber & Wilson, 1995; Nardi, 1996; Parker, 1993). Autonomous agents are, in fact, using symbolic linguistic tools, such as ACL, as mediators in communicative activities.

A number of cognitive models can be used to describe agent's activities in a contextual environment. The situation action model explains the contingent nature of agent activities out of a given situation. It follows that this model focuses on the situated activity without taking into account the organizational structure of multiple agent's relations, the accumulated knowledge or the agent historical values in the knowledge base.

Communicative Activity

One of the most important changes in current agent research is the development of multi-agent coordination and cooperation requiring a range of communicative activities that are *necessarily mediated.* This mediation involves the use of symbolic language tools that requires a common or "translated" ontology. The theoretical approach is centered on mediated linguistic tools and ontology standards that allow different agents to be able to communicate and interact throughout a range of environments. In electronic business environments this becomes an essential feature, as there is no guarantee that the communicative requests of agents encountering other agents can be translated.

Figure 5: Communicative environment

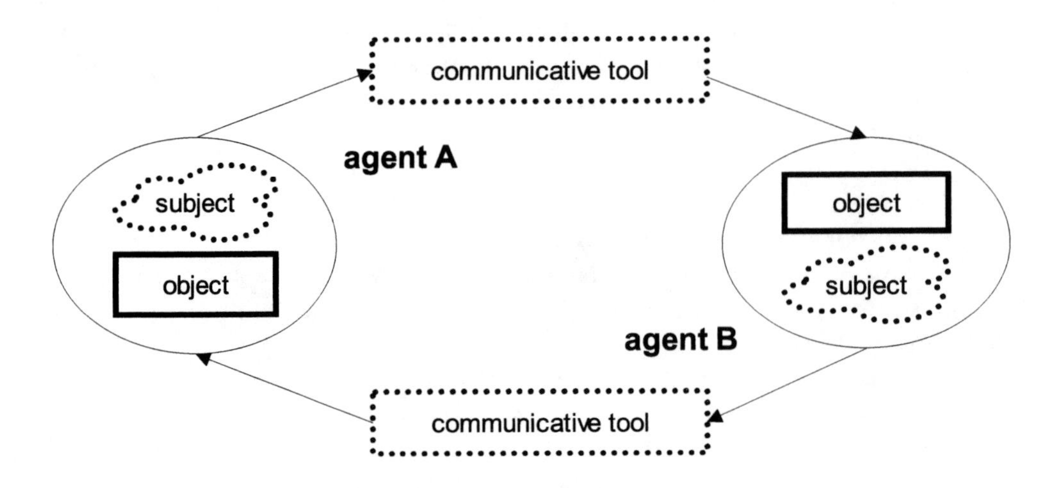

Our communicative environment, Figure 5, shows two agents, in the ovals, with their particular activity model. Some type of agent control language is required to facilitate the communicative activities, thereby enabling the two, or possibly more, agents to communicate.

Therefore, as a result of communicative activities with a number of agents, a cognitive learning activity process can be said to happen in the agent knowledge base. Cognitive scientists and philosophers quite often imply, in a metaphorical way, that speech and language are, in fact, mediating tools. For example, Wittengstein (1958) on his "Philosophical Investigations" relates tools in a toolbox with word generation functionality. Kempson contrasts Austin's (1965) concept of speech-act with Grice's (1989) co-operative principle of communication. In both cases, the underlying concepts are considered in agent communicative activities.

The concept of mediating agent communicative activities is important in modeling cooperative communicative intelligent agent communications. Although agents are software programs, they are conceived in the ALMA model as entities with subjective and objective qualities, as shown in our activity model (Figure 5), with self-awareness and behavior management. Communicative tools necessary to mediate environmental communicative perceptions need to be performed by agents, otherwise an agent will not have a cooperative ontology in place to perform efficient communicative activities.

Agent modeling requires, therefore, an analysis of the context in which the agent performs its activities. Communicating agents must be able to integrate easily within the context of their environment, such as the Internet or wireless networks, where a range of multiple agents interact and cooperate with people, organizations and federated multi-ontology systems. This communication must allow for interactive symbolic *mediated activities* using some form of standardized symbolic language as the communicative tool.

Intelligent Activities

Intelligence, in this case, says that agent B is given decision-making permission by another agent (agent A) to make one or more negotiated decisions on behalf of agent A. From the computer science perspective of intelligence, software agents are considered autonomous, asynchronous, distributed processes with distinct object-related traits, thus they do not act for another agent. From the artificial intelligence (AI) perspective, agents are communicative, intelligent and rational, with the possibility of intentional communication, thus exhibiting *subjective* traits but not *objective* traits. These two perspectives require differing architecture design and modeling approaches. The first incorporates the *object-oriented* characteristics, while the second implies intelligent communication and therefore requires a *subject-oriented* paradigm.

Object-oriented modeling has been well researched, but subject-oriented models are still in their early research stages, and require a multidisciplinary approach with such disciplines as cognitive science and artificial intelligence. Object-oriented modeling, while important for the design of agent encoding/decoding modules and symmetric processes, is inadequate for modeling intelligent agents' communicative activities performed in a contextual environment.

THE ALMA MODEL

As previously described, the unit of analysis in a tool-mediated activity model is a mediated activity in a given contextual environment. The work of an encoding / decoding device is not inferential or creative. It is not inferential, because the symmetrical relation between a message and a signal is quite different from the asymmetric relation of premises to conclusion, i.e., the meaning of a sentence doesn't logically follow from its sound. Also, an encoding/decoding device is rarely creative. It could be dangerous if it were. A creative encoding/decoding will change the process symmetry quite often, generating communication errors. Therefore, encoding/decoding can be just an ancillary process of the creative and inferential language mediated activity of a symbolic linguistic communication in specific contexts (Sperber, 1994). A creative and inferential language mediated activity, in fact, requires the *asymmetric* kind of model that the ALMA model proposes. ALMA proposes to be able to frame a number of agent internal activities, as well as external ones.

Inferential activities are encapsulated *inside* the agent architecture, involving the process of internalized conceptual representations stored into the agent internal knowledge bases. Using the MA model described above, communicative *subjective* self-expressions are *externalized* using mediating linguistic tools and, subsequently, the same agent can *internalize* communicative expressions in an *objective* fashion (Vygotsky, 1978; Vygotsky, 1986).

These activities are extremely important for modeling agents' learning and development processes, and will be investigated thoroughly in our next stage of research. Agents can *externalize* their ontology and knowledge base of data ontogenetically implanted by the programmer at agent creation time. However, the ability of *externalizing* inferential knowledge (*internalized* by agents during the process of communicative mediated activities) implies an agent possesses a number of internal inferential activities mediated by a range of tools encapsulated inside the agent architecture.

The ALMA model is designed to perform activities in an environment, as shown in Figure 5, where the communication process between agents uses an ACL and is based on subject and object communicative *unidirectiona*l interfaces. The interfaces will be designed specifically to provide an agent with a range of communicative activity capabilities — necessary to perform the externalization of internally produced ACL messages and the internalization of ACL messages received from the environment.

Subjective and objective inference engines are to be built inside the agent model. ALMA uses a number of specialized knowledge bases tailored for each of the specific subject and object modules. A knowledge base for high level meta-representations, integrated with the other knowledge bases, will provide a federated KBS environment (Wiederhold, 1994; Gruber, 1993; Guarino & Welty, 2000; Sperber, 2000). A multiple ontology environment capable of storing communicative representations for both internalization and externalization activities will be available, together with an ontology for conceptual meta-representation. The ALMA model can be easily associated with recent cognitive sciences theories, such as Sperber's (Sperber & Wilson, 1995; Sperber, 2000) language meta-representations and relevance theories.

A diagram of the ALMA architecture that is currently being researched is illustrated in Figure 6. The diagram clearly identifies the subject and object modules and associated connections. The ALMA internal modules take into account the current standardization efforts, although certain critical areas need to be built specifically with the theoretical cooperative communicative framework in mind.

The architecture shown in Figure 6 represents the current state of the cooperative communicative intelligent agent conceptual model. The dotted line indicates the proposed boundary of ontology in the model. However, the ontology itself is being researched as being two internal agents, thus agents within an agent. The dotted line also represents the communicative tool that would be incorporated into an intelligent agent. Research continues in looking at meta-representations externalization and in the area of conceptual representation and reflection in information systems (Edmond & Papazoglou, 1998).

Figure 6: ALMA model and architecture

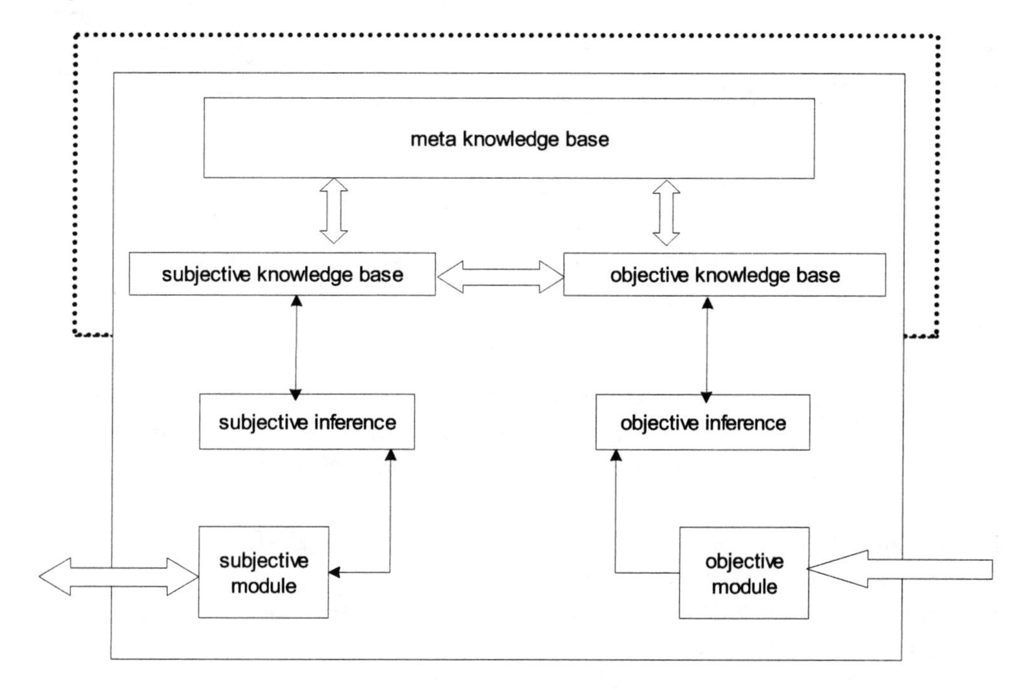

Current research is targeting the subjective and objective modules at the conceptual level. Within the activity model and communicative model, the work by FIPA and OMG, as previously noted, will try to adhere to current ACL standards, while moving the ALMA model into a logical environment.

SCENARIOS

The ALMA concepts can be applied to simple or complex business or individual situations. Although this scenario is not strictly an e-commerce example, it illustrates the ALMA model and architecture and its application. The major difference between simple applications and more complex applications is the number of interactive agents and infrastructures that would be needed to support the cooperative environment and communications. Many of the very simple non-cooperative communicative applications have been done by organizations that create agents.

Simple Scenario

Let us pose a simple scenario with a limited number of variables. A traveller is going from Sydney, to San Francisco, to Seattle, to Shanghai, (via a four-hour stop over in Tokyo, Japan), and back to Sydney. Within each of the three (SF, Seattle,

Shanghai) cities, the traveller has set up at least one meeting. In San Francisco and Seattle, the meetings are two and three hours after arrival of the flight. This trip requires accommodations in San Francisco and Shanghai. A rental car is needed in Seattle, but not the other cities. The traveller has a PDA device and his office has a server that the intelligent agents originate from.

We will use several intelligent agents: one to monitor the flights, one to keep the organizations and home office informed, and another for accommodations and car rental. Other agents at the various organizations are involved, also. In this case, the ontology will need to include two different languages, and possibly three different data formats. The languages to be considered are: English and Chinese. The data formats depend on the receiving devices and servers at the various organizations.

The plane leaves Sydney and runs into headwinds, thereby it will arrive in San Francisco two (2) hours late. This will cause a change to the meeting schedule in San Francisco. The intelligent agent of the traveller (agent A) informs the San Francisco organization of the delay, whereby an agent (agent SF) at the organization where the meeting is to be held checks the meeting participant's schedule, or participant, for a re-booking time. Then agent SF and agent A reschedule the meeting, and all parties are informed by their respective agent. Agent A also needs to inform the accommodations agent that the traveller will be checking in late. Agent A also informs the traveler's home office of the change in schedule.

The plane to Seattle is cancelled 60 minutes before the scheduled flight time. The traveller's agent (agent A) needs to book the traveller on another flight, and inform the car rental company of the change of arrival time and flight. Agent C at the airlines finds what options are available, and agent A makes a decision. Agent B at the car rental company looks for a suitable replacement, and agent A makes the new reservation, and informs the traveller of the car rental and the new flight schedule. Then, agent A informs the home office of the changes.

The plane from Seattle to Shanghai via Tokyo leaves on time. However, problems occur in Tokyo that delay the flight to Shanghai by six hours. All Shanghai meetings need to be rescheduled, and the new schedule needs to be negotiated with the Shanghai people. The accommodations agent needs to be informed of the change in arrival of the traveller.

Complex Scenario

This will actually take less time to explain then the simple scenario above, assuming reader knowledge. We have a truck moving goods from point A to point B. Points A and B are 300 miles apart. Between the pickup, point A, and delivery, point B, the road is a highway (not necessarily a freeway). The variables that the delivery would like to consider are: road conditions, weather conditions, traffic conditions, alternative routes, booking new or additional loads, timing considerations, and mechanical problems of the truck hauling the goods. Before leaving the pickup point, the driver will most likely have some information on weather and road conditions, while traffic conditions can change very quickly. The driver might even have some load information.

An intelligent agent is sent out from a truck. The cooperative communicative intelligent agent is originating from a mobile environment, while checking traffic, road, and weather conditions stored and updated on a continuous basis by other entities in other environments. The agent also needs to communicate with the truck's home base to keep it informed of the truck's situation.

If a problem occurs, such as mechanical, a traffic accident or load top up, the intelligent agent can assess the situation and recommend an alternative route. The route needs to take into account weather, time and load factors of the vehicle. If the truck has two intelligent agents, one for monitoring travel conditions and one for load updates for back filling or forward filling, then they will need to communicate each other's current situation. That is, the two agents will need to communicate the options, while recommending a route that maximizes the load and minimizes the travel time. The load maximizing agent will have booked the additional load at the best price for the given conditions. It may even enter into some type of negotiations on time and other factors. An agent also needs to let the truck's base know where it is and what is going on. Thus, for this situation there could be several agents that are coordinated by a cooperative communicative intelligent agent. There are a multitude of conditions that can affect the hauling of goods. Not all are covered here, such as those that need to be considered in one country or even state, but do not exist in another country or state with that country.

SUMMARY

Our perspective of electronic commerce is quite broad. Although the term ubiquitous is more encompassing, it too fails to draw the reader into the depth or scale that he or she needs to be able to visualize. We have been using the term electronic commerce to include any type of Internet, business-to-employee management, and wireless commerce related to electronic business transactions of various types.

This chapter describes research in the process of a cooperative communicative intelligent agent conceptual model. These concepts have been used in describing a conceptual model with a framework and architecture for cooperative communicative intelligent agents. The cooperative communicative intelligent agents are intended to use an agent communication language (such as ACL or KQML) as a mediating tool for agent communicating activities. The resulting conceptual model is referred to as ALMA (Agent Language Mediated Activity). This model is being developed into a logical model for application to specific e-business models.

REFERENCES

Austin, J. L. (1965). *How To Do Things with Words*. New York: Oxford University Press.

Cohen, P.R. & Levesque, H. J. (1995). Communicative actions for artificial agents. *Proceedings of the First International Conference on Agent Systems.* San Francisco: AAAI Press. *Computerworld.* (2001, March). Retrieved January 9, 2002 from the World Wide Web: http://www.computerworld.com.

eCommerce B2C: U.S. Consumer shopping, buying and demographics. (2001, September). *EMarketer.*

Edmond, D. & Papazoglou, M. P. (1998). (Eds.), *Cooperative Information Systems: Trends and Directions.* Academic Press, pp. 233-262.

FIPA: Agent Communication Language. Retrieved May 3, 2002 from the World Wide Web: http://www.fipa.org/spee/fipa97/FIP.k97.html.

Franklin, S. & Graesser, A. (1996). Is it an agent or just a program? A taxonomy for autonomous agents. *Proceedings of the 3rd International Workshop on Agent Theories, Architectures and Languages*. (ATAL-96). Springer-Verlag.

Fuchi, K. & Yokoi, T. (Eds.), (1994). *Knowledge Building and Knowledge Sharing.* Ohmsha and IOS Press.

Gartner. (2001) Worldwide business-to-business Internet commence to reach $8.5 trillion in 2005. Retrieved February 12, 2002 from the World Wide Web: http://www3.gartener.com/5_about/press_room/pr20010313a.html.

Georgeff, M., Pell, B., Pollack, M., Tambe, M. & Wooldridge, M. (1999). The belief-desire-intention model of agency. *Proceedings of Conference: Agents, Theories, Architectures and Languages (ATAL).*

Grice, Paul. (1989). *Studies in the Way of Words* Cambridge, MA: Harvard University Press.

Gruber, T. R. (1993). A translation approach to portable ontology specifications. *Knowledge Acquisition,* 5,199-220.

Guarino, N. & Welty, C. (2000). *Ontological Analysis of Taxonomic Relationships.*

Kaptelinin, V. (1996). Computer mediated activity. *Context and Consciousness*. Cambridge, MA: MIT Press.

Klusch, M. (2000). *Intelligent Information Agents: Agent Based Information Discovery and Management on the Internet.* Springer-Verlag.

Kutti, K. (1996). Activity theory as a potential framework for HCI research. *Context and Consciousness*. Cambridge, MA: MIT Press.

Laender, A & Storey, V. (Eds.), *Proceedings of ER-2000: The International Conference on Conceptual Modeling.* Springer-Verlag LNCS, October.

Masoud, M. & Jentzsch, R. (2002, December). Submitted *15th Australian Joint Conference on Artificial Intelligence.* Canberra. Australia.

Nardi, B.A. (1996). Studying Context. In B. Nardi (Ed.), *Context and Consciousness.* Cambridge, MA: MIT Press.

Odell, J. (Ed), (2000). *Agent Technology*. (OMG Document 00-09-01, OMG Agents Interest Group, September). OMG Agents Interest Group.

Parker, S.T. (1993). Higher intelligence, propositional language and culture as adaptation for planning. In *Tools, Language and Cognition in Human Evolution.* UK: Cambridge University Press.

Sperber, D. (1994). *What is Intelligence?* Cambridge University Press.

Sperber, D. (Ed). (2000). *Meta-Representations: A Multidisciplinary Perspective*. UK: Oxford University Press.

Sperber, D. & Wilson, D. (1995). *Relevance: Communication and Cognition.* Oxford: Blackwell.

Vygotsky, L.S. (1978). *Minds in Society*. Cambridge, MA: Harvard University Press.

Vygotsky, L.S. (1986). *Thought and Language*. Cambridge, MA: MIT Press.

Wiederhold, G. (1994). Interoperation, mediation and ontology's. *Proceedings of International Workshop on Heterogeneous Cooperative Knowledge Bases.* Tokyo.

Wittgenstein, L. (1958). *The Blue and Brown Book.* New York: Harper and Row.

Chapter XVI

Supporting Mobility and Negotiation in Agent-Based E-Commerce

Ryszard Kowalczyk and Leila Alem
CSIRO Mathematical and Information Sciences, Australia

ABSTRACT

This chapter presents recent advances in agent-based e-commerce, addressing the issues of mobility and negotiation. It reports on selected research efforts, focusing on developing intelligent agents for automating the e-commerce negotiation and coalition formation processes and mobile agents for supporting deployment of intelligent e-commerce agents and enabling mobile e-commerce applications. Issues such as trade-off between decision-making in negotiation and mobility capabilities of the agents are also discussed in this paper.

INTRODUCTION

Electronic commerce offers new channels and business models for buyers and sellers to effectively and efficiently trade goods and services over the Internet. Agent-mediated e-commerce is concerned with providing agent-based solutions for different stages of trading processes in e-commerce, including need identification, product brokering, merchant brokering, contract negotiation and agreement, payment and delivery, and service and evaluation (Bailey & Bakos, 1997; Chavez et al.,

1997; Guttman & Maes, 1998; Gutman et al., 1998). As the market quickly evolves, new advanced dynamic e-commerce (also called negotiated e-commerce, or e-negotiation) solutions emerge to enable mapping more sophisticated and efficient negotiation models in business transactions to e-commerce; in particular, in the contract negotiation and agreement stages of the trading process. It involves the development of e-commerce agents with more intelligent decision-making and learning capabilities in the context of automated contracting that can include comparison shopping, bidding in auctions and contract negotiations. At the same time, the e-commerce environment also becomes more complex and dynamic due to the business trends to trade in several inter-connected marketplaces, and use new wireless communication channels and portable computing devices (e.g., PDAs, mobile phones) in emerging location-aware mobile e-commerce (m-commerce). Here, the mobility aspects of agent technology are predicted to play a significant enabling role.

This chapter presents recent advances in agent-based e-commerce, addressing the issues of mobility and negotiation. It reports on selected research efforts, focusing on developing intelligent agents for automating the e-commerce negotiation and coalition formation processes, and mobile agents for supporting deployment of intelligent e-commerce agents and enabling mobile e-commerce applications. The mobility of e-commerce agents covers advances in location-aware, mobile and networked comparison shopping, mobile auction bidding and mobile contracts negotiation. The negotiation agents are presented in the context of e-commerce negotiation, with incomplete and imprecise information and dynamic coalition formation, where agents negotiate the distribution of the coalition value and the agent level of resources. Furthermore, issues such as trade-off between decision-making in negotiation and mobility capabilities of the agents also are discussed in this paper.

MOBILE E-COMMERCE AGENTS: RECENT ADVANCES

Mobile agents have been recognized as a very prospective technology for both dynamic and mobile e-commerce applications (Sandholm, 2000; Griffel et al., 1997), but the research in that area is still in the very early stages. Although most of the related research considers mobile communication and location-aware computing, there also is growing research on deploying mobile and intelligent agents in advanced e-commerce, including location-aware, mobile and networked comparison shopping, mobile auction bidding and mobile contracting.

Location-Aware Shopping

Agora (Fonseca et al., 2001) is a project conducted at HP Labs to develop a test-bed for applications of agent technology to a mobile shopping mall. A scenario involves mobile shoppers with personal digital assistants (PDA) interacting with store services while in the mall, on the way to the store, or in the store itself. Mall-wide services, such as directories and locators, are available through the PDA connected to the wireless network that provides an URL-based access to the virtual presence of the mall and its services on the Web. Intelligent agents represent both shoppers and the store, and participate in on-line auctions to bid for desired products based on shopper's preferences. A lightweight version of a scenario involving mall infrastructure agents, store agents, a personal shopper assistant and bidding agent, and an English auction agent has been implemented with a multi-agent system Zeus from BT Labs (Zeus) and additional Java-based support software.

Impulse (Impulse, 2000) is an on-going research project at MIT Media Lab that explores a scenario in which the buying and selling agents can run on wireless mobile devices and engage in multi-parameter negotiation for comparison-shopping at the point of purchase. The buyer agent resides on a PDA equipped with a GPS receiver and a wireless Internet connection that enables the URL access and communication with the seller (provider) agents. The agents have been implemented with a Java-based mobile agent system called Hive (Taylor, 2000), also developed at MIT Media Lab.

Mobile Comparison Shopping

An agent-based framework for mobile commerce has been proposed by Mihailescu and Binder (2001). It provides three types of agents: device agents, service agents and courier agents. The device agent is a stationary agent that resides on a mobile device and provides access to wireless services, such as a location-based comparison-shopping. The service agents are owned by service providers and handle service requests from the users. They are heavy-weighted mobile agents operating within the wired network. The courier agents are single-hop, light-weighted mobile agents that can migrate from a service agent to a mobile device in order to establish communication with the user. A test-bed has been developed for a shopping center scenario where consumers can access a web portal wirelessly, via their PDA devices, for services such as product location, product comparison and store location, with the envisaged possibility of negotiation. The test-bed has been implemented with the use of Java-based tools, including Aglets SDK for service agents and KVM SDK for the device agent and courier agents.

Networked Comparison Shopping

MAgNET (Mobile Agents for Networked Electronic Trading) is a mobile, agent-based system prototype developed at the University of California with Java and IBM Aglets SDK to enable buyers to comparison shop for items from different online sellers. In MAgNET, a human buyer creates a mobile shopping aglet that compares quotes for an item from different online sellers by visiting those seller sites, and returns to the buyer with the best offer that it obtains. It can allow the buyer to send a mobile agent to various suppliers to purchase component parts needed to produce a complex product (Dasgupta et al., 1999). It is possible for a supplier to create and dispatch a mobile agent to potential buyers to survey customer responses, determine market values of products, and sell products.

Mobile Auction Bidding

eAuctionHouse (eAuctionSite) is a prototype of the Internet auction server developed as a component of a dynamic e-commerce platform, eMediator (eMediator), at the University of Washington. It supports combination auctions with bidding via quantity-price graphs through the Web browsers, and uses the integrated mobile agent system called Nomad for automated bidding and auction monitoring in selected auctions (Sandholm, 2000). Nomad is based on the Concordia mobile agents system (Concordia) and allows the users to generate mobile agents within the eAuctionHouse through the Web browser and launch them onto the agent dock within the eAuctionHouse site. The agents can be executed locally to participate actively in two auction types on the user's behalf, even when the user is disconnected from the network. In bidding, the agents follow game-theoretical dominant strategies based on the user's reservation price (English auctions) and number of bidders (single-item, single-unit, sealed-bid first-price auctions).

BiddingBot (Fukuta et al., 2001) is a multi-agent system developed at Nagoya Institute of Technology that can support attending, monitoring and bidding in multiple auction sites. It consists of several cooperative bidding agents that have been implemented with a mobile Java-based agent framework called MiLog (Fukuta et al., 2001). In BiddingBot, multiple bidding agents can attend different auctions and bid on behalf of users simultaneously, in order to obtain the items at the best price. BiddingBot's bidding agents can bid according to autonomous and coordinated bidding mechanisms designed for the agents.

Mobile Contracting

Electronic contracting has been investigated as an application niche for mobile agents by Griffel et al. (1997) in the scope of the OSM project (OSM) at University of Hamburg. In particular, a mechanism for contract document circulation has been developed to support the contracting parties engaged in the electronic contract negotiation, with contract documents represented as mobile objects. More specifically, mobile agents are used to circulate the contract data objects between the negotiating participants (who can review and alter the contract terms). A contract-carrying agent also can have a responsibility for and a role in dealing with the contract, while avoiding explicit locking mechanisms. An experimental prototype of a mobile agent system has been built to allow the users to get involved in the contract negotiation process through their Web browsers.

DynamiCS (DynamiCS) is an actor-based framework for mobile negotiation agents proposed by the same group at University of Hamburg (Tu et al., 2000). It involves integration of intelligent decision-making capabilities into mobile agents based on plug-in mechanisms enabling dynamic composition of mobile negotiation agents (Tu et al., 1999). In particular, the DynamiCS framework aims at providing flexibility in integrating negotiation strategies into mobile agents dynamically. It also uses rule management mechanisms to manage actors and coordinate plug-ins' mobility. DynamiCS has been implemented with Java, using Voyager system (Voyager) as the basic mechanism for distribution and mobility.

InterMarket (InterMarket) is a research project proposed to develop an Intelligent Mobile e-Marketplace System at Intershop Research and Fredrich Schiller University based on a mobile agent system, Tracy (Tracy). It aims at enabling mobile access and automated trading in e-marketplaces, based on integration of mobile agents and intelligent decision-making agents offered as an add-on component to a commercial e-marketplace platform. InterMarket proposes stationary (or networked mobile) intelligent trading agents, to automate the users' decision-making and negotiation tasks in e-marketplaces, and mobile agents, to support deployment of the trading agents and provide mobile access and communication to e-marketplaces from mobile devices, such as Personal Digital Assistants (PDA) or mobile phones.

E-NEGOTIATION AND COALITION FORMATION

The increased potential of agent technology in supporting and automating negotiation has been recognized in a wide range of real-world problems, including: group conflict resolution (Nunamaker et al., 1991; Sycara, 1992); business negotiations (Foroughi, 1995); resource allocation and scheduling (Sycare et al., 1991); and e-commerce (Beam & Segev, 1997; Guttman, Moukas & Maes, 1998). The developed approaches also have been the basis for many e-negotiation and coalition formation agent systems for e-commerce.

E-Negotiation

Many existing negotiation agent systems support distributive negotiations based on auctions, or other forms of competitive bidding, where the terms of transaction typically involve a single issue (e.g., price) and/or the agents compete because of their mutually exclusive objectives (Kasbah & AuctionBot). Some systems that can support multi-issue integrative negotiations (e.g., Tete-a-Tete), which may lead to win-win agreements if the agents have mutually non-exclusive objectives, usually provide a varying level of automation support and/or assume a high degree of information sharing (common knowledge) among the agents. They may share common knowledge explicitly (e.g., information about private constraints, preferences and utilities may be disclosed by fully cooperative agents) or implicitly (e.g., agents may have available or assume some information about probability distribution of utilities of other agents). The assumption of common knowledge that allows one to handle some aspects of uncertainty associated with incomplete information (e.g., mixed strategies in game theory) may be difficult to satisfy in the competitive e-commerce environment. Moreover, the existing systems typically assume that all information available to the agents is precisely defined. For example, users are usually required to provide exact and precise information about their private preferences, constraints and objectives (e.g., price < $99, delivery time = 1 day, etc.). Therefore, autonomous negotiation agents that can handle both incomplete and imprecise information may be needed in the real-world negotiation settings.

Fuzzy e-Negotiation Agents (FeNAs) is a prototype system of intelligent agents to support fully autonomous, multi-issue negotiations in the presence of limited common knowledge and imprecise information. The FeNAs consider negotiation as an iterative decision-making process of evaluating the offers, relaxing the preferences and constraints and making the counter-offers, in order to find an agreement that satisfies constraints, preferences and objectives of the parties. The agents use the principles of utility theory and fuzzy constraint-based reasoning during negotiation, i.e., they offer evaluation and counter-offer generation. They negotiate on multiple issues through the exchange of offers on the basis of the information available and negotiation strategies used by each party. The available information can be imprecise, where constraints, preferences and priorities are defined as fuzzy constraints, describing the level of satisfaction of an agent (and its user) with different potential solutions.

The overall objective of an agent is to find a solution that maximizes the agent's utility at the highest possible level of constraint satisfaction, subject to its acceptability by other agents. Depending on the constraints, preferences and objectives of the parties, the FeNAs can support both distributive and integrative negotiations. During negotiation, the agents follow a common protocol of negotiation and individual negotiation strategies. The protocol prescribes the common rules of negotiation: agents can accept or reject offers, send counter-offers or withdraw from negotiation; agents are expected to accept own offers; and negotiation is successful if the final offer satisfies all parties, etc. The private negotiation strategies specify how the individual agents evaluate and generate offers in order to reach a consensus, according to their constraints and objectives. A number of negotiation strategies have been implemented in FeNAs, including the take-it-or-leave-it, no concession, fixed concession, simple concession strategies and their better deal versions.

In the FeNAs negotiation, the set of fuzzy constraints of each party C^j prescribes a fuzzy set of its preferred solutions (individual areas of interest). The possible joint solutions of negotiation (common areas of interest) are prescribed by an intersection of individual areas of interest. In this context, the objective of the FeNAs negotiation is to find a solution within a common area of interest that maximizes constraint satisfaction of the parties.

Figure 1 illustrates a simple example of a fuzzy constraint-based representation of the negotiation problem involving two parties, a and b. $C^a(x)$ and $C^b(x)$ define

Figure 1: Fuzzy constraints of two negotiating parties a and b

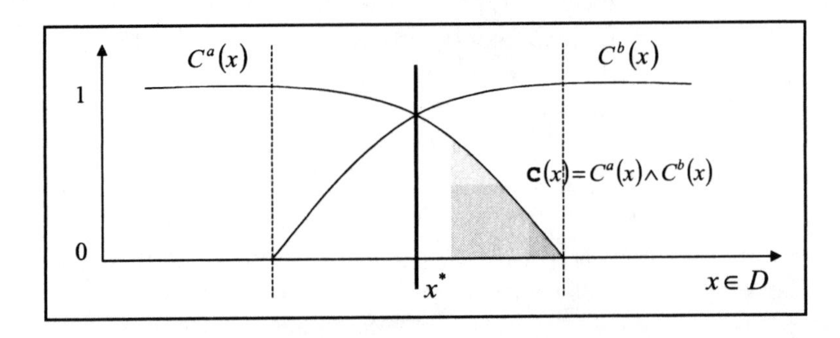

individual areas of interest of the parties a and b, respectively. x* is a solution from an intersection of the individual areas of interest, i.e., the common area of interest, defined by a conjunctive combination $\mathbf{c}(x) = C^a(x) \wedge C^b(x)$, with a maximal joint degree of constraint satisfaction. It should be noted that the common area of interest, i.e., $\mathbf{c}(x)$ is not known to the negotiating parties (agents) a priori. Therefore, the main goal of the FeNAs is to move toward and to explore potential agreements within the common area of interest, in order to find the most satisfactory agreement for the parties.

The FeNAs exchange their preferred solutions in the form of offers according to the individual negotiation strategies (e.g., trade-off and/or concession on a level

Figure 2: Fuzzy constraint-based negotiation

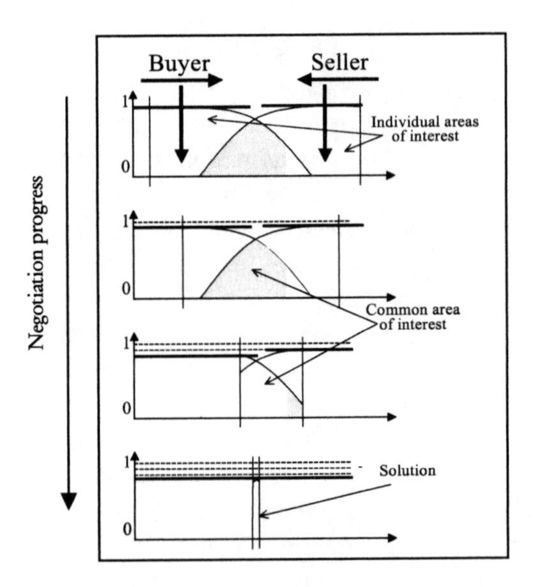

Figure 3: A fuzzy negotiation agent for seller with constraint propagation

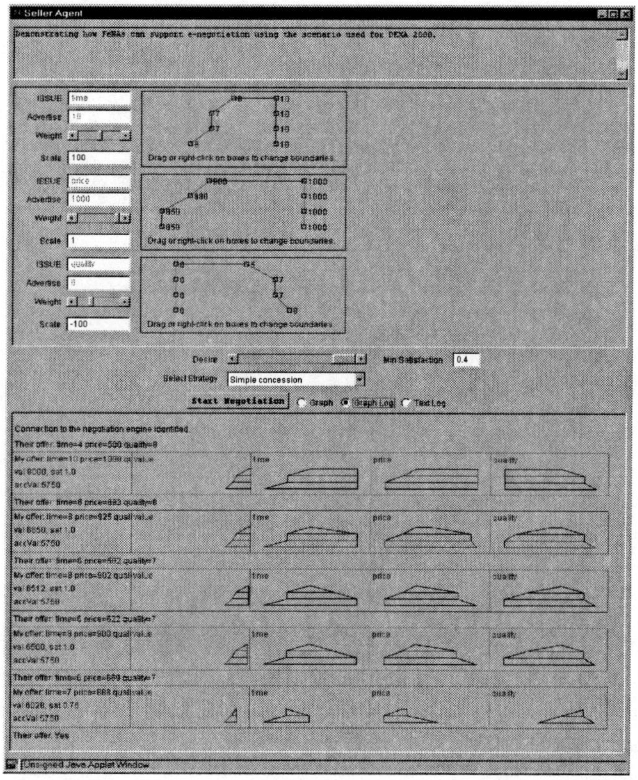

of constraint satisfaction). Typically, each agent starts negotiation by offering the most preferred solution from its individual area of interest, i.e., a solution with the maximal satisfaction degree of the private constraints. If an offer is not acceptable to other agents, they make counter-offers, in order to move them closer to an agreement. It can involve considering alternative solutions (trade-offs) at the same level of constraint satisfaction (if they exist), or making a concession, i.e., offering a solution with a lower degree of constraint satisfaction. The offers already exchanged between the agents constrain the individual areas of interests and the future decisions of the agents. For example, a rational negotiator would not propose an offer with a lower satisfaction value than a satisfaction value of an offer received already from another party). Therefore, the individual areas of interest change (i.e., reduce) when the offers are exchanged during the negotiation process.

The principles of fuzzy constraint propagation, based on the rules of inference in fuzzy logic (Zadeh, 1973; Zadeh, 1978; Dubois et al., 1994; Kowalczyk, 2000), are used in this process. Fuzzy constraint propagation supports searching for a solution by pruning the search space of potential solutions. It also allows the agents to track the changes in their individual areas of interest, i.e., the currently available values and levels of satisfaction of potential alternatives during the negotiation process (see Figure 3).

Dynamic Coalition Formation

Coalition formation is an important method for cooperation among on-line agents. Coalitions among such agents may be mutually beneficial, even if the agents are selfish and try to maximize their own payoffs. As stated by Nwana et al. (1998), coalition formation will be a key issue in electronic commerce. On-line agents that will form a coalition can gain by using the greater market power that coalition provides. In e-commerce, where self-interested agents pursue their own goals, cooperation and coalition formation cannot be taken for granted. It must be pursued and achieved via argumentation and negotiation.

FeNAs-based coalition formation adopts a multi-agent approach to this problem, where we typically create one or more agent, each with its own agenda and preferences, and have the agents electronically negotiate with each other within a predefined set of rules. We are mostly interested in the question of which procedure the agents should use to coordinate their actions, cooperate and form a coalition. Constrains, such as communication cost and limited computational time, are taken into account. This approach, while still primitive, offers the most hope for coalition formation in e-commerce. It is not constrained by Game Theory assumptions, and does not limit itself to cooperative bargaining contexts, as it handles mixed-motive bargaining context, in which the agents compete, as well as cooperate, among each other).

Figure 4: Coalition formation

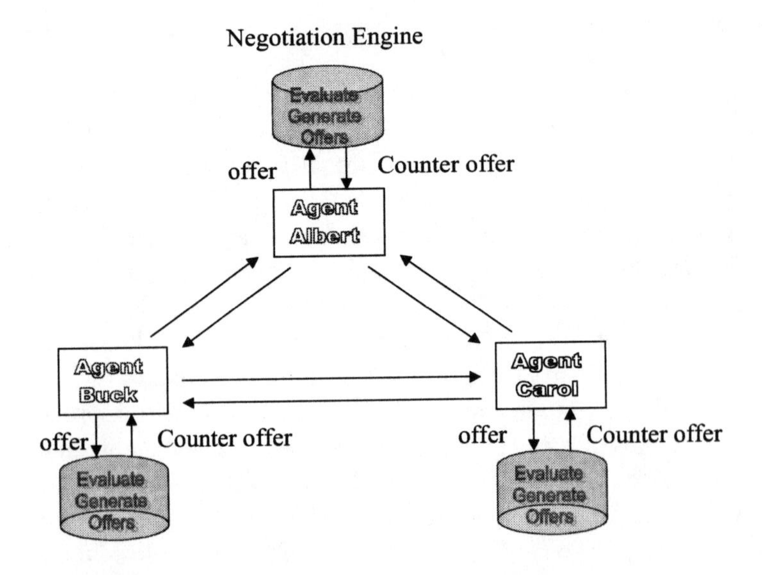

The formation of a coalition among three agents is described as three agents (Albert, Buck, Carol) negotiating competitively, in order to decide which two agents will form a coalition, and to agree on a division, between the coalition members, of the coalition value/utility.

Each agent makes use of a separate negotiation engine to generate and evaluate offers. The negotiation engines have been designed and developed, based on the FeNAs presented in the previous section. Every offer an agent receives is sent to its negotiation engine for evaluation. The engine responds by sending a counter offer: an acceptance offer or a rejection offer. Offer evaluations make use of agent preference values (level of gain and level of activity).

At the end of coalition formation negotiation, one of the following coalition structures is agreed upon:

((Albert, Payoff A), (Buck, Payoff B)) ; Payoff A + Payoff B = Utility (Albert, Buck)

(Albert, Payoff A), (Carol, Payoff C); Payoff A+ Payoff C = Utility (Albert, Carol)

(Buck, Payoff B), (Carol, Payoff C); Payoff B+Payoff C = Utility (Buck, Carol)

The agent's payoff is the agent's personal gain (e.g., the agent portion of the coalition utility), minus the cost incurred in conducting the trading activities (negotiations, etc.):

Payoff = Gain – Cost

The cost of an agent is a linear function of his/her activity level, Cost = b AL. An agent sets its activity level depending on his/her circumstances/ preferences.

AL =1, in this case, the agent has the computational resources and wants to conduct the trading activities on behalf of the coalition members.

AL = 0, in this case, the agent wants to share the computational workload of conducting the trading activities.

AL = -1, in this case, the agent has limited computational resources and does not want to conduct the trading activities, he/she prefers to leave it to the agent representing the coalition.

b is a constant used to balance cost and gain, b = 500. By setting b at 500, an agent not representing the coalition will gain $500, when an agent representing the coalition will lose $500. Such an agent will accept such a loss, provided he/she can get a greater portion of the coalition's utility. A coalition will form once the agents agree on two issues:

- who is getting what: in other words, which portion of the coalition utility each agent is getting; and
- who is doing what: in other words, who will represent the coalition and will conduct the trading activities on behalf of the coalition members.

At any time during the coalition negotiations, each agent has three possible actions: it can accept the offer it received from the agent it was negotiating with; it can make a counter offer to this same agent; and, finally, it can ask a third agent to better the offer received by the agent it was initially negotiating with. Each agent makes use of the coalition formation algorithm presented below.

This algorithm has been used in the experimental Intelligent Trading Agency (ITA) for the user car trading test-bed for three buyers trying competitively to negotiate on which two coalitions will form in order to get a better deal with the car dealer. We have tested this algorithm on a number of scenarios (each with different agent preferences and different negotiation strategies). The following figure shows a screenshot, with some results obtained with one of the scenarios.

Coalition formation has been widely studied among rational agents (Sandholm & Lesser, 1995; Rosenschein & Zlotkin, 1994), and among self-interested agents (Sandholm & Lesser, 1998).

Game Theory most commonly employed by coalition formation researchers usually answers the question of what coalition will form, and what reasons and processes will lead the agents to form a particular coalition. The focus is on understanding why agents form a particular coalition among all possible ones. Game Theory is concerned with how games will be played from both descriptive and

Figure 5: Coalition formation algorithm

```
Begin
  A -> ( B, Payoff B )
  Begin Loop
    If B 'accept'
      End Game [(A, UAB-Payoff B), (B, Payoff B)], (C,0)
    If B 'reject'
      A begins negotiations with C
    If B -> (A, ( B, Payoff B) )                          {better offer}
      Payoff B' = closes offer ~  Payoff B         {B already neg. with C}
      A -> ( B, Payoff B' )


  If B  -> ( A, Payoff A )                               {counter offer}
    If A 'reject'
      A begins negotiation with C
    If A can 'accept'
      A -> (C, (A, Payoff A) )                         {ask C to do better}
      C -> (A, Payoff A' )
      If  Payoff A' > Payoff A
          End Game [(A, Payoff A'), (C, UAB - Payoff A')],(B,0)
      If  Payoff A' <=  Payoff A
          End Game [ (A, Payoff A), (B, UAB - Payoff A) ], (C, 0)
    Else  A -> ( B, Payoff B' )                           {counter offer}
  Repeat Loop
```

Figure 6: Screenshot with results obtained with one of the scenarios

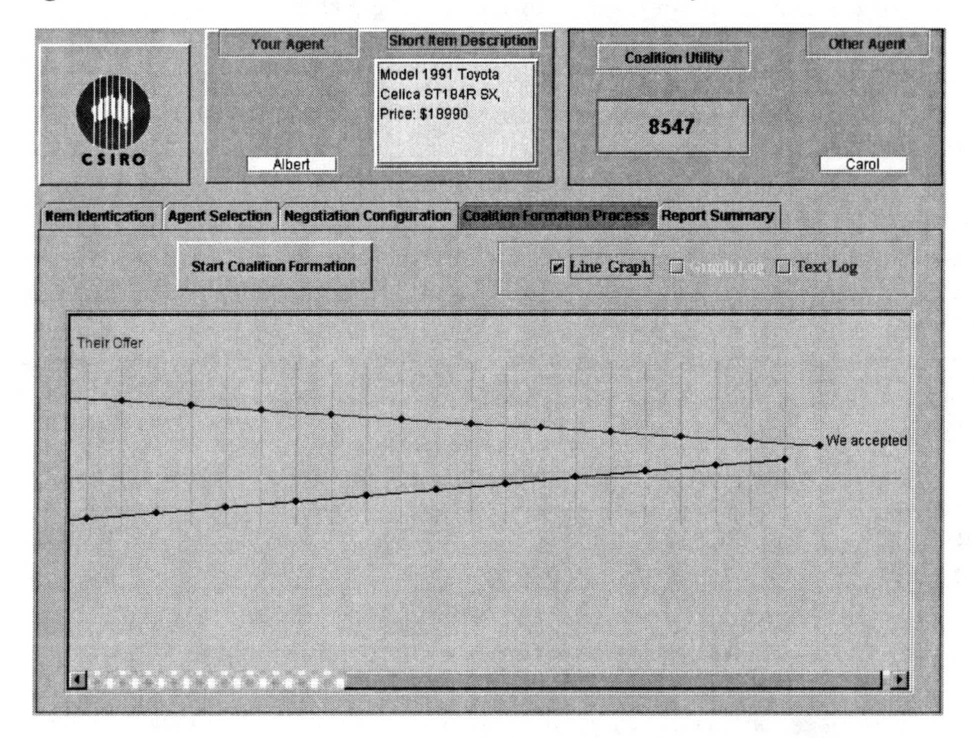

normative points of view. Game solutions consist of strategies in equilibrium (Nash, Perfect, Dominant). Although existing theory is rich in insights, and provides useful benchmarks, it cannot tell us how to program on-line agents for most bargaining contexts of interests. Relevant game-theoric solution techniques, almost invariably, make assumptions (e.g., of shared prior probability estimates, common knowledge of agent's preferences and perfect rationality) that do not apply to real bargaining contexts. The following, more specific, limitations/criticisms are raised by Linhart and Radner, below.

- Common knowledge and rationality assumptions: equilibrium are derived by assuming that players are optimizing against one another. This means that players' beliefs are common knowledge; this could not be assumed in e-trading scenarios.
- Not predictive: game-theorists are not able to predict the outcome of bargaining processes; they focus more on being able to explain a range of observed behavior.
- Non-robustness: results of negotiation depend crucially on the procedure of offers and counteroffers, and at what stage discounting take place. In a real world situation of negotiation, these features are not a given piece of data, but rather, evolve endogenously. Results dependent on them may be too specific to be of any use.

Furthermore, Game Theory has focused almost exclusively on outcomes that, under perfect rationality, constitute equilibria of the game, playing scant attention to processes off the equilibrium path, and the many vagaries of non-rational behavior (Dworman, Kimbrough, & Laing, 1996). Consequently, previous game theory has little to offer, by way of instruction for programming online agents to negotiate effectively in complex real-world contexts. Work in AI, by Rosenschein and Zlotkin, (1994) has aimed at circumventing this problem by seeking to design mechanisms (protocols), under which online agents may negotiate using pre-defined strategies, known a priori to be appropriate. Work by Shehory and Kraus (1999) adjusts the Game Theory concepts to autonomous agents and present different coalition formation procedures. The procedures presented concentrate on widely cooperative problems, such as the postmen problem. While both works are certainly of useful value, their range of potential applicability does not cover all the requirements for electronic commerce.

Remarks

The presented approach allows agents to negotiate autonomously, form coalitions and communicate in the presence of incomplete, imprecise and conflicting information. The agents are rational and self-interested, in the sense that they are concerned with achieving the best outcomes for themselves. They are not interested in social welfare, or outcomes of other agents (as long as they can agree on a solution). The rationality of the agents is bound by the availability of the information and computational resources. It means that the agents try to achieve as good an outcome as possible in both negotiating a deal and forming a coalition. In other words, they do not always have the information and computational resources to obtain the theoretically optimal outcome (according to the game theoretical results).

The solutions presented have been used and tested on the experimental Intelligent Trading Agency for the user car trading test-bed (http://www.cmis.csiro.au/aai/projects). The FeNAs have been tested also, with different negotiation scenarios for document translation services (Kowalczyk & Bui, 1999; Kowalczyk, 2000).

The results of the initial experiments with the fuzzy e-negotiation work indicate that the FeNAs system can handle a variety of e-negotiation problems with incomplete common knowledge and imprecise/soft constraints. In particular, they can provide automation support for multi-issue integrative negotiation, as experimented with in scenarios for car-trading and document translation negotiation.

The results of the initial experiments with the coalition formation work indicate that simple artificial agents formulate effective strategies for negotiating the formation of coalitions in mixed motive and multilateral negotiation contexts, and, therefore, seem appropriate for developing practical applications in electronic commerce.

DISCUSSION: MOBILITY VS. NEGOTIATION SUPPORT

The research efforts presented in the previous sections aim at providing new capabilities for e-commerce solutions with the use of agent technology, including mobile e-commerce, automated negotiation and coalition formation. In particular, the mobility factor adds several aspects to the scope of agent-mediated e-commerce and specific solutions they address, such as support for agents' migration and users' mobility; their implementation and deployment approaches are discussed, briefly, in the remainder of this paper.

The e-commerce application areas with mobile agents considered in this paper include comparison shopping, auction monitoring and bidding, and contract negotiation, (i.e., common application areas in agent-mediated e-commerce). They extend, however, the scope of the previous agent-mediated e-commerce applications to wireless m-commerce (Agora, Impulse, MB, InterMarket) and mobile agent-mediated networked e-commerce (MAgNET, eAuctionHouse, BiddingBot, OSM/DynamiCS).

The intelligent agents aim to provide automation support for decision-making tasks, in particular, for negotiation in e-commerce. The mobile agents extend that support by allowing for participation in several marketplaces in networked e-commerce, and enabling users' mobility and wireless participation. The agents' mobility adds ubiquity power to the participants to the e-commerce game. In particular, it allows the agents to respond quicker to local changes in marketplaces and make trading decisions faster than remote agents or human participants. The agents can move across the network to reduce network traffic and communication latency, and also can perform their trading tasks when the users are disconnected from the network. In addition, some of them enable portable devices that support users' mobility, i.e., they allow the users to access, move and disconnect from the wireless network while the agents perform trading tasks on their behalf.

Most presented systems have been deployed with general-purpose mobile agent systems that have been extended with decision-making (intelligent agent) capabilities. Typically, they have used commercial agent development tools, such as Concordia in eAuctioHouse, IBM Aglets in MB and MAgNET and Voyager in OSM/DynamiCS. Some systems have used prototypical mobile agent systems, such as Hive in Impulse, Tracy in InterMarket and MiLog in BiddingBot. The use of third-party systems provides specific mobility and communication capabilities, and permits implementation of some high-level, application-specific functions. The use of proprietary systems typically allows for more flexible implementation at the expense of the development costs. However, in almost all cases, an additional Java-based software component has been required, even when commercial tools have been used.

Most efforts focus on developing intelligent trading agents to automate the users' decision-making and negotiation tasks, and mobile agents to support deployment of the trading agents and enable agents' mobility and users' access from mobile devices. However, there is a trade-off between decision-making capabilities that make the agents "heavy," and mobility of agents, which requires the agents be "light-weighted" due to the limited network bandwidth and device computational resources. For example, eAuctionHouse deploys mobile agents that are created within the e-marketplace by the users, through the Web browser, for participation in the bidding processes on the users' behalf. InterMarket also deploys intelligent trading agents within the e-marketplace with a mobile agent system. But, it enables mobile access, with the use of mobile communication agents that can reside on the users' personal computers and mobile devices, and move to the e-marketplace to deliver the users' instructions to the trading agents. MAgNET integrates some decision-making capabilities into mobile agents that allow the shopping agents to compare quotes from different seller sites they can visit. DynamiCS proposes plug-in, decision-making capabilities for mobile agents that permit flexible changes to the decision-making capabilities within the agents, keeping them reasonably small. To cope with the limited computational resources of the mobile devices and wireless network bandwidth, some systems, like MB and InterMarket, adopt "light-weighted" mobile communication agents to deliver instructions to "heavier" decision-making agents operating within the networked environment.

CONCLUSION

This paper presents recent advances involving the use of mobile agents and intelligent agents for advanced e-commerce solutions. A number of the selected agent systems have been overviewed, with the aim of providing a representative view of the current research trends in developing intelligent and mobile agent-mediated e-commerce, including location-aware, mobile and networked comparison shopping, auction bidding and contract negotiation.

In general, the considered systems aim to provide new capabilities for advanced e-commerce solutions with the use of agent technology, in particular focusing on integration of the complementary capabilities of the intelligent and mobile agents. The intelligent agents aim at providing automation support for decision-making tasks in e-commerce. The mobile agents extend that support, by allowing for participation in several marketplaces in networked e-commerce, and enabling users' mobility and wireless participation. It extends the scope of the agent-mediated e-commerce to wireless m-commerce and mobile agent-mediated networked e-commerce.

In most cases, a considerable software development effort was required to implement decision-making capabilities and support software enhancing the agent systems used during the development. It is consistent with several views on the early stage of maturity of agent development tools available today. Therefore, there remains a need for agent development tools that can support efficient deployment of both agents' mobility and decision-making (especially negotiation) in e-commerce applications.

REFERENCES

Alem, L., Kowalczyk, R., & Lee, M. (2000). Recent advances in e-negotiation agents (SSGRR'00). Italy. *International Conference on Advances in Infrastructure for Electronic Business, Science and Education on the Internet.*

AuctionBot. (2000). Retrieved from the World Wide Web: http://auction.eecs.umich.edu/.

Bailey, J. & Bakos, Y. (1997). An exploratory study of the emerging role of electronic intermediaries, *International Journal of Electronic Commerce, 1* (3), Spring 1997.

Baruceanu, M. & Lo, W. (2000). A multi-attribute utility theoretic architecture for electronic commerce. Barcelona, Spain: *Proceedings of 4th International Conference on Autonomous Agents,* pp. 239-247.

Beam, C. & Segev, A. (1997). Automated negotiations: A survey of the state of the art. *Wirtschaftsinformatik, 39*(3), 263-268.

Braun, P., Eismann, J., Erfurth, C., & Rossak, W. (2001) Tracy - A prototype of an architected middleware to support mobile agents. Washington, D.C: *Proceedings of the 8th Annual IEEE Conference and Workshop on the Engineering of Computer Based Systems (ECBS),* pp. 255-260.

Chavez, A., Dreilinger, D., Guttman, R., & Maes, P. (1997) A real-life experiment in creating an agent marketplace. *Proceedings of the Second International Conference on the Practical Application of Intelligent Agents and Multi-Agent Technology,* London: Practical Application Company.

Collins, J. & Gini, M. (2000). Exploring decision processes in multi-agent automated contracting (Technical Report, TR 00-53, University of Minnesota, 2000). Reprinted in *IEEE Internet Computing*, pp 61-72.

Dasgupta, P., Narasimhan, N., Moser, L., & Smith, P.M. (1999). MAgNET: Mobile agents for networked electronic trading. *IEEE Transactions on Knowledge and Data Engineering, 24* (6), pp 509-525.

Dubois, D., Fargier, H., & Prade, H. (1994). Propagation and satisfaction of flexible constraints. In R. Yager and L. Zadeh (Eds.), *Fuzzy Sets, Neural Networks and Soft Computing.*

Dworman, G., Kimbrough, S., & Laing, J. (1996). Bargaining by artificial agent in two coalition games: A study in genetic programming for electronic commerce, *(OPIM working paper 96-04-09).* Philadelphia, PA: The Wharton School, University of PA.

EauctionSite. (2002). Retrieved from the World Wide Web: http://ecommerce.cs.wustl.edu/.

Faratin, P., Sierra, C., & Jennings, N. (1998). Negotiation decision functions for autonomous agents. *International Journal of Robotics and Autonomous Systems, 24* (3-4), pp. 159-182.

Faratin, P., Sierra, C., Jennings, N., & Buckle, P. (1999). *Designing Flexible Automated Negotiators: Concessions, Trade-Offs and Issue Changes* (RR-99-031, 1999). Institut d'Investigacio en Intelligencia.

Foroughi, A. (1995). A survey of the use of computer support for negotiation. *Journal of Applied Business Research,* 121-134.

Frank, R. H. (3rd ed.) (1996). *Microeconomics and Behaviour.* McGraw-Hill.

Fukuta, N., Ito, T., Ozono, T. & Shintani, T. (2001) A framework for cooperative mobile agents and its case-study on BiddingBot. *Proceedings of the JSAI 2001 International Workshop on Agent-based Approaches in Economic and Social Complex Systems (AESCS 2001),* pp. 91-98.

Griffel, F., Tuan, M., Munke, M., & da Silva, M. (1997). Electronic contract negotiation as an application niche for mobile agents. *Proceedings of the International IEEE Workshop on Enterprise Distributed Object Computing (EDOC).* Australia.

Guttman, R. & Maes, P. (1998). Agent-mediated integrative negotiation for retail electronic commerce. *Proceedings of the Workshop on Agent-Mediated Electronic Trading.* Minneapolis.

Guttman, R. H., Moukas, A. G., & Maes, P. (1998). Agent-mediated electronic commerce: A survey. *Knowledge Engineering Review*, June 1998.

Impulse (2001). Retrieved from the World Wide Web: http:// agents.www.media.mit.edu/groups/agents/projects/impulse.

ITA. Retrieved from the World Wide Web: http://www.cmis.csiro.au/aai/projects.

Jennings, N., Faratin, P., Lomuscio, A., Parson, S., Sierra, C., & Wooldridge, M. (2001). Automated negotiation: Prospects, methods and challenges. *International Journal of Group Decision and Negotiation. 10*(2), pp. 199-215.

Kasbah. Retrieved from the World Wide Web: http://kasbah.media.mit.edu.

Keeney, R. & Raiffa, H. (1976). *Decisions with Multiple Objectives: Preferences and Value Trade-offs*. New York: John Wiley and Sons.

Kotz, D. & Gray, R. (1999). Mobile agents and the future of the Internet. *ACM Operating Systems Review*, August 1999, pp. 7-13.

Kowalczyk, R. & Bui, V. (2000). On constraint-based reasoning in e-negotiation agents. In F. Dignum and U. Cortés (Eds.), *Agent Mediated Electronic Commerce III, LNAI.* Springer-Verlag, pp. 31 - 46.

Kowalczyk, R. (2002). Fuzzy e-negotiation agents. *Journal of Soft Computing, Special Issue on Fuzzy Logic and the Internet, 6* (5), pp. 337-347.

Kowalczyk, R., Unland, R., & Ulieru, M. (2002). Integrating mobile and intelligent agents in advanced e-commerce: A survey. Erfurt, Germany. *3rd International Symposium on Multi-Agent Systems, Large Complex Systems, and E-Businesses,* pp. 692-709.

Kowalczyk, R., Franczyk, B., Speck, A., Braun, P., Eismann, J., & Rossak. W. (2002). InterMarket - Towards intelligent mobile agent e-marketplaces. Lund, Sweden. *The 9th Annual IEEE Conference and Workshop on the Engineering of Computer based Systems*, pp. 268-275.

Lander, S. & Lesser, V. (1993). Understanding the role of negotiation in distributed search among heterogenous agents. Chambery, France. *Proceedings of the 13th International Joint Conference on Artificial Intelligence,* 438-444.

Lewicki, R., Saunders, D., & Minton, J. (1997). *Essentials of Negotiation*. Irwin.

Lomuscio, A., Wooldridge, M. & Jennings, N. (2000). A classification scheme for negotiation in electronic commerce. In F. Dignum & Sierra, C. (Eds.), *Agent-Mediated Electronic Commerce: A European Perspective*, Springer-Verlag, pp. 19-33.

Ma, M. (1999). Agents in e-commerce. *Communications of the ACM, 42* (3), 79-80.

Maes, P., Guttman, R., & Moukas, A. (1999). Agents that buy and sell. *Communications of the ACM, 42*. (3), pp. 81-91.

Mihailescu, P. & Binder, W. (2001). A mobile agent framework for m-commerce. *GI Jahrestagung,* 2, pp. 959-967.

Nunamaker Jr., J.F., Dennis, A. R., Valacich, J. S., & Vogel, D. R. (1991). Information technology for negotiating groups: Generating options for mutual gain. *Management Science,* October 1991.

Nwana, H., Rosenschein, J., Sandholm, T., Sierra, C., Maes, P., & Guttmann, R. (1998). Agent-mediated electronic commerce: Issues, challenges and some viewpoints. Minneapolis. *2nd International Conference on Autonomous Agents*.

Oliveira, E. & Rocha, A.P. (2000). Agents advanced features for negotiation in electronic commerce and virtual organization formation process. In F. Dignum & C. Sierra (Eds.), *Agent Mediated Electronic Commerce – A European AgentLink Perspective*. Springer-Verlag, 2001, pp. 78-97.

Papaioannou, T. (2000). *Mobile Information Agents for Cyberspace – State of the Art and Visions* (CIA_2000, vo. 1860 LNCS). Springer-Verlag.

Parsons, S. & Jennings, N. R. (1996). Negotiation through argumentation – a preliminary report. Japan. *Proceedings of the 2nd International Conference On Multi-Agent Systems,* 267-274.

Raiffa, H. (1982). *The Art and Science of Negotiation*. Cambridge, MA: Harvard University Press.

Rosenschein, J. & Zlotkin, G. (1994). *Rules of Encounter: Designing Conventions for Automated Negotiation among Computers.* Cambridge, MA: MIT Press.

Sandholm, T. & Huai, Q. (2000). Nomad: Mobile mobile agent system for an Internet-based auction house. *IEEE Internet Computing*, pp. 80-86.

Sandholm, T. & Lesser, V. (1998). Issues in automated negotiation and electronic commerce: Extending the contract net framework. *Reading in Agents*. Morgan Kaufmann Publishers, 66-73.

Sandholm, T. & Lesser, V. (1997). Coalition among computationally bounded agents. *Artificial Intelligence 94*, 99-137.

Shehory, O. & Kraus, S. (1995). Feasible formation of stable coalitions among autonomous agents in general environments. *Computational Intelligence Journal.*

Shehory, O. & Kraus, S. (1999). *Coalition Formation Among Autonomous Agents: Strategies and Complexity*. Preliminary Report, Bar Ilan University: Department of Mathematics and Computer Science.

Smith, H. & Poulter, K. (1999). Share the ontology in XML-based trading architectures. *Communications of the ACM, 42* (3), 1.

Sycara, K. (1992) The PERSUADER. In D. Shapiro (Ed.), *The Encyclopedia of Artificial Intelligence*. New York: John Wiley Sons.

Sycara, K., Roth, S., Sadeh, N., & Fox, M. (1991). Distributed constraint heuristic search. *IEEE Transactions on System, Man, and Cybernetics,* 21 (1991), 1446-1461.

Tete-a-Tete. Retrieved from the World Wide Web: http://ecommerce.media.mit.edu/tete-a-tete/.

Tu, M., Griffel, F., & Lamersdorf, W. (1999). Integration of intelligent and mobile agents for e-commerce – A research agenda. In St. Kirn and M. Petsch (Eds.), *Workshop Intelligente Softwareagenten und betriebswirtschaftliche Anwendungsszenarien, TU Ilmenau, FG Wirtschaftsinformatik* 2. Arbeitsbericht.

Tu, M., Seebode, C., Griffel, F., & Lamersdorf, W. (2000). DynamiCS: An actor-based framework for negotiating mobile agents. *Electronic Commerce Research Journal, 1*(1/2).

Wurman, P. R, Walsh, W. E, & Wellman, M.. P. (1998). Flexible double auctions for electronic commerce: Theory and implementation. *Decision Support Systems,* 24, pp. 17-27.

Zadeh, L.A. (1973). Outline of a new approach to the analysis of complex systems and decision processes. *IEEE Trans. Man. Cybernetics*, 3, pp. 28-44.

Zadeh, L.A. (1978). Fuzzy sets as a basis for a theory of possibility. *Fuzzy Sets and Systems,* 1, pp. 3-28.

Chapter XVII

Deploying Java Mobile Agents in a Project Management Environment

F. Xue and K.Y.R. Li
Monash University, Australia

ABSTRACT

This chapter introduces mobile agent technology *and explains how it can help businesses to implement client-server enterprise computing solutions. A Java mobile agent-based project management system prototype is presented to demonstrate the main features of mobile agents (mobility, functionality, intelligence and autonomy), and how they help to enhance communication processes and facilitate security within the project environment. It suggests a practical way to isolate all host resources from all visiting agents using host agents and exported host functions. It also proposes a communication infrastructure to support intelligent dialogue among agents.*

INTRODUCTION

In the computing world, there are three categories of mobility: mobile hardware, mobile users and mobile software. Mobile devices (e.g., PDA and 3G mobile networks), user mobility concepts (e.g., telecommuting) and mobile software (e.g., mobile agents) recently have attracted a great deal of attention from both industry and academia. In this chapter, mobile agent technology and how it can help businesses implement client-server enterprise solutions is examined. Harrison et. al. (1995) concluded that alternative solutions, such as Remote Procedure Calls and Sockets, can perform as well as mobile agents, except in relation to real time interaction with the server. Mobile agents, however, provide better support to mobile clients, including users and hardware devices. Mobile agent technology tolerates unreliable network services, supports transient connected devices and can conserve network bandwidth.

Mobile agent technology, however, is not without its critics. Security, negotiation, intelligence and virus-like behaviors are some of the major causes of concern. This chapter presents a mobile-software based prototype designed to operate within a project management environment. The prototype, though incomplete, highlights the issues associated with mobile agent technology and presents a practical approach to addressing the problem.

INTERNET AND MOBILE AGENT TECHNOLOGY

Business transactions require human activity, such as information collection and analysis, and human interaction, such as negotiation. With the rapid growth of Internet technology throughout the past few years, e-business has started to transform the ways in which we conduct our business. Concepts such as e-shopping carts and e-shops are helping us to implement Business to Consumer (B2C) business. Such working models often require a buyer to visit vendors' Web sites. The data collected is analyzed, and a transaction completed, often without much interactive negotiation. Mobile agent technology now exists to provide an alternative way to implement e-business. The new technology enables automation and negotiation and is receiving a lot of attention from both researchers and industrialists. The new approach can provide enterprise-computing solutions rather than relying on traditional message-based architecture.

WHAT IS MOBILE AGENT TECHNOLOGY?

A mobile agent can migrate from machine to machine in a heterogeneous network under its own control. It is capable of roaming wide area networks (WANS) and the World Wide Web (WWW); interacting with foreign hosts; collaborating with other agents; gathering information on behalf of its owner; and coming 'back home,' having performed the duties pre-defined by its user.

Most mobile agents possess following basic properties:

- Autonomy;

- Object-orientation;
- Mobility;
- Functionality; and
- Reactivity/Intelligence.

Technically, a mobile agent is an object-oriented code that moves itself among various computers in one, or multiple, systems. It conducts designated tasks defined in its internal codes and, during self-execution, can decide when and where to move (Sahuguet, 1997).

In terms of Object-Oriented Programming, a real application consists of lots of functional objects, each an abstraction of related attributes and methods. These objects and attributes remain in local memory. An agent, however, is a special *object* that can linger in the memory of a foreign machine and perform functions — or interact with another agent — for which it was programmed. Agents are autonomous. They can remain unattended for a long time and be activated automatically when pre-defined conditions occur.

Agents can co-operate or communicate. This occurs when one agent makes the location of some of its internal objects and methods known to other agents. By doing this, an agent exchanges data or information with other agents, without necessarily giving away all of its information. Agents also can work together, communicating with each other while performing functions in the same way that a team of humans communicates when completing a task.

However, mobility is neither a necessity, nor sufficient, condition for agent-hood.

An agent can be executed on one computer system (stationary agent), or be executed on different systems at different times (mobile agent).

WHY MOBILE AGENTS?

Mobile agent technology should be seen as an alternative approach to traditional client-server architecture, as well as, a better solution for distributed systems. In the case of the management of distributed resources, a comparison between a client-server solution and a mobile agent-based approach demonstrates that mobile agent technology offers a number of advantages.

The system employing mobile agents will have at least three exciting advantages:

1. There is less data communication between hosts, thus the overall network load can be reduced significantly (Lange & Oshima, 1998);
2. Local execution performed by an agent is between 1,000 and 100,000 times faster than remote messaging (ObjectSpace, 2001); and
3. Problems caused by network latency, unreliable services and disconnected devices can be overcome (Harrison et al., 1995).

Other advantages may include flexibility and scalability of the system, load balancing and on-demand services. These advantages derive from the way in which mobile agents treat distribution problems by using local interaction and mobile logic.

According to Harrison et al. (1995), "While none of the individual advantages of mobile agents... is overwhelmingly strong, we believe that the aggregate advantage of mobile agents is overwhelmingly strong."

WHEN TO USE MOBILE AGENTS?

Although it is clear mobile agent technology brings us many advantages, it is still necessary to identify when such a technology should be used. Mobile agent technology may not be the most suitable choice for all circumstances. The test performed by ObjectSpace (2001) revealed that mobile agent technology can be best applied in the following circumstances.

- If a system is required to send a number of messages to three or more destinations, an agent can be dispatched to visit each remote machine, in turn. However, if there are just two sites, it may not be beneficial to use the agent approach, as the overall I/O communication will not be reduced significantly. Thus, the decision to utilize an agent is dependent on the number of hosts involved.
- The applicability of a mobile agent is determined by whether the task the system performs is independent of its host. For example, an agent can be launched from a PDA before it is switched off. The agent will continue its execution in the network independently.
- If system functions can be partitioned into several programs to run in parallel, each program might be represented by agents who could communicate with one another to achieve the final goal.
- Multiple agents can be created to reside in remote hosts for periodic monitoring functions.

In addition, bandwidth cost, level of autonomy and complexity of agent execution environment require considerations before adopting mobile agent technology (Wang et al., 2001).

EXAMPLES OF MOBILE APPLICATIONS

Agent applications include electronic commerce, group collaboration, workflow automation, active messaging, event monitoring, information gathering, and distributed simulation and network management. The following three examples demonstrate the use of mobile agents in business.

Agent-Based Web Information Retrieval System

An agent is sent from the client and remains as a temporary resident in the remote host. At pre-determined times, or when the agent senses certain triggering events in the server, the agent will obtain the information from the server and return to its owner's machine.

Agent-Based Web Search Engine

A mobile agent is launched to the Web, by its owner, with instructions to purchase a particular commodity. The agent will visit the machine that provides an index directory service first, to determine which shops to visit. The agent then visits all the selected shops and obtains the price of the item it intends to purchase from each shop. The agent then makes the decision to return to the shop that provided the best price and has the stock available to fill the order. The agent places the order and travels back to report to its owner.

The Personal Adaptive Web Sentinel (PAWS)

PAWS is an agent-based personal library of information systems. Based on the owner's preferences, and the index services that are available (such as Alta Vista and Yahoo), the agent automatically updates existing information, adds new information and purges old data from the personal library.

PAWS, itself, is a set of small communicating modules in which the kernel is responsible for controlling the other modules and for the communication with external agents (Falk & Jasson, 1996).

Figure 1: An agent-based information retrieval system

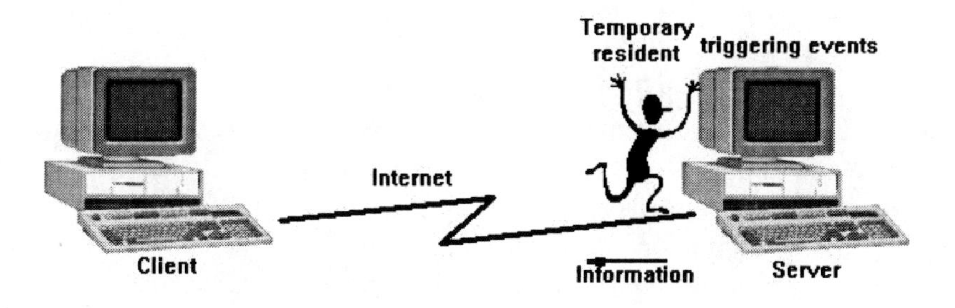

Figure 2: Agent-based shopper assistant

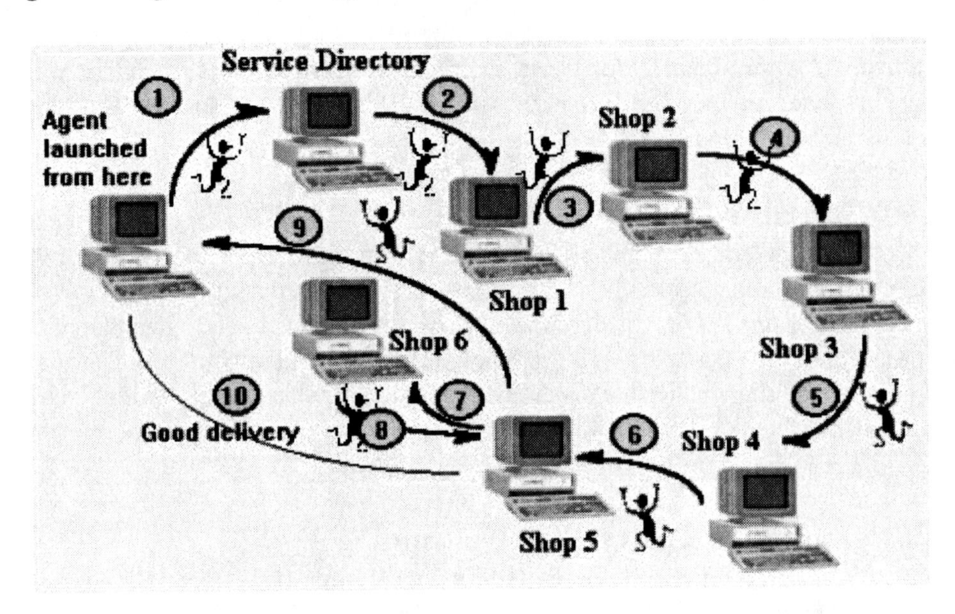

ISSUES RELATED TO MOBILE AGENTS

Although some agent development platforms have been released by research organizations, such as Aglets, from IBM, and Concordia, from Mitsubishi, mobile agent technology has not been able to capture its share of the commercial development world. The following three questions need to be answered.

1. How to establish and manage a security environment for agents?
2. How to cope with remote reference efficiently while agents are moving?
3. How to establish a common local execution and communication environment on all computing devices for receiving visiting agents released from all platforms?

This chapter focuses mainly on the first question.

To be useful, a mobile agent needs to interact with its host system. This creates security concerns for its host. The host system must have security measures to safeguard itself against malicious agents. The very nature of mobile agent technology promotes virus-like behavior. For example, denial of service caused by a mobile agent's ability to replicate and, thereby, overwhelm the CPU of a host server, could pose a major problem (Nguyuen et al., 2002). Agents must also be safeguarded against malicious agents and malicious hosts. Encryption, PKI and digital certificates can address this issue, to a certain extent.

Mobile software agents are required to move across heterogeneous networks, platforms and machine architectures. The use of a common programming language, such as Java, can help address the issue of heterogeneity. The byte-code verifiers of the Java virtual machine also can check for forbidden code sequences and prevent them from loading.

Beside mobility, a mobile agent exhibits intelligence and functionality. Intelligence is embedded in an agent, defining how it should interact and negotiate with remote hosts or other agents. Most researchers define functionality as the tasks the visiting agent needs to perform at the remote host. This idea creates security concerns.

To examine the above issues, and how they possibly may be addressed, and to demonstrate the inter-relationship between intelligence and functionality, a Java-based mobile agent prototype was developed. An Intranet project management mobile agent system was developed for the following reasons.

1. Kotz and Gary (1999) believe that mobile agent technology will appear first in the relatively safe Intranet environment.
2. The success of project management relies on effective communication among stakeholders. Emerging technology such as mobile devices, wireless LAN and 3G mobile net are effective means of enhancing communication processes. Mobile agents provide the appropriate technology to support these new devices and network architectures.

A MOBILE AGENT-ASSISTED PROJECT MANAGEMENT SYSTEM PROTOTYPE

This prototype was designed to construct an agent that could be launched by a project team member onto the network. The agent will then travel, based on the itinerary specified by the member, to visit servers at each site. The mobile agent will interact with a stationary agent at the remote machine to receive permission to obtain data from the local project database. The stationary agent (host agent) obtains function names from the visiting mobile agent. According to the name of the function and the access rights of the incoming agent, the host agent performs the requested function. Depending on the functions performed and the status of the project received, different responses may occur. Upon completion of the task, the host agent forwards the results to the visiting agent. The latter then returns home and presents the information received to its owner.

The Development Environment

There are many agent tools available from different vendors such as IBM and ObjectSpace. The table below lists some of the agent software packages on the market.

Aglets is a Java-based mobile agent system. The Aglet carries with it its program code, as well as, its state (data) while moving from one machine to another. Therefore, an Aglet agent can halt its execution on one machine, dispatch itself to another remote machine and resume its previous process there. It is safe for a computer to host *intrusted* Aglets, as it has a built-in security mechanism (Lange & Oshima, 1998; Gilbert & Janca, 1997; Lange & Oshima, 1996). An Aglets agent only allows the sending of string-encoded messages to stationary agents at formerly known addresses (Harrison et al., 1995).

Concordia is a full-featured framework for the development and management of mobile agent applications on computer networks. The application can process data, even if the user is disconnected from the network. Unfortunately, it does not support multicast, remote agent creation and garbage collection (Mitsubishi Electric Information Technology Centre, 2002).

Voyager® is a Java MA object request broker (ORB). It comes with distributed services, such as directory and publish subscribe multicasting. Voyager® uses regular Java message syntax to construct remote objects, send them messages and move them among applications. In this way, agents can act independently, on the behalf of a client, even if the client is disconnected or unavailable. This approach is particularly valuable in workflow, or resource, automation. It supports proxy updating, name space and garbage collection. Voyager® ORB is intended for

Table 1: Agent packages

Product	URLs	Language
AgentBuilder®	www.Agentbuilder.com	Java
Agentalk	www.ked.ntt.co.jp/csl/msrg/topics	LISP
Aglets	www.trl.ibm.co.jp/aglets/	Java
Concordia	www.meitca.com/HSL/Projects/Concordia/	Java
DirectIA SDK	http://www.animaths.com/	C++
IGEN™	http://www.cognitiveAgent.com/	C/C++
Intelligent Agent Factory	http://www.bitpix.com/	Java
JACK Intelligent Agents	http://www.Agentsoftware.com.au/jack.html	JACK
Microsoft Agent	msdn.microsoft.com/msAgent/default.asp	Active X
Network Query Language	http://www.nqli.com/	
Pathwalker	http://www.fujitsu.co.jp/hypertext/flab/free/paw/	Java
Voyager	http://www.Agentbuilder.com/AgentTools/Voyager	Java

business-application programmers and B2B component builders (ObjectSpace, 2001).

Voyager® ORB was employed as the agent development environment for this project.

Project Prototype Implementation

The project prototype was implemented and tested on Windows platform. The experiment was conducted using two PCs, one a client running Windows 2000 Professional and the other, a server running Windows 2000 Advanced Server.

MSProject2000 was used to create a project; its data was exported to a Microsoft Access database, which was then uploaded to the server.

To manage agents, accounts and functions, a database containing five tables exists on the server.

- The first table contains all account information, including user names, passwords and the user's group.
- The second table stores all registered agent's IDs and their keys.
- The third and the fourth tables together define the information of all functions available. The information contains function names, function identifiers, the required input parameters and the output parameters.
- The fifth table specifies what functions are available to which membership groups. For example, a financial function, like *viewfixedcost,* is available only to those agents created by financial group members.

An agent is created by a user after he/she supplies a user name and a password (see Figure 4). The user also can specify the IP address of the target machine (the server) if none exists (see Figure 5).

On the server, a stationary agent (host agent) is designed to respond to any foreign agents arriving at the machine. Upon their arrival, the host agent will determine whether they are friendly. The incoming agents have to supply an agent identifier for access. The host agent will compare the ID supplied with the relevant record in the registered agent table to identify the agent.

After the agent is identified, the host will check the user account table against the password supplied to identify the user and the membership group.

If the right password is supplied without function ID, it indicates that the incoming agent is requesting a list of functions to bring back home. Based on the user's membership group, the identifiers of all functions available to the group will be given to the incoming agent, as well as, the function name, input parameters and output parameters, in a simple text string format. For example, if the user provides the right password and has financial access privileges, the host agent will supply the incoming agent with all functions available to the financial group. The visiting agent can return home to its owner, who can select which function to perform.

Figure 3: Snap shot of the database tables

Username	Password	IsLogged	GroupType
FIAFEI	000000	1	1
FIARAY	000000	1	1
OPRFEI	000000	1	2

Agent' Identifier	Description	isArrive	isDeparted	Data
130.194.74.113:6000	113: 6000	0	0	
130.194.74.113:7000	113: 7000	0	0	
130.194.74.113:7001	113: 7001	0	0	
130.194.74.113:8000	113: 8000	0	0	
130.194.74.113:9000	113: 9000	0	0	
130.194.74.117:6000	117: 6000	0	0	
130.194.74.21:6500	021: 6500	0	0	

Inputs : Table

FunctionID	FieldName	Allias	FieldType	DataLength
000001	PROJ_ID	Project ID	1	255
000001	TASK_ID	Task ID	4	11
000002	PROJ_ID	Project ID	1	255
000003	PROJ_ID	Project ID	1	255

Outputs : Table

FunctionID	FieldName	Allias	Field	DataLength	Priority
000001	PROJ_NAME	Project Name	1	255	1
000001	TASK_ACT_COST	Actural Cost	6	22	4
000001	TASK_COST	Task Cost	6	22	3
000001	TASK_FIXED_COST	Fixed Cost	6	22	5
[illegible]	[illegible]	[illegible]	[illegible]	[illegible]	[illegible]

FunctionID	Description	GroupType
000001	View Task Cost	1
000002	View Project Cos	1
000003	Amend Task Cos	1
000004	Amend Project C	1
000005	View Task Perce	2

Group	GroupName
1	FINANCE
2	OPERATION

If the correct password is supplied with the function ID, the host agent will search the database to validate the function ID and to ensure that the user has the right to request the function. If accepted, the host agent will manipulate the data on the project database, on behalf of the visiting agent. Some functions require parameters. For example, the function *ViewTtask Cost* requires *Project ID* and *Task ID* as input parameters. In this case, before the host agent manipulates the project data, it will check the function table to identify all the parameters the function requires and retrieve their values from a text-based buffer of the visiting agent. After the host agent finishes the requested task, the results, such as *Task Cost*, *ActualCosts* and *Fixed Costs,* will be exported to the buffer of the incoming agent. The field name for each output parameter also will be delivered.

Other functions implemented include *Updating Activities Status* and *View Percentage of Work Content Completion.* These functions are available only to users with *Operation/Track* access rights.

Client View of the System

Figures 4 and 5 show the user interface at the client computer for dispatching an agent.

Figure 6 shows the agent returns after failing the "agent ID and registration" process at the server.

Figure 7 displays the ***View Task*** function, as selected, after the agent returns with a list of functions. All available functions are listed in the combo box Figure 7 also shows the input parameters required when *View Task Cost* function is selected.

Figure 8 illustrates the values for the input parameters required and the results obtained by the agent from the host.

The user interface unpacks the text message received from the host agent into the output text area.

Figure 4: Dispatch screen

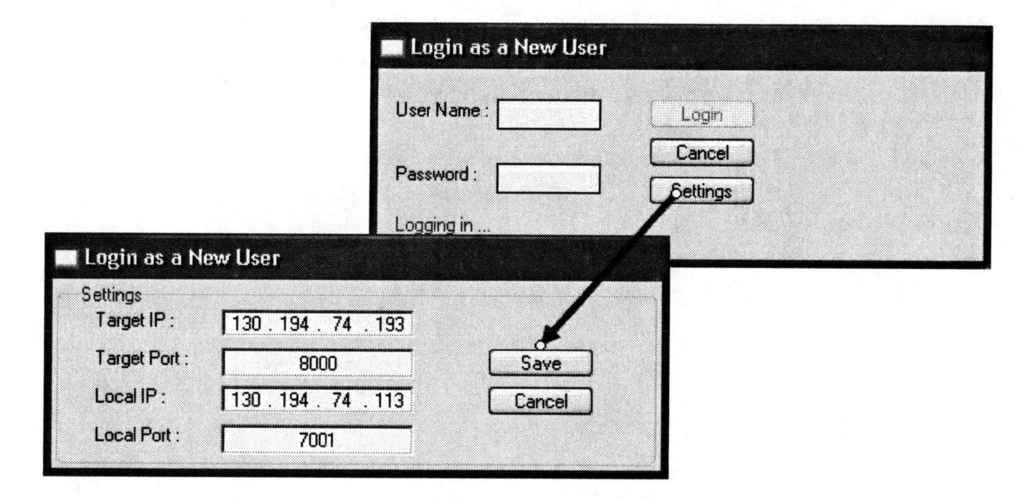

Figure 5: Setting screen

Figure 6: Wrong password supplied

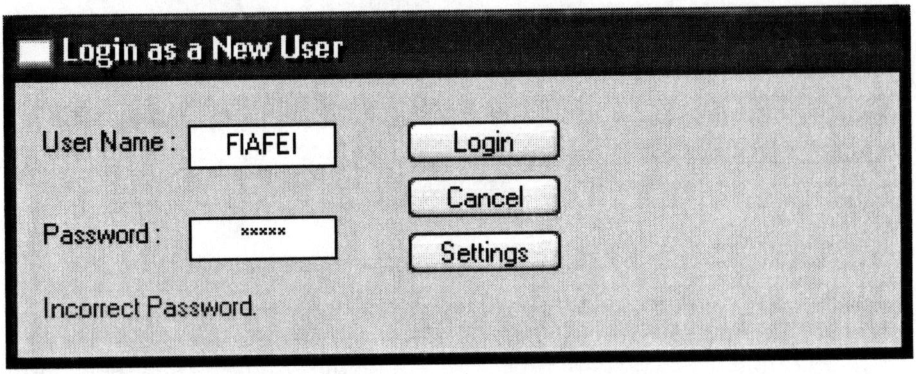

Features of the Prototype

The project prototype, though incomplete, can demonstrate the mobility, functionality and autonomous features of mobile agents. Another important aspect the prototype demonstrated is that the system provides security. Only friendly agents can interact with the host agent. The host agent will prevent any incoming agents from accessing data that they do not have a right to obtain.

Figure 7: Agent Returning with lists of functions

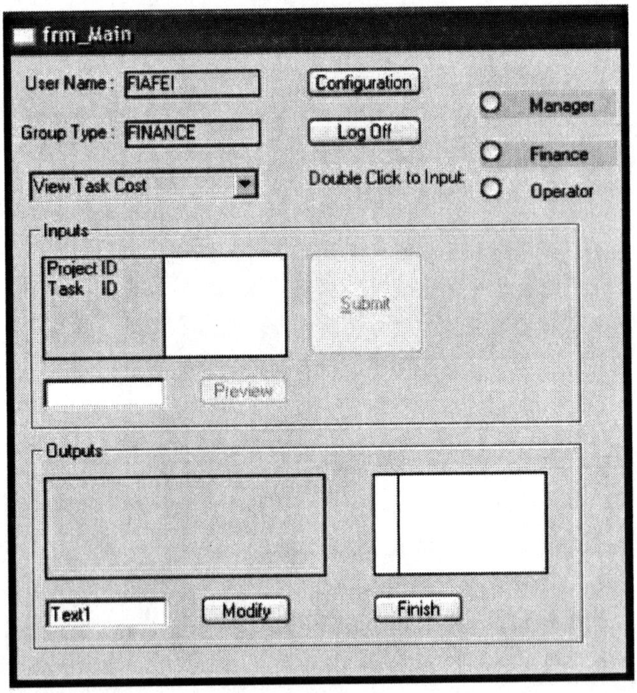

Figure 8: Inputs and returned information

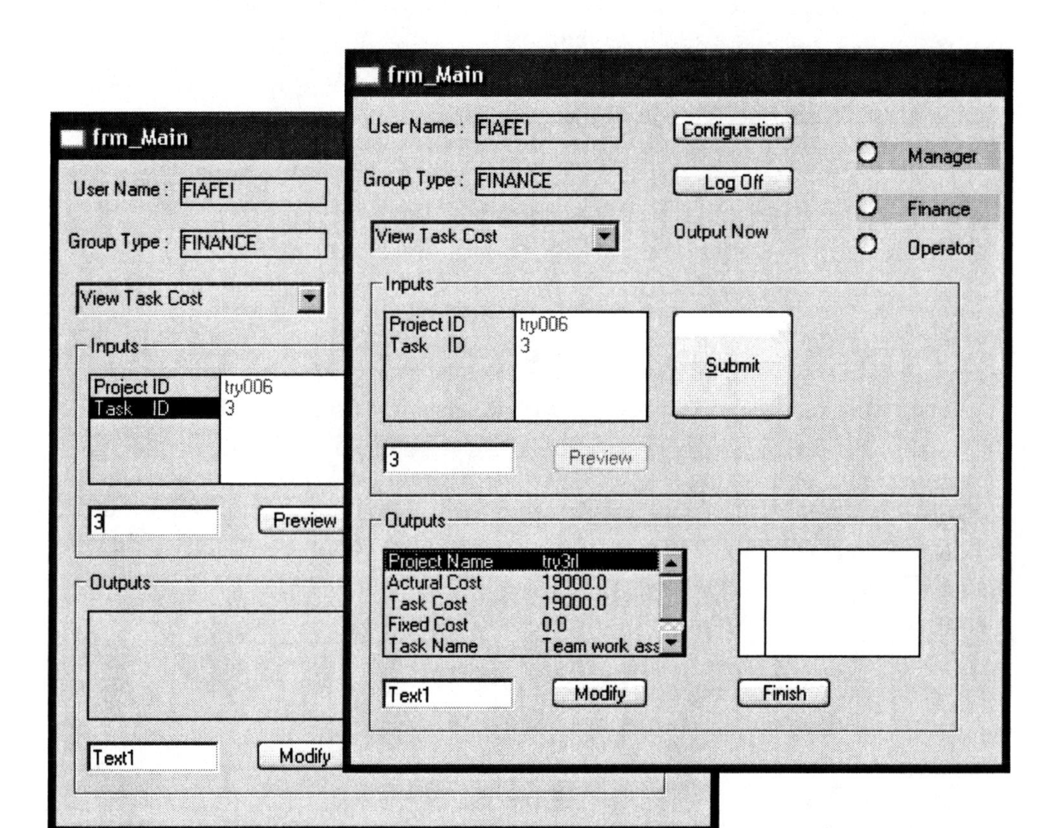

The system also displays an infrastructure in which the host agent receives text-based instructions from the incoming agent, and the host agent returns the required information in a simple string format, after the request has been processed.

The infrastructure utilizes a text-based communication system to facilitate negotiations between the guest agent and the host agent. The host agent exposes its functions to the guest agent, who receives a list of authorized functions, together with the required parameters for each function. The user controlling the incoming agent can select a function from the list and then dispatch the agent back to the server. The agent informs the host agent of the selected function ID and its parameter values. The host agent then performs the requested processes to retrieve data or invoke system actions.

The prototype also exhibits the use of a single intelligent interface for users to "program" their agents and to display the returned data (see Figure 8). The combo box allows the user, before dispatching the agent, to select the function and to enter the values for all required parameters. The output text area displays all messages received when the agent returns home. Both the parameter names and the corresponding values in the output text area are generated dynamically from the text message received from the host agent.

The above communication protocol makes possible a dialogue between the guest agent and the host agent. The guest agent now has sufficient intelligence to instruct the host agent about what it wants, and the latter is able to understand the request and activate the internal function calls.

CONCLUSION AND FUTURE RESEARCH

The mobility, functionality and autonomous features of mobile agents make them powerful emerging technology for enhanced enterprise computing. The ability of mobile agents to communicate and negotiate with one another mimics how humans interact with each other in the real world. Mobile agents *bring back* human interaction and negotiation to our cyberworld. The project prototype has demonstrated that they can be used to facilitate communication and security services within a project management environment. The prototype also highlights the power of mobile agents to provide security and control access. The prototype provides the foundation for building an agent-based, fully automatic alert system for project management. Under such a system, a mobile agent can be deployed as a virtual inspector to visit or reside on servers at remote sites.

Besides Voyager® ORB, there are many other agent development environments. Research should be directed at comparing various environments under the following criteria:

1. stability and security;
2. commercial extent;
3. platform compatibility; and
4. convenience of use

To provide maximum security, the prototype can be extended as follows:

- the system has a front machine functioning as an agent chatting room where the mobile agent resides;
- a firewall is set up to isolate all enterprise resources from the front machine;
- the incoming agent can exchange instructions and data with the host agent in the front machine only;
- the host agent can move through the firewall and is responsible for all enterprise data manipulation; and
- if needed, an account agent can be established in the front machine to manage and track foreign agent logging in activities.

For the system to be more useful, it should be extended so that an agent can be instructed to perform a sequence of requests and understand how to proceed from one request to the next. For example, in the *View Project Status* function, the guest agent would send a new request to the host agent to retrieve, from the project database, notes explaining the reasons for poor performance (if it found the project was behind schedule). If the status return indicated the project was on time, the agent would return home.

REFERENCES

Anders Falk, A. & Jonsson, I. (1996). PAWS: An agent for WWW-retrieval and filtering. Media Lab, Ericsson Telecom. Retrieved March 6, 2000 from the World Wide Web: http://www.fek.su.se/forskar/program/imorg/dok/1996-08-28/erimedlab/paws/NewPAAM.doc.html.

Gilbert, D. & Janca, P. (1997). IBM Intelligent Agents. (White Paper, IBM Corporation).

Harrison, C.G., Chess, D. M., & Kershenbaum, A. (1995). *Mobile Agents: Are they a Good Idea?* (IBM Research Report, RC 19887 (88465). New York: T.J. Watson Research Center.

Kotz, D. & Gray, R. (1999). Mobile agents and the future of the Internet. *ACM Operating System Review, 33*(3) 7-13. Retrieved March 6, 2000 from the World Wide Web: http://www.cs.dartmouth.edu/~dfk/papers/kotz:future2/.

Lange, D. B. & Chang, D.T. (1996). IBM Aglets Workbench. (White Paper, IBM Corporation).

Lange, D. B. & Oshima, M. (1998). *Programming and Deploying Java Mobile Agents with Aglets.* Addison-Wesley.

Mitsubishi Electric Information Technology Centre. (2002). Mobile agent computing: a white paper. Retrieved July 21, 2002 from the World Wide Web: http://www.concordiagents.com/Whatisit.htm.

Nguyen, A., Stewat, I., & Yang, X. (2002). A mobile agent model application for e-commence. Retrieved March 6, 2002, from the World Wide Web: http://ausweb.scv.edu.au/aw01/papers/referred/nguyen2/paper.html.

Objectspace. (2001). Voyager ORB 4 Documentation. Retrieved November 20, 2001 from the World Wide Web: http://support.objectspace.com/doc/index.html.

Sahuguet, A. (1997). About agents and database, Retrieved March 6, 2000 from the World Wide Web: http://www.cis.upenn.edu/~sahuguet/Agents/Agents_DB.pdf.

Wang , Y. H., Chung, C.M., Cheng, A. C., & Wang, W. N. (2001). Mobile agents over e_business. TTaiwan: TamKang University (Conference paper, the International Workshop on Agent Technologies for Internet Applications).

Chapter XVIII

Factors Influencing Users' Adoption of Mobile Computing

Wenli Zhu
Microsoft Corporation, USA

Fiona Fui-Hoon Nah and Fan Zhao
University of Nebraska–Lincoln, USA

ABSTRACT

This chapter introduces a model that identifies factors influencing users' adoption of mobile computing. It extends the Technology Acceptance Model (TAM) by identifying system and user characteristics that affect the perceived usefulness and perceived ease of use of mobile computing, which are two key antecedents in TAM. Furthermore, it incorporates two additional constructs, trust and enjoyment, as determinants in the model, and proposes specific factors that influence these two constructs. The long-term goals of this work are to gain an increased understanding of adoption issues in mobile computing, and to explain how specific HCI design issues may affect adoption by users.

INTRODUCTION

The emergence of mobile computing, combined with the increased popularity of the Internet, is changing our daily lives. The increasing use of small portable computers, wireless networks and satellites unfolds the new technology of mobile computing, which allows transmission of data to computers that are *not* physically linked to a network. As a result, people can communicate "on the move."

Mobile devices with new input and output methods and form factors are dramatically different from traditional desktop computers (Rodden et al., 1998). These technological changes make increasing demands on both the quality of user interface and the functionality of mobile devices (Johnson, 1998). The questions of interest are: 1) what factors influence users' adoption of mobile computing; 2) how does the design of mobile devices and interface affect user adoption; and 3) to what degree do specific factors such as trust and enjoyment (in using mobile devices) play a role in adoption?

The objectives of this chapter are: 1) to review literatures on technology adoption and the various design dimensions of mobile devices; and 2) to propose a model for adoption of mobile computing.

BACKGROUND

Technology Adoption Models

In reviewing the literature on IT adoption, we found the Technology Acceptance Model (TAM) proposed by Davis (1989) to be most relevant to this research. TAM was derived from the Theory of Reasoned Action (TRA) proposed by Fishbein and Ajzen (1975). According to Davis (1989), perceived usefulness and perceived ease of use are two key factors influencing people's attitude toward IT usage intention and actual IT usage. Perceived usefulness is defined as "the degree to which a person believes that using a particular system would enhance his or her job performance" (Davis, 1989, p. 320). Perceived ease of use is defined as "the degree to which a person believes that using a particular system would be free of effort" (Davis, 1989, p. 320). Davis and his colleagues (Davis, 1989; Davis et al., 1989; Davis et al., 1992) demonstrated that perceived usefulness had a strong direct effect on usage intentions, whereas perceived ease of use affected usage intentions indirectly via perceived usefulness. They also demonstrated that TAM had a higher explanatory power than TRA for predicting (word processing) software usage (Davis et al., 1989).

In an extension to TAM, Davis and his colleagues (Davis et al., 1992) examined the impact of enjoyment on usage intentions. They reported two studies concerning the relative effects of usefulness and enjoyment on intentions to use and usage of computers. As expected, they found that usefulness had a strong effect on usage intentions. In addition, enjoyment had a significant effect on intentions. A positive relationship between usefulness and enjoyment was also observed.

Based on these results, Davis and his colleagues (Davis et al., 1992) argued that further research is needed to examine the role of additional constructs, such as: availability of a particular software application; ease of learning; social normative influence; system design characteristics; system familiarity or experience; top management support; user involvement; and task characteristics, etc.

After its initial evaluation with e-mail, word processing and graphics applications, TAM was extended to other applications, including voice-mail (Adams et al., 1992), spreadsheets (Mathieson, 1991), Database Management Systems (Szajna, 1994), Group Support Systems (Chin & Gopal, 1995) and adaptive technology for the physically-challenged (Goette, 1995). TAM also was studied in different contexts to identify cultural differences (Straub, 1994) and gender differences (Gefen & Straub, 1997; Venkatesh & Morris, 2000) in its application.

In this research, we extend the TAM model to study human-computer factors leading to users' adoption of mobile computing. We review, summarize and identify system and user characteristics that influence users' adoption of mobile computing.

Design Dimensions of Mobile Devices: Input, Output, Navigation and Bandwidth

The degree of user interaction with mobile devices tends to be limited, due to small screen size, lack of a standard-sized keyboard and mouse, limited graphical content, and low bandwidth. According to Jones et al. (1999), the effectiveness in completing a task will drop by 50% when small screen devices are used. Small screen size poses limitations on the design and use of a mobile device.

There are numerous input styles available for mobile devices. The use of a miniature keyboard is common, but may not be suitable for all situations. For example, Kristoffersen and Ljungberg (1999) found that field workers had difficulties using a keyboard when both hands were required to operate the keyboard or when there was no flat surface on which to place the keyboard. In such cases, a pen-based interface provides a more ergonomic solution. The soft keyboard is a popular device that provides more flexibility to users. It is becoming a mainstream technology as small mobile computers, such as the Palm Pilot, and touch screens of all sizes increase in popularity and affordability. Furthermore, some forms of handwriting recognition for entering data have been introduced to provide a natural substitute for other input devices. In circumstances such as driving, visual concentration cannot be switched to wireless devices. Speech interface, then, has its advantages: it's natural, easy to learn, hands-free, and eyes-free (Graham & Carter, 1999).

The output limitations of wireless devices are mainly caused by the limited screen size. Besides using short messages and simple graphics for output display, some audio "display" systems are being researched, for example, non-speech sound systems, which utilize non-speech sound to increase usability without the need for more screen space (Brewster et al., 1998).

There are two kinds of navigational information that can be provided to users: navigational breadth, which shows available options on the same hierarchical level; and navigational depth, which shows the current location in the hierarchy (Nielsen, 1998). Cockburn and Jones (1996) emphasized four guidelines for Web site navigation design: use simple and natural dialogue; speak the user's language; minimize memory load; and be consistent. These four guidelines also apply to the navigation design of mobile devices. On a mobile device, navigation is based mainly on menus. Menu systems must meet the challenge of providing sufficient breadth and depth of available options to users. Information must not be hidden too deeply beneath many layers. To improve navigation, Marsden and Jones (1998) suggested using better categorizations, less key-press actions, and better visualization. Scrolling is another concern for navigation. Scrolling provides the opportunity to display a greater number of options. Although users may accept scrolling as a necessary feature of mobile devices (i.e., to overcome the problem of limited screen size), scrolling up and down (or sideways, if the device allows) is cumbersome. Scrolling beyond two pages causes even more trouble for the users. Therefore, scrolling should be kept to a minimum, if necessary at all (Jones et al., 1999).

Limitations on bandwidth create frustrations and put pressure on users who transfer large amount of data and pay by the second. Currently, limitations on bandwidth only allow the transmission of 160 characters for expressive messaging to mobile platforms, and, typically, in low-resolution monochrome. In the future, bandwidth limitations will be addressed through 3G technology.

Factors Influencing Usefulness of Mobile Devices

For a mobile device to be perceived as useful, it has to: 1) offer services that satisfy users' needs; 2) match the users' degree of mobility; 3) be compatible with other devices that the users might also be using, such as a desktop computer; 4) be accessible throughout a wide coverage area; and 5) provide stable and reliable access.

First, the functionality provided by mobile devices and the service offerings by mobile service vendors must fit the users' needs and requirements. For example, a stock broker who needs instant access to real-time stock quotes will not find a service that provides only delayed stock quotes very useful. Second, users' degree of mobility also influences how they perceive the usefulness of a mobile device. In general, the more mobile the user is, the greater the user's need for mobile computing. Users who stay at a fixed location most of the time, and can get the information they need via traditional means, will not find a mobile device and service as useful as users who are always "on the run." Third, compatibility is a key factor in technology adoption (Moore & Benbasat, 1991, 1996), and we argue that it is even more so in the context of mobile technology. It is important for mobile devices to be compatible with existing computing devices (such as desktop computers) in order to facilitate data synchronization and transfer. Lastly, to maximize the benefits of mobile computing and to achieve "any time, any place" access, two factors – degree of network coverage and reliability of access – are important. The closer "any time, any place" coverage is achieved, the more useful the mobile device and the higher the likelihood of widespread adoption.

Trust in Mobile Computing

Trust in mobile computing is affected by two main components: mobile technology and mobile vendor (Siau & Shen, forthcoming). Unlike wired networks, wireless communications suffer in bandwidth, stability of connection and predictability in functions. As noted by Siau and Shen (forthcoming), the lack of security and instability of mobile communications undermine users' trust in mobile technology. In other words, security and usability (i.e., ease of use and usefulness) of mobile devices may influence users' trust in mobile technology, which may then influence users' intention to use a mobile device and service.

Privacy policies of the mobile vendor also influence trust. The stricter the privacy policies and adherence to these policies, the greater the degree of trust exhibited by users. Other factors that influence trust in the context of mobile computing are familiarity, reputation and integrity of the mobile vendor (Siau and Shen, forthcoming). Ratnasingham and Kumar (2000) identified three levels of trust: competence trust, predictability trust and goodwill trust. In this case, competence trust refers to mobile vendors' skills and technical knowledge in providing mobile services. Predictability trust refers to trust in mobile vendors' consistent behaviors that provide cues and knowledge for users (including potential users) to make predictions and judgments based on past experiences. Goodwill trust refers to trust in mobile vendors' care, concern, honesty and benevolence. Familiarity with a mobile vendor increases one's confidence in the quality of services provided, thus increasing perceived competence trust. Reputation of the vendor increases predictability trust. Integrity, on the other hand, refers to honesty and benevolence of the mobile vendor. Hence, it is related to goodwill trust.

Enjoyment of Using Mobile Devices

Enjoyment is related to the concept of flow (Csikzentmihalyi & Csikzentmihalyi, 1988), which is the feeling or sensation of enjoyable experiences and the process of optimal experience. Researchers have suggested that flow is a useful construct for understanding interactions with computers (Csikszentmihalyi, 1990; Ghani, 1991; Ghani & Deshpande, 1993; Webster et al., 1993). Flow in human-computer interaction is related to the following characteristics (Brigish, 1993): easy to use, fun to use, fast, personalizable, comprehensiveness, and highly visual and browsable.

Three determinants of flow that might be relevant to mobile computing are: 1) perceived congruence of skills and challenges; 2) focused attention; and 3) interactivity

(Hoffman & Novak, 1996). Challenges and skills are the universal preconditions of flow. For flow to occur, it is necessary to achieve a balance between the level of challenges faced in using a mobile device and the level of skills possessed by the user (Csikszentmihalyi & Csikszentmihalyi, 1988). If the user interface is not understandable to the user (i.e., high challenge with respect to the user's skills), enjoyment or flow is not likely to take place. Once congruence between skills and challenges is achieved, flow may be initiated. It is important to note that, in order to sustain this flow state, congruence should always be present. During the flow state, the user experiences enjoyable feelings. If challenges are higher than skills (e.g., if a mobile device is difficult to use or its usage is difficult to understand), the user will exit the activity (i.e., adoption will not occur). Focused attention is also necessary to induce flow. Focused attention is defined as the "centering of attention on a limited stimulus field" (Csikszentmihalyi, 1977, p. 40). According to Csikszentmihalyi and Csikszentmihalyi (1988), when one is in the state of flow, one is so engrossed in the activity that one simply does not have enough attention left to think about anything else. Next, interactivity is the availability of immediate feedback between entities. This exchange of information and feedback is in the form of a sensory dialogue. It is important for an activity to be interactive, in order to induce or maintain flow. In human computer interaction, interactivity can be thought of as an activity where the user requests some action to be performed and the computer responds promptly to that request by taking the appropriate action or displaying the appropriate results to the user. Hence, response time must be acceptable to the user for flow to occur (Nah & Kim, 2000). According to Nielsen (1993), a response time of one second is about the limit for users' flow of thought to stay uninterrupted.

DEVELOPMENT OF A MODEL FOR ADOPTION OF MOBILE COMPUTING

Model Description

Based on the above-reviewed literature and discussions, we propose the following model for the adoption of mobile computing (as shown in Figure 1).

Hypotheses

The following factors are hypothesized to influence users' perceived ease of use.

1. Input device: the different types of input devices will affect the ease of use of a mobile device.
2. Output device: the different screen size and the use (or lack of use) of sound or speech will affect the ease of use of a mobile device.

Figure 1: Proposed model for users' adoption of mobile computing

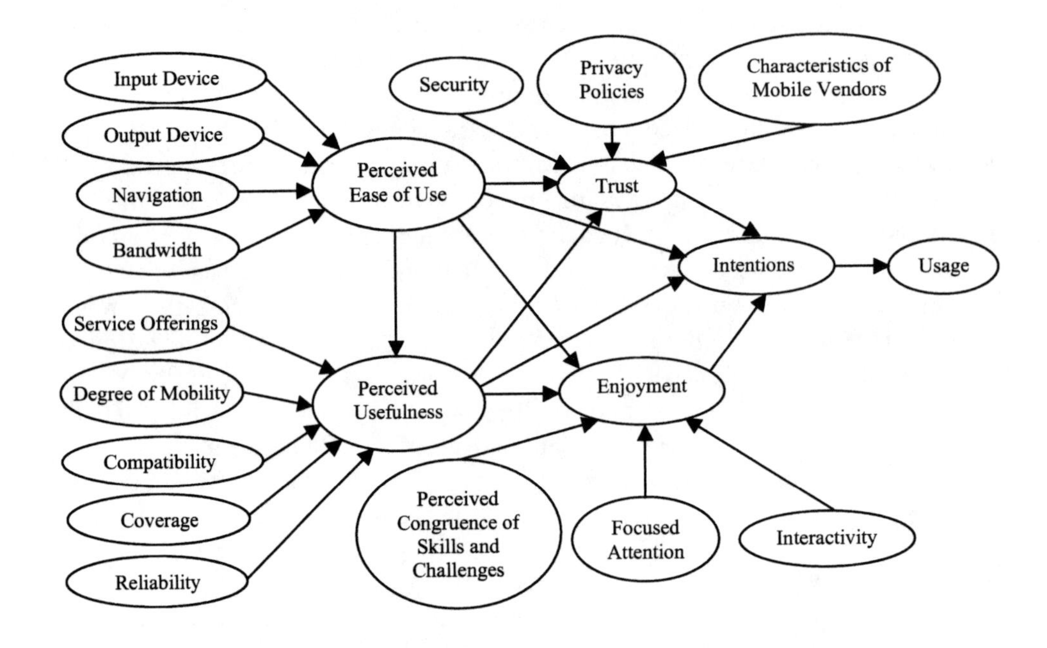

3. Navigation: navigation is the key in helping users find the information they want in the shortest amount of time. Therefore, it is a key element in determining the ease of use of a mobile device.
4. Bandwidth: bandwidth affects the degree of multi-media interaction that is possible, which, in turn, affects users' perceived ease of use.

The following factors are hypothesized to influence users' perceived usefulness.

1. Service offerings: the quality and the variety of services provided by the service providers and how well they support users' needs and requirements will directly affect users' perception of the usefulness of mobile computing.
2. Degree of mobility: we hypothesize that the more mobile the user, the more valuable mobile computing is to the user.
3. Compatibility: mobile devices must work well with users' existing computing devices, such as a desktop computer, in order to synchronize and transfer data. If users cannot transfer information back and forth between the mobile device and other computing device, the usefulness of the mobile device will be limited.
4. Coverage: the greater the area of coverage, the better "any place" access can be achieved, and hence, the more useful the mobile device is to users who are always "on the move."

5. Reliability: stable and uninterrupted access is necessary to achieve "any time" access. A mobile device that does not provide reliable and uninterrupted access is not likely to be as popular or useful to the user.

Based on the literature review, we propose that trust and enjoyment will be two other key constructs that will affect users' intention to use mobile computing. More specifically, we hypothesize that the following factors will influence users' trust.

1. Security: security is a necessary and important factor of trust. The more security provided on the mobile devices, the higher the level of trust.
2. Privacy policies: privacy is an important component influencing users' trust. The stricter the privacy policies, the greater the likelihood of gaining users' trust.
3. Characteristics of mobile vendors: familiarity, reputation and integrity of mobile vendors have been highlighted as three important characteristics that influence trust. The higher the familiarity, reputation and integrity of the vendor, the higher the level of trust exhibited by the users.
4. Perceived ease of use: we hypothesize that the user-friendliness and ease of use of the mobile device will influence users' perception of trust. The higher the perceived ease of use, the greater the level of trust.
5. Perceived usefulness: we hypothesize that the perceived usefulness of the mobile device will influence users' perception of trust. The greater the perceived usefulness of the mobile device, the greater the level of trust.

The following factors are hypothesized to influence enjoyment.

1. Perceived congruence of skills and challenges: congruence between skills and challenges is necessary for flow or enjoyment to occur. If the mobile device is too difficult or challenging to use (relative to the user's skill level), then the user will not enjoy using the mobile device.
2. Focused attention: if the mobile device is fun to operate, then focused attention may result. Focused attention increases enjoyment in interacting with the mobile device.
3. Interactivity: availability of immediate feedback can induce and maintain flow or enjoyment. The responsiveness of the mobile device to users' requests can influence the level of enjoyment.
4. Perceived ease of use: ease of use is hypothesized to influence enjoyment in using a mobile device.
5. Perceived usefulness: usefulness of a mobile device is also hypothesized to affect the level of enjoyment.

We hypothesize that perceived ease of use, perceived usefulness, trust and enjoyment will affect intentions to use mobile devices, which will then influence actual usage.

Proposed Study

The proposed model will be tested, verified and enhanced through interviews with users and non-users of mobile computing. The survey approach may be employed to test the model. Experimental studies are being planned to gain a more in-depth understanding of design issues in human-computer interface of mobile devices.

FUTURE TRENDS AND CONCLUSION

Mobile computing is a fairly new area of research. From a design perspective, guidelines will need to be developed for designing the content, input-output methods, and navigation for mobile devices, and for understanding the appropriateness of using different input/output devices for different task types and scenarios. From a technological point of view, improvements and developments are needed in the following areas – security, bandwidth, reliability/stability of connection, coverage, compatibility, functionality or service offerings, and multi-media interaction – before widespread adoption is likely to take place. With regard to mobile vendors, the establishment of trust is crucial. Security control implemented by the vendor, as well as the vendor's privacy policies and perceived characteristics (e.g., reputation), are important in trust formation.

In this chapter, we have identified both system design and user factors influencing individuals' usage of mobile computing. Human-computer interaction (HCI), which is related to system design, is a very important area in mobile computing. Our long-term goals are not only to gain an increased understanding of adoption issues in mobile computing, but also to explain how specific HCI design issues can affect the likelihood of adoption. As a result, specific guidelines can be proposed to improve the ease of use and usefulness of mobile computing, as well as, to increase users' trust and enjoyment in using mobile computing. Users' adoption of mobile computing can, therefore, be accelerated.

REFERENCES

Adams, D. A., Nelson, R. R., & Todd, P. A. (1992). Perceived usefulness, ease of use and usage of information technology: A replication. *MIS Quarterly, 16*(2), 227-247.

Brewster, S., Leplatre, G., & Crease, M. (1998). Using non-speech sounds in mobile computing devices. Retrieved from the World Wide Web: http://www.dcs.gla.ac.uk/~stephen/papers/mobile98.pdf.

Brigish, A. (1993, September). The electronic marketplace: Evolving toward 1:1 marketing. *Electronic Marketplace Report (formerly Electronic Directory & Classified Report), 7*(9) 6-7.

Chin, W. & Gopal, A. (1995). Adoption intention in GSS: Relative importance of beliefs. *Data Base, 26*(2&3), 42-63.

Cockburn, A. & Jones, S. (1996). Which way now? Analysing and easing inadequacies in WWW navigation. *International Journal of Human-Computer Studies,* 45, 105-129.

Csikszentmihalyi, M. (1977). *Beyond Boredom and Anxiety.* San Francisco, CA: Jossey-Bass.

Csikszentmihalyi, M. (1990). *Finding Flow: The Psychology of Engagement with Everyday Life.* New York, NY: BasicBooks (a division of HarperCollins Publishers).

Csikszentmihalyi, M. & Csikszentmihalyi, I. (1988). *Optimal Experience: Psychological Studies of Flow in Consciousness.* Cambridge, MA: Cambridge University Press.

Davis, F. D. (1989). Perceived usefulness, perceived ease of use and user acceptance of information technology. *MIS Quarterly, 13*(3), 319-339.

Davis, F. D., Bagozzi, R. P., & Warshaw, P. R. (1989). User acceptance of computer technology: A comparison of two theoretical models. *Management Science, 35*(8), 982-1002.

Davis, F. D., Bagozzi, R. P., & Warshaw, P. R. (1992). Extrinsic and intrinsic motivation to use computers in the workplace. *Journal of Applied Social Psychology, 22*(14), 1111-1132.

Fishbein, M. & Ajzen, I. (1975). *Beliefs, Attitude, Intention, and Behavior: An Introduction to Theory and Research.* Reading, MA: Addison Wesley.

Gefen, D. & Straub, D. W. (1997). Gender differences in the perception and use of e-mail: An Extension to the technology acceptance model. *MIS Quarterly, 21*(4), 389-400.

Ghani, J. A. (1991). Flow in human-computer interactions: Test of a model. In J. Carey (Ed.), *Human Factors in Management Information Systems: An Organizational Perspective.* Norwood, NJ: Ablex.

Ghani J. A. & Deshpande, S. P. (1993). Task characteristics and the experience of optimal flow in human-computer interaction. *Journal of Psychology, 128*(4), 381-391.

Goette, T. (1995). *Determining Factors in the Successful Use of Adaptive Technology by Individuals with Disabilities: A Field Study.* (Unpublished doctoral dissertation). GA: Georgia State University.

Graham, R. & Carter, C. (1999). Comparison of speech input and manual control of in-car devices while on-the-move. Retrieved from the World Wide Web: http://www.cs.strath.ac.uk/~mdd/research/pt2000/abstracts/carter.html.

Hoffman, D. L. & Novak, T. P. (1996). Marketing in hypermedia computer-mediated environments: Conceptual foundations. *Journal of Marketing,* 60, 50-68.

Johnson, P. (1998). Usability and mobility: interactions on the move. Retrieved from the World Wide Web: http://www.cs.colorado.edu/~palen/chi_workshop/papers/Johnson.pdf.

Jones, M., Marsden, G., Mohd-Nasir, N., & Boone, K. (1999). Improving Web interaction on small displays. Retrieved from the World Wide Web: http://www.handheld.mdx.ac.uk/www8/www8.pdf.

Kristoffersen, S. & Ljungberg, F. (1999). *Making Place to Make IT Work: Empirical Explorations of HCI for Mobile CSCW.* (Proceedings of the International ACM SIGGROUP Conference on Supporting Group Work), 276-285.

Marsden, G. & Jones, M. (1998). Ubiquitous computing and cellular handset interfaces – Are menus the best way forward. Retrieved from the World Wide Web: http://people.cs.uct.ac.za/~gaz/papers/menus2.pdf.

Mathieson, K. (1991). Predicting user intentions: Comparing the technology acceptance model with the theory of planned behavior. *Information Systems Research, 2*(3), 173-191.

Moore, G. C. & Benbasat, I. (1991). Development of an instrument to measure the perceptions of adopting an information technology innovation. *Information Systems Research, 2*(3), 192-222.

Moore, G. C. & Benbasat, I. (1996). Integrating diffusion of innovations and theory of reasoned action models to predict the utilization of information technology by end-users. In K. Kautz & J. Pries-Heje (Eds.), *Diffusion and Adoption of Information Technology* (pp. 132-146). New York: Chapman-Hall.

Nah, F.H. & Kim, K. (2000). World Wide Wait. In M. Khosrowpour (Ed.), *Managing Web-enabled Technologies in Organizations: A Global Perspective* (pp. 146-161). Hershey, PA: Idea Group Publishing.

Nielsen, J. (1993). *Usability Engineering*. San Francisco, CA: Morgan Kaufmann.

Nielsen, J. (1998). Sun's new Web design. Retrieved from the World Wide Web: http://www.sun.com/980113/sunonnet/.

Ratnasingham, P. & Kumar, K. (2000). *Trading Partner Trust in Electronic Commerce Participation*. (Proceedings of International Conference on Information Systems), 544-552.

Rodden, T., Chervest, K., Davies, N., & Dix, A. (1998). Exploiting context in HCI design for mobile systems. Retrieved from the World Wide Web: http://www.comp.lancs.ac.uk/computing/users/tam/workshop/submission.html.

Siau, K. & Shen, Z. (forthcoming). Building customer trust in mobile commerce. *Communications of the ACM.*

Straub, D. W. (1994). The effect of culture on IT diffusion: E-mail and FAX in Japan and the U.S. *Information Systems Research, 5*(1), 23-47.

Szajna, B. (1994). Software evaluation and choice: Predictive validation of the technology acceptance instrument. *MIS Quarterly, 17*(4), 319-324.

Venkatesh, V. & Morris, M. G. (2000). Why don't men ever stop to ask for directions? Gender, social influence, and their role in technology acceptance and usage behavior. *MIS Quarterly, 24*(1), 115-139.

Webster, J., Trevino, L. K., & Ryan, L. (1993). The dimensionality and correlates of flow in human computer-interactions. *Computers in Human Behavior,* 9, 411-426.

Chapter XIX

Mobile Computing Business Factors and Operating Systems

Julie R. Mariga
Purdue University, USA

ABSTRACT

This chapter introduces the enormous impact of mobile computing on both companies and individuals. Companies face many issues related to mobile computing. For example: which devices will be supported by the organization? which devices will fulfill the business objectives? which form factor will win? which features and networks will future devices offer? which operating systems will they run? what will all this cost? what are the security issues involved? what are the business drivers? This chapter will discuss the major business drivers in the mobile computing field, and provide an analysis of the top two operating systems that are currently running the majority of mobile devices. These platforms are the 1) Palm operating system (OS), and 2) Microsoft Windows CE operating system. The chapter will analyze the strengths and weaknesses of each operating system and discuss market share and future growth.

INTRODUCTION

Mobile computing, defined as a generalization of all mobile computing devices, including Personal Digital Assistants (PDAs, e.g., Palm Pilots, Pocket PCs), smart phones, and other wireless communication devices, will continue with dramatic changes throughout the next five years. There are a number of reasons for change, but two main factors are the convergence of next-generation handhelds and high-speed wireless technology. The operating systems found in today's handhelds will provide the foundation for future devices and applications. The two main operating systems for PDA's are the Palm OS and the Windows CE OS. Which operating system should companies or individuals implement? It depends on a number of items. One important issue to consider is what application(s) are needed by the user(s). Once this question is answered, it may help to eliminate some operating systems and devices. Another important item to consider is portability. Portability of applications is important because devices change rapidly and, if applications are portable, they can be reused on new devices, without having to be rewritten. If applications are developed in a language that allows for portability, such as Java, then these can be deployed to a wide range of devices, including handhelds that support various operating systems, embedded Linux devices and pure Java devices. Another important issue to consider in selecting an OS is what type of development tools is available, as well as the number and strength of the programmers, so they can create and maintain applications. Currently, the Palm OS supports the largest number of packaged applications. Many of these applications, however are better suited for individual, rather than business, use.

BACKGROUND

According to Jones (2001), there are four main factors driving the mobile business phenomenon. They are: 1) Economics, 2) Business Need, 3) Social Trends, and 4) Technology. Economics includes the falling prices of mobile airtime and the inexpensive cost of devices. Jones (2001) states that, during the next five years, costs will continue to decrease, allowing new mobile applications to be developed and reducing Bluetooth chip sets' cost to under $5, which will enable electronic devices to be networked together. Business needs include organizations needing new types of mobile applications to increase customer service and enable better supply chain management. In many countries, mobile devices have become a lifestyle accessory, mainly among younger adults. As young adults continue to want more functionality from their devices and applications, there will be a mix among the mobile technology and entertainment and fashion. New core technologies, such as WAP, i-mode, Bluetooth and 3G networks, are enabling a new generation of mobile applications. As these four factors continue to evolve, they will continue to push the growth of the mobile business arena.

The main differences between the Palm OS and the Pocket PC OS are discussed in the next section of the chapter.

OPERATING SYSTEMS

Palm Operating System

The Palm operating system was developed specifically for use with Palm Pilots. As these devices began to proliferate, 3Com (who has since spun off its Palm division as a separate company) licensed the Palm OS to other handheld device manufactures and developers. As a result, the Palm OS currently maintains a large market share in the handheld device market. Through the use of the Palm OS, software developers can build data applications for use with Palm devices, which can be implemented via wireless or synchronized data access to corporate data. According to Dulaney (2001), for IS organizations setting standards for PDAs, Palm OS devices will have the broadest appeal and application support. The Palm OS has the following strengths:

1) The large number of partners working with Palm;
2) The number of applications available to run on top of the Palm OS; and
3) The current amount of market share owned by Palm.

There are weaknesses with Palm and the Palm OS. Some of them are:

1) The core OS functionality is limited (as compared to Windows CE); and
2) Palm (as a company) is undergoing major changes, and some question their leadership and future business directions.

There is an ongoing industry debate about the future of Palm and Microsoft. Palm supporters are becoming concerned that the Palm market share is stagnating, while Windows CE market share is increasing. "The conventional wisdom is that Microsoft is gaining a lot of momentum. But Palm is still the market leader," said Alexander Hinds, the president and CEO of Blue Nomad, makers of the Wordsmith word processing application for the Palm OS. (Costello, 2001). According to Costello (2001), one of the reasons for the sense of impending doom ascribed to Palm (by many) is Microsoft's push into the enterprise market, an area Microsoft traditionally has dominated. Many industry experts believe that Windows CE will post large gains in the market in the next year, and that Palm will need to work more closely with third-party developers, who are already strong in the enterprise, in order to counter that move. However, many of the vendors supporting the Palm OS are confident in the future, and believe that the release of Palm 5.0 will be a key in sustaining the Palm market share. Many observers tout the functionality of the Windows CE platform, with its native support for full-motion video, digital music and high-resolution color screens as major selling points for corporate customers. This situation hasn't been helped by the length of time Palm has let lapse between major upgrades to its operating system. Palm OS 5.0, a major upgrade that will mark the platform's transition to the more powerful StrongARM processor, is set for sometime in 2002.

Windows CE Operating System

The Microsoft operating system, Windows CE, has been renamed the Pocket PC. The latest release is Pocket PC 2002. The new version comes with bundled software, but most industry analysts see the new release as an evolutionary upgrade, not a major change. The OS now looks very similar to the Windows XP operating system. With Pocket PC 2002, memory management is handled much more effectively. This is an important improvement, because most of the applications that run on top of the OS require extensive hardware requirements, as compared to applications that run on the Palm OS. Both Pocket Word and Excel come bundled with the OS, but it still does not include Pocket PowerPoint. Another application that comes bundled is MSN Messenger, which works just like the desktop version. Many users will enjoy this, as messaging is a popular application on PDAs. Another bundled application is Windows Media Player 8, which supports streaming video. Other important features include a new terminal client that provides access to Windows NT and other servers. It also includes virtual private network (VPN) access support for connecting to an intranet. The new version also supports Windows 2000 level password security. This is an important feature because, if the handheld device is lost or stolen, the data stays protected. Pocket PC 2002 also improves as a personal organizer. A few problems still exist; for example, Pocket Internet Explorer does not handle frames and pop-up windows correctly. Another problem occurs when transferring between Pocket Word and the desktop version of Word. Some of the formatting in the document can be lost. The Pocket PC OS has the following strengths:

1) Utilizes the Microsoft infrastructure;
2) Overall performance of the OS; and
3) Many developers are familiar with the Microsoft development environment.

There are weaknesses with Pocket PC OS. Some of them are:

1) The anti-Microsoft sentiment; and
2) The inability of the OS to work with Java applications; this must be done through the use of third party tools.

FUTURE TRENDS

The market for mobile devices and applications will continue to grow throughout the next ten years. There is still a lot of speculation as to what companies will be the leaders, but, according to data from Stanley (2001), the breakdown for PDA sales both worldwide and in the United States is shown in Table 1.

Table 1

Worldwide PDA Unit Sales (in millions):	**2001**	**2005**	**2007**
Windows CE	4,205	15,960	24,355
Palm OS	7,595	14,035	16,395
Other OS	4,575	13,530	20,620
Worldwide PDAs Total	16,375	43,525	61,370

USA PDA Unit Sales (in millions):	**2001**	**2005**	**2007**
Windows CE	1,430	5,565	8,445
Palm OS	4,780	8,390	9,710
Other OS	430	1,725	2,955
USA PDAs Total	6,640	15,680	21,110

CONCLUSION

Mobile computing platforms are going to continue to evolve in the next five to ten years, but companies must start making decisions and putting together business plans now. Below is a list of recommendations that both individuals and organizations should start implementing.

Recommendations

- Organizations should select mobile technologies that match their business goals.

- IT leaders should plan to manage the purchase and deployment of handheld devices in the same manner that PCs are purchased and deployed.
- Organizations should support personal devices for business purposes.
- Companies need to build flexible and extendible architectures that can support multiple devices and applications.
- Companies must create policies in the types of devices, operating systems and applications that will be supported. Adopt standardized synchronization and development tools.
- Technology, mobile economics, social trends and business needs will combine to drive the evolution of mobile e-commerce.
- Currently there are not any "killer" mobile applications. Users have different needs and value applications differently.
- Companies developing and operating mobile applications will see risks, and it will require a new skill set, as well as new attitudes, in management.
- Windows CE should be strongly considered for vertical market applications.

REFERENCES

Biggs, M. (2001). Analysis: It's time for a handheld strategy. Retrieved November 16, 2001 from the World Wide Web: http://www.mbizcentral.com/story/MBZ20011115S0009.

Costello, S. (2001). Comdex–Despite Microsoft, developers stick with Palm - for now. Retrieved November 16, 2001 from the World Wide Web: http://www.nwfusion.com/news/2001/1116palmms.html.

Dulaney, K. (2001). *Outfitting the Frontline: Phones, PDAs, and Strategies for Use.* Location: Gartner Group Symposium 2001.

Duwe, C. (2001). Microsoft Pocket PC 2002. Retrieved September 6, 2001 from the World Wide Web: http://www.zdnet.com/filters/printerfriendly/0,6061,2810757-3,00.html.

Egan, B. (2001). *Mobile and Wireless Computing: The Next User Revolution.* Location: Gartner Group Symposium 2001.

Jones, N. (2001). *Mobile Commerce Business Scenario.* Location: Gartner Group Symposium 2001.

Microsoft muscles in on global PDA shipments. (2001). *Mbusiness, 16* (November).

Schwartz, E. (2001). Opinion: The pocket PC steamroller. Retrieved September 21, 2001 from the World Wide Web: http://www.mbizcentral.com/story/MBZ20010921S0011.

Shaw, K. (2002). Looking into Palm's future. Retrieved January 3, 2002 from the World Wide Web: http://www.nwfusion.com/newsletters/mobile/2001/01142706.html.

Stanley, D. (2001). Palm's up? Palm's out? Retrieved December 17, 2001 from the World Wide Web: http://www.unstrung.com/server/display.php3?id=725&cat_id=2.

About the Authors

Julie Mariga is an Associate Professor of Computer Information Systems and Technology at Purdue University in West Lafayette, Indiana, USA. Her areas of interest include collaborative computing, mobile computing and increasing under-represented groups in the Information Technology field. Professor Mariga is an award-winning teacher at Purdue University. She has won a university-level teaching award, two school-level teaching awards, and three department-level teaching awards. Professor Mariga co-authored the textbook *Client/Server Information Systems: A Business Oriented Approach* (1999), is a contributing author to *The Internet Encyclopedia* (2002), both of which are published by John Wiley & Sons, Inc. She also has written 11 referred conference articles. Professor Mariga is a member of various professional organizations, including Information Resources Management Association (IRMA), where she served as track chair for Mobile Computing and Commerce, as well as Minorities in Information Technology, and International Association of Computer Information Systems (IACIS).

* * *

Latif Al-Hakim is an independent consultant specializing in information systems design. Currently, he is the Lecturer of Logistics and Operations Management in the Department of Economics and Resources Management Faculty of Business at the University of Southern Queensland, Australia. His experience spans 34 years in industry, research and development organizations and universities. Dr. Al-Hakim received his first degree in Mechanical Engineering from Basrah University in 1968. His master's degree (1977) in Industrial and Systems Engineering and doctorate (1983) in Management Science were awarded from the University of Wales, UK. Dr. Al-Hakim has held various academic appointments and lectured on a wide variety of interdisciplinary Management and Industrial Engineering topics. He has published extensively in facilities planning and systems modeling. Research papers have appeared in various international journals and have been cited in other research and postgraduate work. His current research interest is on supply chain management. He has supervised several DBA and masters students in topics related to his current interest. Starting with designing of information and quality systems, Dr. Al-Hakim's involvement with the industry continued in the form of consultancy and technical advising. He has consulted in the automotive, aerospace, house appliance, metals, plastics, clothing, food and service industries. In addition to teaching and consulting, he has conducted technology transfer training courses and seminars in various fields of advanced manufacturing.

Leila Alem is a Senior Research Scientist at CSIRO Mathematical and Information Sciences, Australia. Her research spans a variety of HCI/cognitive psychology/AI research issues, as focused through investigations into systems that support human learning. Her specific interests are in the areas of learner-centered design, models of skill development, dynamic and individualized instructional planning and learner and student modeling. She is currently exploring the area of online negotiation agents' systems in e-commerce, specifically, exploring the means for supporting the formation of a coalition among those agents. She has published in several international conferences and journals, and has an active participation in the international research community (member of conference program committees, initiator of workshops, and reviewer of PhD and master's theses).

Akhilesh Bajaj is Assistant Professor of Information Systems Management at the H. John Heinz III School of Public Policy and Management at Carnegie Mellon University, USA. He received a BTech in Chemical Engineering from the Indian Institute of Technology, Bombay, a master's degree in Business Administration from Cornell University, and a doctorate in Management Information Systems (minor in Computer Science) from the University of Arizona. He has published more than 25 referred articles in journals, books and academic proceedings. Dr. Bajaj's research deals with the construction and testing of tools and methodologies that facilitate the construction of large organizational systems, as well as studying the decision models of the actual consumers of these information systems. He is on the editorial board of the *Journal of Database Management*. He teaches graduate level courses on basic and advanced database systems, as well as enterprise wide systems. Dr. Bajaj's information is available at http://bajaj.heinz.cmu.edu/MyInfo/.

Ulrike Baumöl is Program Manager of the post-graduate program, Executive Master of Business Engineering, and Assistant Professor at the University of St. Gallen, Switzerland. She received her master's degree in Business Administration (1992) from the University of Dortmund, Germany. As a research assistant at the University of Dortmund from 1992 through 1998, she received her doctoral degree, with a dissertation on software target costing (1998).

Pratyush Bharati is an Assistant Professor in the Management Science and Information Systems Department of the College of Management at the University of Massachusetts, USA. He received his PhD from Rensselaer Polytechnic Institute. His present research interests are in management of IT for service quality, diffusion of e-commerce technologies in small and medium sized firms and Web-based decision support systems. His research has been published in several international journals including *Communications of the ACM* and *Decision Support Systems*. He is a member of the Association of Computing Machinery and Association of Information Systems.

Pak Yuen P. Chan received his Masters of Commerce (Honours) degree in Information Systems from the University of Canterbury. He currently works as a Programmer for the Applied Technology Center, Hong Kong University of Science and Technology.

Abhijit Chaudhury is a Professor of Management Information Systems at Bryant College, USA. He received his bachelor's and master's degrees in Engineering. He has a PhD in Information Systems from Purdue University. He is a very active researcher, with around 40 papers and presentations completed in the U.S. in the last 10 years. Dr. Chaudhury has taught at the University of Texas, Austin, the Babson College, the University of Massachusetts and the Bentley College. His research has been published in several journals including *Information Systems Research*, *Communications of ACM*, several transactions of *IEEE*, and the *Journal of Management Information Systems*. He is an author of a textbook entitled *E-Commerce and E-Business Infrastructure*, published by McGraw-Hill Publishers. Another two books of his on mobile-commerce and business modeling have been published by Kluwer Publishers.

Qiyang Chen is Associate Professor of Management Information Systems in the School of Business at Montclair State University, USA. He received his doctorate in Information Systems from the University of Maryland in Baltimore, a master's degree from China Space Academy and a bachelor's degree from the National University of Defense Technology in China. His research interests, publications and industry consulting activities are in the areas of strategic issues in MIS, database design and data modeling, human-computer interaction, and soft computing.

Bishwajit Choudhary is a Management Consultant with experiences from European companies in banking, consulting, telecom and professional services sectors. He has worked with technology forecasting, industry analyses, vendor assessment, realizing business cases, developing new product concepts for range of emerging solutions in banking, digital security, electronic-IDs, mobile-commerce and smart cards. He has successfully concluded several Pan-European partnerships, high-profile sales contracts and spearheaded development of new business units during his professional career. Currently, he serves as Advisor (Business Development & Strategy) at Norwegian Banks' Payments Central (NBPC) in Oslo.

Jong-hoon Chun received his bachelor's degree in Computer Science from University of Denver, Colorado, USA, in 1986 and master's and doctorate degrees in Computer Science from Northwestern University, Evanston, IL, in 1988 and 1992, respectively. He taught at the University of Oklahoma from 1992 to 1995. He is currently an Associate Professor at the Department of Computer Science and Engineering, Myung-Ji University, Korea, and CEO of CoreLogiX, Ltd., Korea.

Renzo Gobbin received his Masters in Applied Science from the University of Canberra and his Bachelor of Arts from Australian National University, Australia. He is currently a multi-disciplinary PhD by research student at the University of Canberra in the School of Information Sciences and Engineering. His area of research is in the areas of artificial intelligence, intelligent agents communication, philosophy and the application of intelligent agents technology to e-commerce and e-Learning.

Markus Greunz has been a Research Assistant and doctoral student at the Media and Communications Management Institute of the University of St. Gallen, Switzerland, since November, 1999. He is member of the research group, Media Platforms and Management (MPM) with a research focus on new methods and technologies for the design and implementation of media platforms. He holds a master's degree in Management Information Systems from Johannes Keppler University, Linz, Austria, with a focus in the fields of software engineering, multimedia application development and knowledge, and process management. Prior to joining the doctoral program at the institute he was a software developer at the Ars Electronica FutureLab, Linz, Austria.

Ric Jentzsch is a Senior Lecturer at the University of Canberra in the School of Information Sciences and Engineering, Australia. He has a PhD from University of New South Wales, Australia and a Masters in Computer Information Systems from Colorado State University, USA. His research interest is in the area of diffusion and application of e-business for small to medium enterprises (SME), enabling technologies for small to medium enterprises (SME), intelligent agents – use, business models, applications, and frameworks for e-business.

Boris Jukic, PhD, is an Assistant Professor of Management Information Systems at George Mason University, USA. Dr. Jukic earned a bachelor's degree in Computer Science and Electrical Engineering from University of Zagreb, Croatia in 1991; a master's degree in Business Administration from Grand Valley State University in 1994; and a doctorate in Management Science and Information

Systems from the University of Texas at Austin in 1998. His research interests include application of economic theory in various areas of Information Technology, ranging form computer network resource management to the application of new database technologies in electronic commerce. His research has appeared in various engineering, management and economics publications and conference proceedings.

Nenad Jukic, PhD, is an Assistant Professor of Information Systems at Loyola University, Chicago, USA, where he also serves as a coordinator for graduate courses and programs in the areas of Data Warehousing, Business Intelligence, and e-Commerce. Dr. Juki was born and raised in Croatia, where he received a bachelor's degree in Electrical Engineering and Computer Science from the University of Zagreb in 1991. He received a master's degree (1993) and his doctorate (1997) in Computer Science form the University of Alabama. Dr. Juki conducts active research in various information technology related areas, including database management, e-business, data warehousing, and data mining. His work was published in a number of journal and conference publications. Aside from academic work, his engagements include work for U.S. military and government agencies, as well as consulting for corporations that vary from start-up to Fortune 500 companies.

Dongkyu Kim received his bachelor's and master's degrees in Computer Science and Statistics from Seoul National University, Seoul, Korea in 1995 and 1997, respectively. He has completed the doctoral program in Computer Science and Engineering at Seoul National University and oversees the development of the Content Management System at CoreLogiX, Ltd.

Jeuk Kim received a bachelor's degree in Oceanography in 2000, and the master's of science degree in 2002 from Seoul National University, Seoul, Korea. He works for Daewoo Information Systems Co., Ltd.

Kiryoong Kim received his bachelor's degree in Statistics from Seoul National University, Seoul, Korea, in 2001. He is in the master's course at the School of Computer Science and Engineering, Seoul National University. His current research interests include the electronic catalog and information systems.

Ryszard Kowalczyk is a Principal Research Scientist at CSIRO Mathematical and Information Sciences, Australia. He has several years of basic research and industrial R&D experience in the area of Intelligent Systems involving knowledge-based systems, fuzzy systems, evolutionary systems, and intelligent agent systems for decision support and automation in mission critical enterprise solutions. After a number of years with a corporate R&D division of a major Australian corporation, Dr. Kowalczyk joined the Commonwealth Scientific and Industrial Research Organization (CSIRO) in 1996, where he led the Applied Artificial Intelligence R&D

Group in carrying out research on intelligent software agents for e-commerce, e-business and e-service delivery. He also spent a year with INTERSHOP Research, where he led research in intelligent software agents for e-negotiation in dynamic e-commerce. Dr. Kowalczyk has a number of patents in the area of intelligent agents, and several publications, including four edited books and more than seventy research papers in books, journals and international conference proceedings. He has been a regular reviewer for *IEEE Transactions on Systems, Man and Cybernetics, IEEE Transactions on Fuzzy Systems, International Journal on Information Sciences, International Journal on Knowledge and Information Systems, ACS Journal of Research and Practice in Information Technology, Journal of Applied Systems Studies, Computational Intelligence Journal,* the *International Journal of Computers and Electrical Engineering,* and the *International Journal of Applied Artificial Intelligence.* He has been a member of and has organized a steering and program committee for several international conferences and workshops.

Ighoon Lee received his bachelor's and master's degrees in Computer Science and Statistics from the University of Seoul, Korea in 1996 and 1998. He has finished the doctoral program in Computer Science and Engineering at Seoul National University.

In Lee is an Assistant Professor in the Department of Information Management and Decision Sciences at the College of Business and Technology, Western Illinois University, USA. He received his master's degree in Business Administration from the University of Texas at Austin and his doctorate from the University of Illinois at Urbana-Champaign. He is a member of IRMA, INFORMS, DSI, and ACM. His current research interests include e-commerce development methodology, agent-oriented enterprise modeling, and intelligent manufacturing systems.

Sang-goo Lee received the bachelor's degree in Computer Science from Seoul National University, Seoul, Korea, in 1985, and master's and doctorate degrees in Computer Science from Northwestern University, Evanston, IL, in 1987 and 1990, respectively. He taught at the University of Minnesota from 1989 to 1990. He is currently an Associate Professor at the School of Computer Science and Engineering, Seoul National University, Korea.

Markus Lenz is a Research Assistant and doctoral student in the research area of Business Media at the Media and Communications Management Institute of the University of St. Gallen. He joined the institute in March 2000 and is a member of the Competence Center Electronic Markets. He has a master's degree in Business Administration, with a major in Strategic Management and Marketing, from Ludwig-Maximilians-University in Munich, Germany. Prior to joining the doctoral program, he worked for the Indo-German Chamber of Commerce in Bombay, India and as a consultant for Simon-Kucher & Partners in Bonn, as well as for CSC Ploenzke in Cologne.

K. Y. R. Li, MSc, PhD, CEng, MIEE, MACS, is a Senior Lecturer at Monash University, Australia. His current research interests include mobile agents, automatic essay scoring using LSA, virtual project office, Web-based project evaluation using AHP and culture issues of e-business/V-business. His aim is to develop cost-effective business solutions utilizing emerging technologies. For the last few years, he has initiated a number of industrial collaborative research programs. Some of the successful multimedia intranet-based developments are: an induction program for Bristol Myers Squibb, competency testing for detective training at the Victoria Police and a just-in-time training program for nurses at South Health Network in Melbourne. Besides teaching and research, he also consults.

Chang Liu is Assistant Professor of Management Information Systems at Northern Illinois University, USA. He received his doctorate of business administration from Mississippi State University in 1997. His research interests are electronic commerce, Internet computing and telecommunications. His research works have been published in *Information & Management*, *International Journal of Electronic Commerce and Business Media*, *Journal of Computer Information Systems*, *Mid-American Journal of Business*, and *Journal Informatics Education Research*. Dr. Liu teaches database and electronic commerce courses.

Brian Mackie, PhD, is Assistant Professor of Management Information Systems at Northern Illinois University, USA. He received his doctorate in business administration from the University of Iowa in 1999. His research interests are database, networks and electronic collaboration. Dr. Mackie teaches Database and Operating Environments.

Jack Marchewka is an Associate Professor and the Barsema Professor of Management Information Systems in the Department of Operations Management and Information Systems (OMIS) at Northern Illinois University, USA. In addition, he is also the Director of the Business Information Technology Transfer Center (BITTC) His current research interests include IT project management, electronic commerce and knowledge management. His articles have appeared in the *Information Resources Management Journal*, *Information Technology and People*, and the *Journal of International Information Management.*

Annette M. Mills is a Senior Lecturer in Information Systems with the Department of Accounting, Finance and Information Systems at the University of Canterbury (New Zealand). She holds a PhD in Information Systems from the University of Waikato. Her research interests include technology adoption, IS education, electronic commerce, and IT in developing countries.

Fiona Fui-Hoon Nah is an Assistant Professor of Management Information Systems at University of Nebraska–Lincoln, USA. Previously, she was a member of the faculty at Purdue University. She received her doctorate in Management Information Systems from University of British Columbia in 1997. Dr. Nah serves on the editorial boards of *Information Resources Management Journal, Journal of Database Management, Journal of Global Information Management*, and *Annals of Cases on Information Technology*. Her research works have appeared in *Communications of the ACM, Journal of Information Technology, Journal of Information Technology Cases and Applications* and *Journal of Software Maintenance*, among others. Her research interests include human-computer interaction, trust and adoption issues in mobile commerce, group decision support, and enterprise resource planning.

Manoj Parameswaran, PhD, is an Assistant Professor of Information Systems at the Decision and Information Technologies Department, R.H. Smith School of Business, University of Maryland, USA. He earned a doctorate in Management Science and Information Systems from the University of Texas at Austin in 1999. He holds a joint appointment with the Institute for Systems Research, and is a research associate at the Center for Research in Electronic Commerce at the University of Texas, Austin. His research and teaching interests include logistics of digital products, electronic markets, telecommunications, and industrial organization in the digital economy. He has worked on network resource allocation for unicast and multicast distribution of multimedia products. His article on economic issues of electronic commerce was featured in *The Financial Times*.

Sang-uk Park received a bachelor's degree in Computer Science and Engineering from Myung-Ji University, Korea, in 1999. He is a Technical Staff Member of CoreLogiX, Ltd., and is developing the Content Management System.

Bernhard Rumpe currently holds a fellowship position at the Technische Universität Munich, Germany, awarded from the Bavarian Government. His main interests are the methods and techniques for the development of business systems. These include the impact of new technologies and application areas, such as e-commerce, to companies, as well as the methodical and technical issues of realizing these systems. He contributed, in various publications, to the UML standardization, as well as to development and enhancement of software engineering processes. He is an author/ editor of eight books, among them the *UML Adaptation for Frameworks (UML-F)*. He is Editor-in-Chief of the new Springer *International Journal on Software and Systems Modeling* (www.sosym.org).

Ada Scupola is an Assistant Professor at the Department of Social Sciences, Roskilde University, Denmark. She holds a PhD in Business Administration from the same department, an MBA from the University of Maryland, College Park, USA, and a master's degree in Information Systems from the University of Bari, Italy. Her research interests are adoption and diffusion of IT (with particular focus on SMEs), business value of IT, strategic management of IT and electronic commerce. Her research has appeared in many books, international journals, such as the *Journal of Information Sciences* and the *Journal of Global Information Technology Management*, and in international conferences such as HICCS. She is on the editorial review board of the *Journal of Information Technology in Organizations*.

Abhishek Sharma is a graduate student at Loyola University of Chicago Graduate School of Business, USA. He received a bachelor's degree in Technology (1997) and a master's degree in Business Administration (2002) from the same university. His main areas of research interest include wireless devices and their impact on various business perspectives, Internet-based peer-to-peer networking systems, and database/data warehousing systems.

Thomas Stiffel is Research Assistant and doctoral student at the Institute of Information Management, University of St. Gallen, Switzerland. He received his master's degree in Industrial Engineering and Management (1999) from University of Karlsruhe, Germany. His research focus is on business model innovation and transformation in the new economy.

John Wang is a Professor in the Department of Information and Decision Sciences at Montclair State University (MSU), USA. Having received a scholarship award, he came to the USA and completed his doctorate in Operations Research at Temple University in 1990. He worked as an assistant professor at Beijing University of Sciences & Technology, China, for two years. In the fall of 1992, he transferred to MSU. Dr. Wang received his tenure in 1997, and was promoted to full professor in 2000 for his outstanding and extraordinary contributions. Dr. Wang has published 72 papers in referred journals and conference proceedings, as well as two research books. He has been an active member of five renowned professional organizations. He has served as session chairman and track chair seventeen times on the most prestigious international and national conferences. His research activities and articles have been well received, enabling him to build a reputation with other significant professionals in his field. He was invited to serve as a referee for *Operations Research* (a flagship journal) and *IEEE Transactions on Control Systems Technology* (a very prestigious journal). He has also developed several computer software programs based on his research findings. Dr. Wang is a highly accomplished and well-established operations research scholar. His current research interests include optimization, nonlinear programming, and manufacturing systems engineering. A long-term goal is to study the synergy of operations research and cybernetics.

Merrill Warkentin is a Professor of Management Information Systems in the College of Business & Industry at Mississippi State University, USA. He has authored more than 100 articles, chapters and books. His research, primarily in E-

Commerce, virtual teams, system security, and expert systems, has appeared in such journals as *MIS Quarterly, Decision Sciences, Information Systems Journal, Decision Support Systems, Communications of the AIS, Electronic Markets, Information Systems Management, Journal of Knowledge Engineering & Technology, Journal of Electronic Commerce Research, Logistics Information Management, ACM Applied Computing Review, Expert Systems* and *Journal of Computer Information Systems*. Professor Warkentin is a co-author of *Electronic Commerce 2002: A Managerial Perspective (2e)* (Prentice Hall, 2002) and editor of *Business-to-Business Electronic Commerce: Challenges and Solutions* (Idea Group Publishing, 2002). He is currently an Associate Editor of *Information Resources Management Journal* and the *Journal of Electronic Commerce in Organizations*, and is a Guest Editor of two special issues of the *Journal of End User Computing*. Dr. Warkentin has served as a consultant to numerous companies and organizations, and has been a featured speaker at more than one hundred industry association meetings, executive development seminars, and academic conferences. He has been a lecturer at the Army Logistics Management College and, since 1996, he has served as National Distinguished Lecturer for the Association for Computing Machinery (ACM). Professor Warkentin received bachelor's and master's degrees, and his doctorate, from the University of Nebraska-Lincoln. He can be reached at mwarkentin@acm.org and www.MISProfessor.com.

Robert Winter is Professor and Director of the Institute of Information Management, University of St. Gallen, Switzerland. He is also Director of St. Gallen's post-graduate program, Executive Master of Business Engineering. He received master's degrees in Business Administration (1984) and Business Education (1986) from Goethe University, Frankfurt, Germany. As a research assistant with Goethe University from 1984 through 1994, he received a doctorate for his work in the field of multi-stage production planning (1989) and venia legendi for his work on formal semantics of conceptual information systems design (1994). His research interests include business engineering / information systems development, information systems architectures, and information logistics (particularly data warehousing and enterprise application integration).

F. Xue holds a bachelor's degree in Computer Science. He earned a master's degree in Information Technology at the University of Newcastle in 2000. He has worked as a professional system engineer specialized in Client-Server system developments. His professional qualification includes MCSE plus Internet Certificate. He completed the project for his master's degree in Business Systems Research at Monash University in 2001. He currently lectures at the School of Multimedia Systems of Monash University, Australia, with a major research interest in mobile agent technology for business.

Fan Zhao is a doctoral student in the Management Information Systems program at the University of Nebraska–Lincoln, USA. He received his master's degree in Textile Engineering from Georgia Institute of Technology in 2000, where he carried out research in developing digital imaging applications for the textile industry. His current research interests are in human-computer interaction and enterprise resource planning.

Wenli Zhu is a Usability Engineer at Microsoft Corporation, USA. She received her bachelor's degree in Management Information Systems from Tsinghua University, Beijing, China, in 1989. She received her master's degree and doctorate in Industrial Engineering (Human Factors) from Purdue University, West Lafayette, Indiana, USA, in 1991 and 1995, respectively. Dr. Zhu has worked in the software industry for more than eight years, mainly in designing and evaluating human-computer interfaces. Her research interests include human-computer interaction, usability engineering, technology adoption and software product development methodology.

Index

A

B

C

J

L

M

N

O

P

R

S

T

U

V

W